# INTERNATIONAL SEMINAR ON
# NUCLEAR WAR AND PLANETARY EMERGENCIES
## 36th Session:

Energy: Global Nuclear Power Future; Global Monitoring of the Planet Proliferation: Nuclear Weapons; AIDS and Infectious Diseases: Avian Flu — Global Health; Climatology: Global Warming/Aerosols and Satellites; Pollution: Plastic Contaminants in Water; Information Security: Relevance of Cyber Security; Limits of Development: Development of Sustainability; Defence Against Cosmic Objects; WFS General Meeting: Cultural Emergy-Focus: Terrorism; Permanent Monitoring Panel Reports; Limits of Development Permanent Monitoring Panel Meeting; World Energy Monitoring Workshop

## THE SCIENCE AND CULTURE SERIES
### Nuclear Strategy and Peace Technology

Series Editor: Antonino Zichichi

1981 — International Seminar on Nuclear War — 1st Session: The World-wide Implications of Nuclear War

1982 — International Seminar on Nuclear War — 2nd Session: How to Avoid a Nuclear War

1983 — International Seminar on Nuclear War — 3rd Session: The Technical Basis for Peace

1984 — International Seminar on Nuclear War — 4th Session: The Nuclear Winter and the New Defence Systems: Problems and Perspectives

1985 — International Seminar on Nuclear War — 5th Session: SDI, Computer Simulation, New Proposals to Stop the Arms Race

1986 — International Seminar on Nuclear War — 6th Session: International Cooperation: The Alternatives

1987 — International Seminar on Nuclear War — 7th Session: The Great Projects for Scientific Collaboration East-West-North-South

1988 — International Seminar on Nuclear War — 8th Session: The New Threats: Space and Chemical Weapons — What Can be Done with the Retired I.N.F. Missiles-Laser Technology

1989 — International Seminar on Nuclear War — 9th Session: The New Emergencies

1990 — International Seminar on Nuclear War — 10th Session: The New Role of Science

1991 — International Seminar on Nuclear War — 11th Session: Planetary Emergencies

1991 — International Seminar on Nuclear War — 12th Session: Science Confronted with War (unpublished)

1991 — International Seminar on Nuclear War and Planetary Emergencies — 13th Session: Satellite Monitoring of the Global Environment (unpublished)

1992 — International Seminar on Nuclear War and Planetary Emergencies — 14th Session: Innovative Technologies for Cleaning the Environment

1992 — International Seminar on Nuclear War and Planetary Emergencies — 15th Session (1st Seminar after Rio): Science and Technology to Save the Earth (unpublished)

1992 — International Seminar on Nuclear War and Planetary Emergencies — 16th Session (2nd Seminar after Rio): Proliferation of Weapons for Mass Destruction and Cooperation on Defence Systems

1993 — International Seminar on Planetary Emergencies — 17th Workshop: The Collision of an Asteroid or Comet with the Earth (unpublished)

1993 — International Seminar on Nuclear War and Planetary Emergencies — 18th Session (4th Seminar after Rio): Global Stability Through Disarmament

1994 — International Seminar on Nuclear War and Planetary Emergencies — 19th Session (5th Seminar after Rio): Science after the Cold War

1995 — International Seminar on Nuclear War and Planetary Emergencies — 20th Session (6th Seminar after Rio): The Role of Science in the Third Millennium

1996 — International Seminar on Nuclear War and Planetary Emergencies — 21st Session (7th Seminar after Rio): New Epidemics, Second Cold War, Decommissioning, Terrorism and Proliferation

1997 — International Seminar on Nuclear War and Planetary Emergencies — 22nd Session (8th Seminar after Rio): Nuclear Submarine Decontamination, Chemical Stockpiled Weapons, New Epidemics, Cloning of Genes, New Military Threats, Global Planetary Changes, Cosmic Objects & Energy

1998 — International Seminar on Nuclear War and Planetary Emergencies — 23rd Session (9th Seminar after Rio): Medicine & Biotechnologies, Proliferation & Weapons of Mass Destruction, Climatology & El Nino, Desertification, Defence Against Cosmic Objects, Water & Pollution, Food, Energy, Limits of Development, The Role of Permanent Monitoring Panels

1999 — International Seminar on Nuclear War and Planetary Emergencies — 24th Session: HIV/AIDS Vaccine Needs, Biotechnology, Neuropathologies, Development Sustainability — Focus Africa, Climate and Weather Predictions, Energy, Water, Weapons of Mass Destruction, The Role of Permanent Monitoring Panels, HIV Think Tank Workshop, Fertility Problems Workshop

2000 — International Seminar on Nuclear War and Planetary Emergencies — 25th Session: Water — Pollution, Biotechnology — Transgenic Plant Vaccine, Energy, Black Sea Pollution, Aids — Mother–Infant HIV Transmission, Transmissible Spongiform Encephalopathy, Limits of Development — Megacities, Missile Proliferation and Defense, Information Security, Cosmic Objects, Desertification, Carbon Sequestration and Sustainability, Climatic Changes, Global Monitoring of Planet, Mathematics and Democracy, Science and Journalism, Permanent Monitoring Panel Reports, Water for Megacities Workshop, Black Sea Workshop, Transgenic Plants Workshop, Research Resources Workshop, Mother–Infant HIV Transmission Workshop, Sequestration and Desertification Workshop, Focus Africa Workshop

2001 — International Seminar on Nuclear War and Planetary Emergencies — 26th Session: AIDS and Infectious Diseases — Medication or Vaccination for Developing Countries; Missile Proliferation and Defense; Tchernobyl — Mathematics and Democracy; Transmissible Spongiform Encephalopathy; Floods and Extreme Weather Events — Coastal Zone Problems; Science and Technology for Developing Countries; Water — Transboundary Water Conflicts; Climatic Changes — Global Monitoring of the Planet; Information Security; Pollution in the Caspian Sea; Permanent Monitoring Panels Reports; Transmissible Spongiform Encephalopathy Workshop; AIDS and Infectious Diseases Workshop; Pollution Workshop

2002 — International Seminar on Nuclear War and Planetary Emergencies — 27th Session: Society and Structures: Historical Perspectives — Culture and Ideology; National and Regional Geopolitical Issues; Globalization — Economy and Culture; Human Rights — Freedom and Democracy Debate; Confrontations and Countermeasures: Present and Future Confrontations; Psychology of Terrorism; Defensive Countermeasures; Preventive Countermeasures; General Debate; Science and Technology: Emergencies; Pollution, Climate — Greenhouse Effect; Desertification, Water Pollution, Algal Bloom; Brain and Behaviour Diseases; The Cultural Emergency: General Debate and Conclusions; Permanent Monitoring Panel Reports; Information Security Workshop; Kangaroo Mother's Care Workshop; Brain and Behaviour Diseases Workshop

2003 — International Seminar on Nuclear War and Planetary Emergencies — 29th Session: Society and Structures: Culture and Ideology — Equity — Territorial and Economics — Psychology — Tools and Countermeasures — Worldwide Stability — Risk Analysis for Terrorism — The Asymmetric Threat — America's New "Exceptionalism" — Militant Islamist Groups Motives and Mindsets — Analysing the New Approach — The Psychology of Crowds — Cultural Relativism — Economic and Socio-economic Causes and Consequences — The Problems of American Foreign Policy — Understanding Biological Risk Chemical Threats and Responses — Bioterrorism — Nuclear Survivial Criticalities — Responding to the Threats — National Security and Scientific Openness — Working Groups Reports and Recommendations

2004 — International Seminar on Nuclear War and Planetary Emergencies — 30th Session: Anniversary Celebrations: The Pontifical Academy of Sciences 400th — The 'Ettore Majorana' Foundation and Centre for Scientific Culture 40th — H.H. John Paul II Apostolate 25th — Climate/Global Warming: The Cosmic Ray Effect; Effects on Species and Biodiversity; Human Effects; Paleoclimate Implications; Evidence for Global Warming — Pollution: Endocrine Disrupting Chemicals; Hazardous Material; Legacy Wastes and Radioactive Waste Management in USA, Europe; Southeast Asia and Japan — The Cultural Planetary Emergency: Role of the Media; Intolerance; Terrorism; Iraqi Perspective; Open Forum Debate — AIDS and Infectious Diseases: Ethics in Medicine; AIDS Vaccine Strategies — Water: Water Conflicts in the Middle East — Energy: Developing Countries; Mitigation of Greenhouse Warming — Permanent Monitoring Panels Reports — Workshops: Long-Term Stewardship of Hazardous Material; AIDS Vaccine Strategies and Ethics

2004 — International Seminar on Nuclear War and Planetary Emergencies — 31st Session: Multidisciplinary Global Approach of Governments and International Structures: Societal Response — Scientific Contributions to Policy — Economics — Human Rights — Communication — Conflict Resolution — Cross-Disciplinary Responses to CBRN Threats: Chemical and Biological Terrorism — Co-Operation Between Russia and the West — Asymmetrical Conflicts — CBW Impact — Cross-Disciplinary Challenges to Emergnecy Management, Media Information and Communication: Role of Media in Global Emergencies — Emergency Responders — Working Groups' Reports and Recommendations

2005 — International Seminar on Nuclear War and Planetary Emergencies — 32nd Session: Limits of Development: Migration and Cyberspace; in Europe; Synoptic European Overview; From and Within Asia; Globalization — Climate: Global Warming; a Chronology; Simple Climate Models; Energy and Electricity Considerations — T. S. E.: CJD and Blood Transfusion; BSE in North America; Gerstmann-Straussler-Scheinker Disease — The Cultural Emergency: Innovations in Communications and IT — Cosmic Objects: Impact Hazard; Close Approaches; Asteroid Deflection; Risk Assessment and Hazard Reduction; Hayabusa and Follow Up — Aids and Infectious Diseases: Ethics in Medicine; International Co-operation; Laboratory Biosecurity Guidelines; Georgian Legislation; Biosecurity Norms and International Organizations, Legal Measures Against Biocrimes — Water and Pollution: Cycle Overview; Beyond Cost and Price; Requirements in Rural Iran; Isotope Techniques; Clean and Reliable Water for the 21st Century — Permanent Monitoring Panels Reports — Workshops: Global Biosecurity; Cosmic Objects

2006 — International Seminar on Nuclear War and Planetary Emergencies — 34th Session: Energy: Nuclear and Renewable Energy; Energy Technologies for the 21st Century; Repositories Development; Nuclear Power in Europe and in Asia; The Future of Nuclear Fusion — Climate: Global Warming; Celestial Climate Driver; Natural and Anthropogenic Contributions; Climate Data and Comparison with Models; Understanding Common Climate Claims — AIDS and Infectious Diseases: New Threats from Infectious Agents — SARS Epidemic; Vaccines Development; Transmissible Spongiform Encephalopathies Update — Limits of Development: International Points of View on Migration — Pollution: Science and Technology; Subsurface Laser Drilling — Desertification: A Global Perspective; Integrated Approaches — Disarmament and Cultural Emergencies: A WFS Achievement in China; Non-Proliferation — Permanent Monitoring Panel Reports — Workshops: Energy; Information Security; Building Resilence Associated with the Third Meeting on Terrorism

2007 — International Seminar on Nuclear War and Planetary Emergencies — 36th Session: Energy: Global Nuclear Power Future; Global Monitoring of the Planet Proliferation: Nuclear Weapons; AIDS and Infectious Diseases: Avian Flu — Global Health; Climatology: Global Warming/Aerosols and Satellites; Pollution: Plastic Contaminants in Water; Information Security: Relevance of Cyber Security; Limits of Development: Development of Sustainability; Defence Against Cosmic Objects; WFS General Meeting: Cultural Emergy-Focus: Terrorism; Permanent Monitoring Panel Reports; Limits of Development Permanent Monitoring Panel Meeting; World Energy Monitoring Workshop.

# THE SCIENCE AND CULTURE SERIES
## Nuclear Strategy and Peace Technology

# INTERNATIONAL SEMINAR ON NUCLEAR WAR AND PLANETARY EMERGENCIES
## 36th Session:

Energy: Global Nuclear Power Future; Global Monitoring of the Planet Proliferation: Nuclear Weapons; AIDS and Infectious Diseases: Avian Flu — Global Health; Climatology: Global Warming/Aerosols and Satellites; Pollution: Plastic Contaminants in Water; Information Security: Relevance of Cyber Security; Limits of Development: Development of Sustainability; Defence Against Cosmic Objects; WFS General Meeting: Cultural Emergy-Focus: Terrorism; Permanent Monitoring Panel Reports; Limits of Development Permanent Monitoring Panel Meeting; World Energy Monitoring Workshop

"E. Majorana" Centre for Scientific Culture
Erice, Italy, 19–24 Aug 2006

Series Editor and Chairman: A. Zichichi

Edited by R. Ragaini

World Scientific

NEW JERSEY • LONDON • SINGAPORE • BEIJING • SHANGHAI • HONG KONG • TAIPEI • CHENNAI

*Published by*

World Scientific Publishing Co. Pte. Ltd.
5 Toh Tuck Link, Singapore 596224
*USA office:* 27 Warren Street, Suite 401-402, Hackensack, NJ 07601
*UK office:* 57 Shelton Street, Covent Garden, London WC2H 9HE

**British Library Cataloguing-in-Publication Data**
A catalogue record for this book is available from the British Library.

**INTERNATIONAL SEMINAR ON NUCLEAR WAR AND PLANETARY EMERGENCIES — 36TH SESSION: ENERGY: GLOBAL NUCLEAR POWER FUTURE; GLOBAL MONITORING OF THE PLANET PROLIFERATION: NUCLEAR WEAPONS; AIDS AND INFECTIOUS DISEASES: AVIAN FLU—GLOBAL HEALTH; CLIMATOLOGY: GLOBAL WARMING/AEROSOLS AND SATELLITES; POLLUTION: PLASTIC CONTAMINANTS IN WATER; INFORMATION SECURITY: RELEVANCE OF CYBER SECURITY; LIMITS OF DEVELOPMENT: DEVELOPMENT OF SUSTAINABILITY; DEFENCE AGAINST COSMIC OBJECTS; WFS GENERAL MEETING: CULTURAL ENERGY-FOCUS: TERRORISM; PERMANENT MONITORING PANEL REPORTS; LIMITS OF DEVELOPMENT PERMANENT MONITORING PANEL MEETING; WORLD ENERGY MONITORING WORKSHOP**



ISBN-13 978-981-270-922-6
ISBN-10 981-270-922-3

Printed in Singapore by World Scientific Printers (S) Pte Ltd

# CONTENTS

## 9. Defence Against Cosmic Objects



## 10. WFS General Meeting Cultural Emergency Focus: Terrorism



## 11. Permanent Monitoring Panel Reports

# 1. OPENING SESSION

## THE INTERNATIONAL SEMINARS ON PLANETARY EMERGENCIES AND ASSOCIATED MEETINGS 36th SESSION

ANTONINO ZICHICHI
CERN, Geneva, Switzerland and
University of Bologna, Italy

I welcome you all to this 36th Session of the International Seminars on Nuclear War and Planetary Emergencies and declare the Session to be open.

Here we are again, one hundred scientists from 26 countries and 77 institutions, gathered in Erice to analyse a series of crucial multidisciplinary scientific issues. But first of all, let me remind you of the reason for which we have been meeting in Erice every August for the last 25 years, and that is to ensure the contribution of the scientific community to the fight to overcome the 63 Planetary Emergencies, subdivided into 15 classes, as determined by the World Federation of Scientists 33 years ago, summarized as follows:

| | **SUMMARY** | Number of P.E. |
|---|---|---|
| I | *Water* | =4 |
| II | *Soil* | =3 |
| III | *Food* | =5 |
| IV | *Energy* | =5 |
| V | *Pollution* | =6 |
| VI | *Limits of Development* | =3 |
| VII | *Climatic Changes* | =1 |
| VIII | *Global Monitoring of the Planet* | =6 |
| IX | *New Military Threats in the Multipolar World* | =3 |
| X | *Science and Technology for Developing Countries to Avoid a North-South Environmental Holocaust* | =1 |
| XI | *The Problem of Organ Substitution* | =5 |
| XII | *Infectious Diseases* | =7 |
| XIII | *Cultural Pollution* | =7 |
| XIV | *Common Defence Against Cosmic Objects* | =2 |
| XV | *The Huge Military Investments* | =5 |
| | **Total** | **=63** |

The main topics of this Session are:

On Sunday 20 August: **The Global Energy Crisis**—co-chaired by Richard Garwin.

In view of its impact on the progress and well-being of both Industrialised and Developing Countries, this Emergency has been one of our most important concerns.

Last year, we reviewed the Nuclear and Renewable Energies aspect, and came to the conclusion that we needed to investigate in more detail the nuclear energy production schemes.

This year, we have programmed a full day session, chaired by Professor Richard Garwin of the Thomas J. Watson Research Center. Jacques Bouchard of the French Atomic Energy Commission and Steve Fetter of the School of Public Policy, University of Maryland will give us *Two Views on the Global Nuclear Energy Power Future*.

Jacques Bouchard, Kazuaki Matsui of the Institute of Applied Energy of Tokyo and Phillip Finck of the Argonne National Laboratory will then analyse the *Generation IV Concepts*.

Comments on *PUREX and MOX Burning* will be provided by Jacques Bouchard, Kazuaki Matsui and Richard Garwin, while Philipp Finck will talk on *Global Nuclear Energy Partnership*

Finally, Charles McCombie of the Arius Association of Baden, Switzerland will report on the delicate question of *Repositories around the World,* Steve Fetter on the *Resources for the Long Term* and Carmen Difiglio of the U.S. Department of Energy will speak of *Coal with Sequestration, the Principal Competitor.*

On Monday 21 August, three important topics:

1. **Nuclear Proliferation**—co-chaired by Richard Wilson of Harvard University:
   **Ramamurti Rajaraman** of the School of Physical Sciences, Jawaharlal Nehru University will speak on *The Implications of the Indo-US Nuclear Agreement for S. Asia and the World;*
   Followed by **Alexander Konovalov** of the Moscow State Institute of International Relations on *The Problem of Nuclear Proliferation in a New Security Environment,* and **Kazuaki Matsui** of the Tokyo Institute of Applied Energy on *Proliferation and the Nuclear Fuel Cycle Issues in Japan;.*
   **Joachim Krause** of the Christian-Albrechts-University of Kiel will pose the question *How Serious is the Crisis of the International Nuclear Non-Proliferation Regime?*
   **Christian Bühlmann** of the R&D services of the Military Federal Department of Switzerland will explain a people's rational behaviour in *Rogue State Helvetia? Switzerland and the Atomic Bomb 1945-1988;*
   **Ahmad Kamal** will speak on *The Demise of the Nuclear Non-Proliferation Treaty,* followed by Roland Timerbaev of the Center for Policy Studies in Russia on *Nuclear Proliferation: Current State and Prospects.* **Richard Garwin** will conclude by providing his views and comments on the problem.
2. **Avian Flu** Update and **Global Health**
   **Albert Osterhaus** of the Erasmus Medical Center in Rotterdam will provide us with the latest findings on the battle waged by the international community of scientists against Avian Flu;
   **Ahmad Kamal** will give us an authoritative report on the need to bring a Change in Global Health.
3. **Global Warming** – co-chaired by William Sprigg
   Global Warming is a recurring topic in our Seminars. We identified early on that it was crucially important to ensure that the models being used to evaluate Global Warming trends are scientifically accurate. This year, we are looking at the evaluation of aerosols and their impact on climate change.
   **Lawrence Friedl** of the NASA Applied Sciences Program will speak on *Air Quality Forecasting and International Policy.* **Sundar Christopher** *on Satellite Remote*

*Sensing of Aerosol Climate Effects.* **Graeme Stephens** of Colorado State University on *Connections between Aerosol, Atmospheric Radiation and Hydrological Processes.* **Gregory Carmichael** of the University of Iowa on *How Good are our Global and Regional Models?* Finally, **Christopher Essex** of University of Western Ontario will tell us how he rates *Fundamental Science in Climate Forecasting with Models.*

On Tuesday 22 August, again three main topics:

1. **Water Pollution**—Co-chaired by L. Everett and F. vom Saal
   This year, Water Pollution is focused on Plastic Contaminants in Water.
   We'll hear testimonies from **Charles Moore** of the Algalita Marine Research Foundation on *Global Ocean Contamination*, **Shanna H. Swan** of the Rochester Center for Reproductive Epidemiology on *Health Effects Of Plastic Additives on Humans,* Frederick vom Saal of the University of Missouri on *Consequences for Animals,* and **Jean-François Debroux** of the San Francisco Kennedy/Jenks Consultants on *the Problem of Remediation.*
2. **Information Security**—co-chaired by Henning Wegener
   Information Security is fast turning out to be a major concern worldwide. This year, we focus on the *Relevance of Cyber Insecurity.*
   After an introduction by Henning Wegener, **Pradeep Khosla** of the Carnegie Institute of Technology will speak on the *Challenge Presented by New Digital Networks,*
   **Udo Helmbrecht** of the Bonn Federal Office for the Security of Information Technologies about the *Economic Dimension of Cyber Insecurity,*
   **William Barletta** of Lawrence Berkeley National Laboratory on the *Evolving Face of Cyber Conflict* and, finally, **Jody Westby** of Global Cyber Risk, Washington, on *Bridging the Legal Divide in Information Security.*
3. **Defence against Cosmic Objects**—co-chaired by Walter Huebner
   After an overview of the recent research activities by Walter Huebner, we will hear **John Zinn** of Los Alamos National Laboratory speak on the *Two Fireballs over New Mexico and their Modeling,* **Michael Belton** of the Belton Space Exploration Initiatives in Tucson on the *Scientific Results of the Deep Impact Mission* and **Hajime Yano** of the University for Advanced Studies in Japan on the *Hayabusa Mission to the Itokawa Asteroid.*

On Wednesday 23 August, **Cultural Emergency** and **PMP Reports**

We will first hear **Ahmad Kamal** present a global report of the Cultural Emergency focused on Terrorism, and then the yearly reports from various PMPs before concluding.

And now I shall give the floor to our co-chairman, **Professor T.D. Lee**, who will report on how the WFS had a major impact on defeating a manifestation of Cultural Emergency in the People's Republic of China.

# A CASE OF CULTURAL EMERGENCY IN CHINA

TSUNG-DAO LEE
Department of Physics, Columbia University, New York, USA

ABSTRACT

I would like to show you the cover of the new and important volume by Professor Zichichi. We are very happy to inform you that it is now available in print in Chinese.

This book speaks out against superstition, and it is therefore important to China as China is a feudal, ancient country where a law has recently been passed against superstition. Nino's book, translated into Chinese, is both extremely timely and useful.

Of course, as we all know, the bond between China and Italy dates back to Marco Polo. The intercultural exchange between pasta and the Chinese noodles is also well known.

Some twelve years ago, China celebrated the formulation of natural law by Galileo by designing a postcard, and also celebrated Enrico Fermi, my teacher, for the launching of nuclear power in 1942.

So the bond between Italy and China has existed for almost a millennium and I think it is worthwhile to note that they are joined again today against the falsification of truth by superstition.

## OPENING SESSION OF THE INTERNATIONAL SEMINAR AT THE ETTORE MAJORANA CENTRE FOR SCIENTIFIC CULTURE

SENATOR ANTONIO D'ALI
President of the Regional Province of Trapani, Italy

No one in the world can ignore the importance of the multi-decadal activities of the Ettore Majorana Centre for Scientific Culture, and the historic milestones for promoting peace and the dialogue between people and countries that its scientists, from the world over, have recorded between these same millenary walls. Your activities have brought Erice to reclaim its central role in a territory, which historically and also presently in a wider strategic vision of the European Union, is bound to become the link between the people of the Mediterranean basin (East and West), integrated inside the large Euro Mediterranean area of co-operation (the new Mediterranean co-operative programme) and the co-operation with the Southern Mediterranean Basin (Mediterranean Basin Co-operation Programme). These programmes will mainly develop projects for economy, co-operative actions and investment on innovations, sea and transport policies, and cultural and other forms of co-operation for an available set of funds of nearly one billion Euros.

Erice, a long time ago already a nexus of important traffic and exchanges, is now reclaiming this central role, and we wish for it to be not only geographic but also mainly cultural.

Once again, from you, the scientific community and Professor Zichichi, comes a new challenge that we willingly accept and implement. A challenge which in the first place is cultural: to put in place a new contemporary "humanitarian" concept, that brings together Science and its technological application to better serve the needs of Humanity.

The contents of Professor Zichichi's Project, "A Project for Humanity", elaborated according to the guidelines of the World Federation of Scientists, was accepted in its entirety by the administration which I preside and the community of citizens of this Province. We hope that this Project will constitute a new approach to Development, based on the analysis of the problems affecting Humanity.

The Province of Trapani, and I hope yourselves too, intends to promote the new approach indicated in the Project elaborated by Professor Zichichi which, based on the scientific analysis of scientists and experts the world over, leads to solutions for the betterment of the quality of life that can be presented to governments who have the political responsibility for human progress: from results of global research that have a bearing on the future of Humanity, to others of a more particular or sectorial nature that could resolve day-to-day problems of existence facing individuals.

I believe that the mission of Science, which is to produce the building elements of the human progress structure, and that of the politicians, which is to decide their use, must increase the occasions during which they confront ideas and enter into dialogue, so as to ensure that the "structure" fits the aspiration for serenity, peace and well-being of the whole of Humanity and that of each individual.

This is why we are grateful to all of you, and more particularly to Professor Zichichi, who demonstrate to the whole world, through the Erice Meetings and your daily work, how extraordinary and indispensable are the contributions of Science to the building of a better future.

# 2. ENERGY

# FOCUS: GLOBAL NUCLEAR POWER FUTURE

# INTRODUCTION, STAGE SETTING AND RULES OF THE GAME

RICHARD L. GARWIN
Thomas J. Watson Research Center, IBM Research Division
Yorktown Heights, USA

I want to use my 15 minutes simply to set the stage and to indicate what is expected of us and of the participants who will not be making prepared speeches. For those of you unfamiliar with Erice, the introduction you have just had from Professors Zichichi and Lee provides a good background. I have long been a participant and am a founding director of the center here, and I believe that the nuclear weapon discussions in Erice played an important role in improving understanding with Soviet scientists and among those of the West, including within the U.S. community.

In looking toward global nuclear energy futures, one needs a way, in principle, to reflect many benefits and hazards as costs. Not only is there the cost of the approach to be taken, but the *opportunity cost* of the approaches not taken—essentially frozen out.

Our purpose here is not to issue a statement, and certainly not in any way to discuss or arrive at one today. Our purpose is not to reach consensus, although that would be nice, if unexpected. But our purpose certainly is to have a better understanding by all participants of technical matters and, particularly, of the approaches advocated or analyzed by the others in their presentations. So the question is: what coefficients can one use to reflect as costs and *opportunity costs* of various uncertain hazards and benefits? For instance:

- Greenhouse gas emissions of nuclear power—perhaps 1% those of fossil fuel plants.
- Contribution to the proliferation of nuclear weapons.
- The risk of costly accidents—deaths and denial of productive use of tens of thousands of square kilometers of land.

There are the specific approaches to be taken, including such details as micro-encapsulated fuel as will be presented in the talk by Dr. Slabber and which has been deployed in a number of operating reactors, especially those at Peach Bottom and Fort St. Vrain in the United States. These have quite different fuel and waste forms compared with the metal-oxide ceramics that are used in light-water reactors—LWRs, which are by far the most common reactors in the world.

Then there is the general question of reactors that burn $^{235}U$, with perhaps some *in situ* breeding such as takes place in CANDU and even in LWRs, as contrasted with a frank breeder reactor that makes at least as much fuel as it burns, but obviously needs much reprocessing in order to remove the fission products, and many cycles of re-use of refabricated fuel.

This is an excellent time to have such a discussion in view of the several proposals that essentially involve "leased fuel," so that fuel of low enrichment (up to 19.9% $^{235}U$) is provided by a certified provider, in competition with others, and the spent fuel (not belonging to the using nation) is returned either to the nation that supplies the fresh LEU fuel or to another nation that perhaps bids competitively to dispose of the spent fuel.

To my mind, nothing could speed up the deployment of nuclear power more than the availability of competitive, commercially-mined geologic repositories, certified by IAEA to accept IAEA-certified spent fuel forms or reprocessed waste.

On the other hand, the U.S. government in its evolving proposals for the Global Nuclear Energy Partnership, first expressed by President George W. Bush in February 2006, proposes to begin by reprocessing all U.S. domestic spent fuel, with the assertion that this will reduce the proliferation hazard and will ultimately save space in the Yucca Mountain repository by a factor of ten or several hundreds. I volunteer here my view, expressed in my testimony of April 6, 2006, to the Science Committee of the U.S. House of Representatives, that as then conceived, the U.S. GNEP program was likely to impede the resurgence of nuclear power in the United States, and it was badly formulated. The correct approach would be to have competitive conceptual designs of the essential element of GNEP—the Advanced Burner Reactor (essentially a fast-neutron breeder-type reactor)—*together with* its specific fuel form and associated reprocessing. The three are tightly coupled—reactor, fuel, and reprocessing—and *that* is what needs emphasis and funds now. Recent modifications of the GNEP proposal only move it further in the wrong direction.

I am sure there will be much discussion of this GNEP proposal at our session.

Well, right now I want to get off the stage and to make way for two substantive views of the Global Nuclear Power Future, as presented by Jacques Bouchard of the CEA, France, and by Steve Fetter, from the University of Maryland. We will invite questions and comments from the audience at appropriate times, but I urge you to be very brief in your question or comment, and not to make a speech. Our purpose is primarily to get the speakers to consider the analyses of others at this session and to leave us with analysis and not a sales pitch.

In particular, I hope that in our work in Erice we are guided by Albert Einstein's prescription, inscribed in stone at the National Academy of Sciences in Washington, D.C., that *"The right to search for truth implies also a duty; one must not conceal any part of what one has recognized to be true."*

# THE FUTURE OF NUCLEAR ENERGY

JACQUES BOUCHARD
French Atomic Energy Commission, Paris, France

## INTRODUCTION

The world energy needs will still increase over the next two or three decades. To limit the use of fossil fuel and the dangerous consequences of the greenhouse effect, it will be necessary to develop every possibility of using renewable sources as well as nuclear energy. Facing the challenge of a strong increase of the worldwide demand for nuclear energy, we have a rather large industrial offer of Gen III reactors,which rely on the large experience with light water reactors and bring new improvements in safety. The back end of the fuel cycle for these Gen III reactors should still be improved to answer the public concern on waste management and proliferation risks.

We have also the beginning of an international cooperation for the development of Gen IV systems, the aim of which is the sustainability of nuclear energy production. In addition to economy and safety, an efficient use of resources—waste minimization and security, including proliferation resistance—are important criteria for Gen IV systems. Diversity will also be a necessity when considering an important growth of the share of nuclear fission in the world energy mix. Besides a large part of electricity production, direct uses of the heat produced with nuclear energy should be considered in industrial applications. In particular, great attention is given to the possibility of a massive production of hydrogen extracted from water using nuclear energy to provide a solution for the energy supply of transportation.

## ENERGY NEEDS

We are still in a world with growing energy needs for two main reasons: the large increase of population resulting from an improved life expectancy and a higher consumption per capita in developing countries including the most populated ones, China and India in particular. Even a drastic reduction of consumption in industrialized countries, which represent today 80% of the world energy demand, will not compensate the growing needs of developing countries.

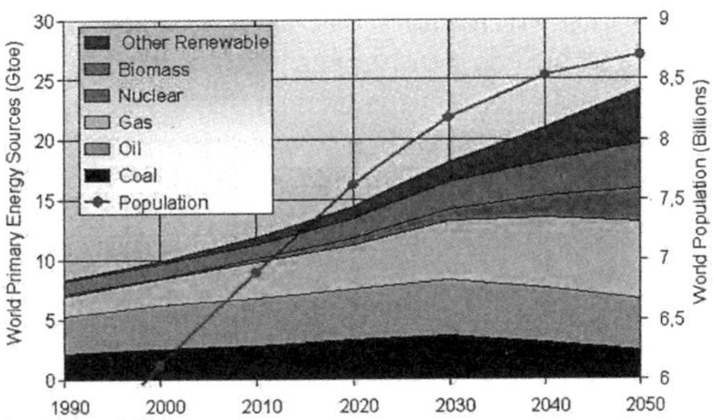

Source IEA : Energy to 2050 - Scenarios for a Sustainable Future

Specialized institutions such as the International Energy Agency of the Organization for Economic Co-operation and Development (OECD) or the World Energy Council have issued plausible scenarios with a world population growing to 9–10 billionsby 2050 and a world energy demand at the same time at least twice the present value.

For many reasons - resources, geopolitics and climate consequences -

fossil fuels, which amount to 87% of the present energy balance, cannot or should not follow such a growing demand. A larger use of renewable sources is foreseen in many countries but will not be sufficient to satisfy the needs, in particular the mass production for industry or large urban areas.

Thus the share of nuclear energy should be increasing and today common predictions give a world nuclear capacity in the range of 1000-1500 GWe as compared to 360 GWe in 2005. In an International Ministerial Conference organized by the International Atomic Energy Agency (IAEA) and held in Paris last year, many countries, and not only the largest ones, indicated their interest to adopt or to increase the use of nuclear energy, (China and India are planning a nuclear capacity of 250 GWe each by 2050). Many delegates mentioned the need to improve political and financial mechanisms in order to allow the implementation of nuclear energy production in developing countries.

GENERATION III NUCLEAR PLANTS

There is a large industrial offer of Generation III reactors which can be deployed in the near term. The table shows those which have been identified by the Generation IV International Forum (GIF) in 2002 on the basis of industrial proposals and with a credible prospect of deployment by 2015 at the latest. It is not a complete list as it includes only offers from member countries of GIF at the time.

| *Generation III reactors identified as 'Near Term Deployment' by the Generation IV Forum* |
|---|
| Advanced Pressurized Water Reactors |
| AP 600, AP 1000, APR1400, APWR+, EPR |
| Advanced Boiling Water Reactors |
| ABWR II, ESBWR, HC-BWR, SWR-1000 |
| Advanced Heavy Water Reactors |
| ACR-700 (Advanced CANDU Reactor 700) |
| Small and middle range power integrated Reactors |
| CAREM, IMR, IRIS, SMART |
| High Temperature, Gas Cooled, Modular Reactors |
| GT-MHR, PBMR |

This new generation of reactors has been designed taking advantage of the large experience acquired in the operation of Gen II plants and of the lessons learnt from important accidents, in particular the Three Mile Island event in 1979. Therefore, light water reactors are still dominating and the main objective for the design of Gen III reactors has been to make new improvements in safety while retaining economic competitiveness.

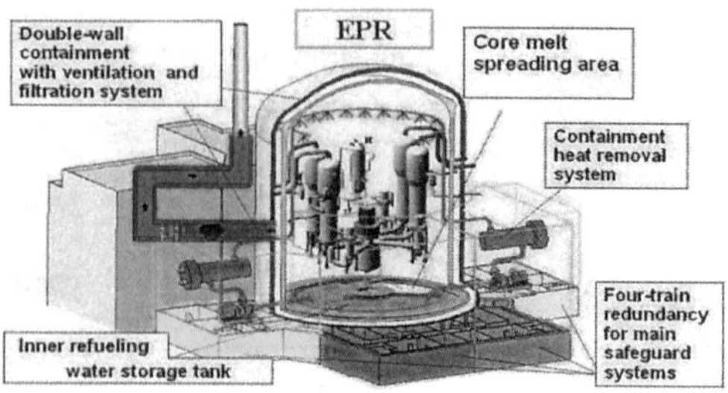


Different approaches have been studied according to countries and companies, for instance, small versus large reactors or passive versus active safety systems, and they are still competing in the industrial offer. At the end, in addition to a further reduction of accident probabilities, it is mainly the mitigation of very low probability severe accidents which constitute a major step characterizing the Gen III design.

An illustration of this new generation of industrial plants is given by EPR, the 1600 MWe plant offered by AREVA NP, Inc., and which is under construction in two countries, Finland and France. The figure shows the main innovative features of the EPR design for what concerns safety. One can notice in particular the four train redundancy for main safeguards systems which contributes to the accident probability reduction, and the core melt spreading area, aimed at avoiding radioactive dispersion in case of the most severe accident for a light water reactor, the core meltdown.

Figure 3 shows the foreseen implementation of the new EDF reactor besides two existing 1300 MWe units on the Flamanville site.

FROM GENERATION III TO GENERATION IV

Two main options are still considered for the backend of LWR's fuel cycle, the once through policy or reprocessing and recycling. The once through policy could be a transitory solution but is not compatible with a sustainable development of nuclear energy. Spent fuel reprocessing, followed by plutonium recycling in MOX fuels reduce the waste volume and long term toxicity, while allowing a better use of valuable materials.

The recycling industry is mature. For France only, there have been more than 20,000 tons of spent LWR fuel reprocessed at the La Hague plant and more than 1,000 tons of MOX fuel fabricated in the MELOX plant at Marcoule. 32 light water reactors in Europe are partly loaded with MOX fuels.

The reprocessing and recycling policy is also economically attractive. The most recent studies based on the French experience show an overall cost of the back end operations which is only a few percent of the kWh cost. With the recent increase of uranium prices (a factor three in two years), the cost comparison of reprocessing and recycling with the once through policy could even end up in favor of the former.

Light water reactors have many advantages but they cannot satisfy alone the sustainable development of nuclear energy. Their efficiency is limited because of an output temperature which does not exceed 300°C and moreover they cannot burn more than 1% of the natural uranium which is used for their fuel fabrication, even if one includes the benefit of plutonium recycling.

A sustainable development of nuclear energy requires the use of fast neutron systems to make full use of natural uranium and to burn a maximum of long lived elements, thus reducing the environmental impact of radioactive waste.

The gas or liquid metal cooling of reactors, either thermal or fast neutron systems, allows higher output temperatures and better efficiency. It helps to save resources and it opens the possibility of other industrial applications of nuclear energy besides electricity production.

To bring new concepts of reactors to industrial maturity, along with energy production cost competitive with light water reactors, requires a strong program of research and development and will take at least two or three decades. During this time, there will be a large market for Generation III systems.

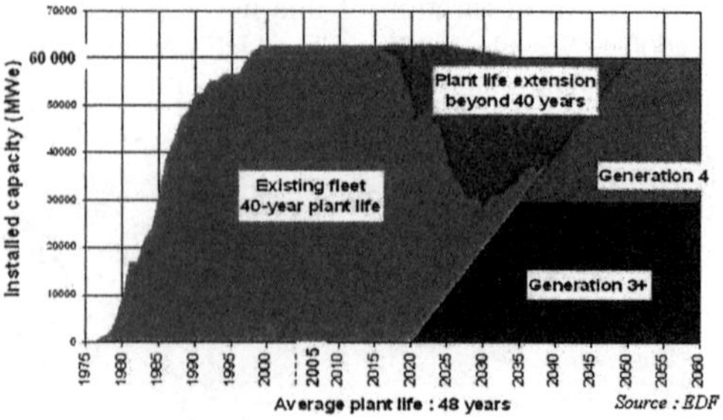


The transition from Generation III to Generation IV should be organized for both reactors and fuel cycle plants. As an example, the figure to the right shows a scenario considered by the French utility EDF for the renewal of its existing fleet of PWRs.

GENERATION IV

The main objectives for the development of Generation IV systems are the following:

SUSTAINABILITY
Saving of natural resources;
Waste minimization;
Security, non proliferation, physical protection.

DIVERSITY
High temperature heat for industrial processes;
Hydrogen as an energy carrier;
Drinking water.

AND: economy and safety still better than for Generation III systems.

The Generation IV International Forum (GIF) was created in 2000 from a U.S. DOE initiative with the aim of developing a collaborative R&D program to bring to industrial maturity a few systems fulfilling the Gen IV criteria. Nine countries signed a charter in 2001 and were joined a year later by Switzerland and Euratom on behalf of the European Union.

An intergovernmental

framework agreement has already been signed by eight of the members, and the forum has recently decided to welcome Russia and China as new members.

The first task of GIF has been to identify innovative concepts of systems (reactor and fuel cycle) with technological breakthroughs. Experts of the member countries have selected the six concepts presented on the figure below among more than one hundred proposals.

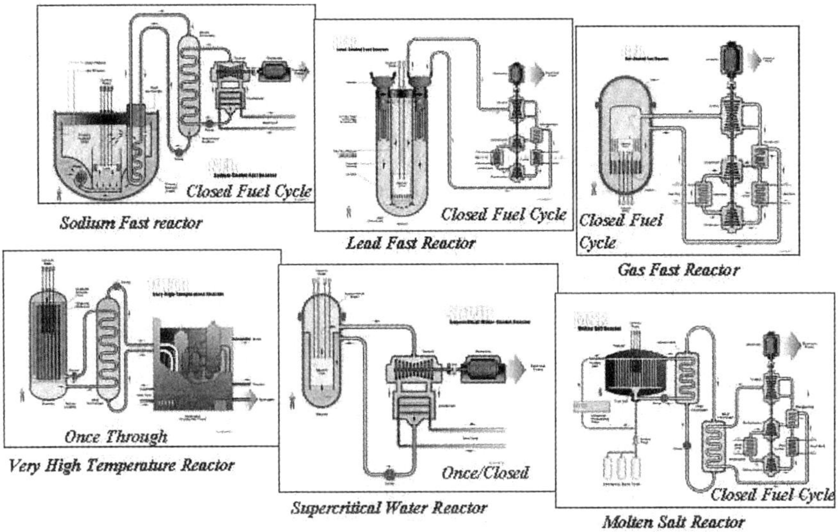


Four among the six selected concepts are based on fast neutron reactors and a closed fuel cycle. As mentioned before, the use of fast reactors is a necessity to satisfy the sustainability criteria.

Saving uranium resources is a first objective. There is plenty of uranium on earth but only a small fraction can be recovered at a cost which allows the competitiveness of nuclear energy. In this respect, the figure to the right shows the necessity of implementing fast reactors without waiting too long. Another reason to burn all the uranium is to avoid the waste management of very large amounts of depleted uranium, a by-product of the necessary enrichment for LWR fuels, which can be burned only in fast reactors.

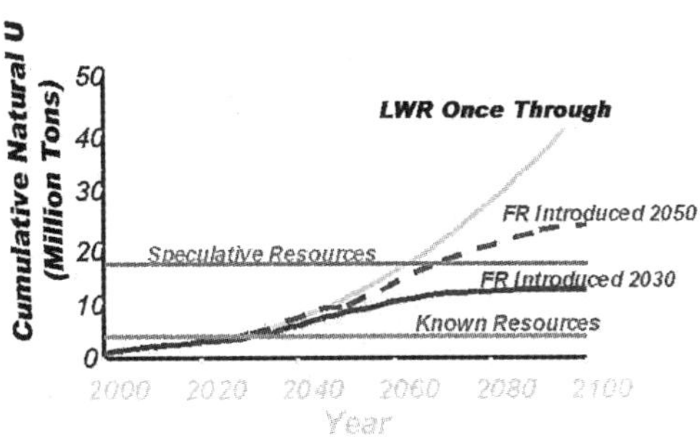


The most important part of radioactive waste coming from the use of nuclear power is contained in the spent fuels. As shown in the following figure, direct disposal of the spent fuel leads to waste which will remain very active for hundreds of thousands of

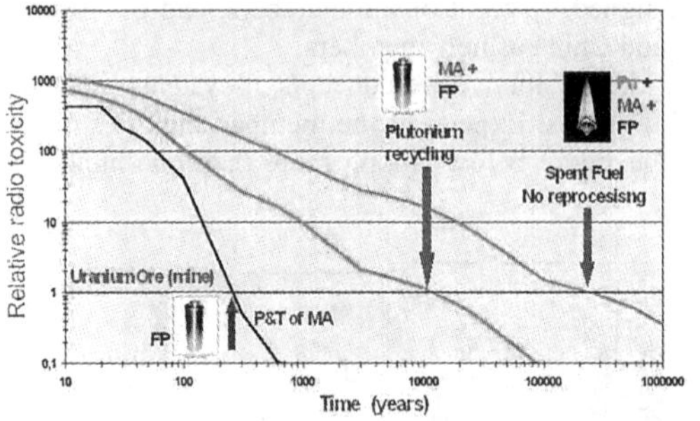


years. With the present technology for reprocessing and recycling, uranium and plutonium are separated from the waste and can be burned in reactors. The remaining waste, containing fission products and other actinides, will be active for a shorter time, in the range of ten thousand years. The objective for Generation IV systems is to burn all the actinides, including plutonium but also neptunium, americium and curium. This way, the waste will be limited to fission products (the actual ashes of fission) and will decay in a few hundred years to a level of activity comparable to the one of uranium ore extracted for fuel fabrication.

The Global Actinide Recycling, a process which is under development in several member countries of the Generation IV International Forum, will not only allow a better efficiency in the use of natural uranium and a minimization of radioactive waste, but it will also improve the proliferation resistance of the back end of the fuel cycle.

As compared to the once-through policy which, by burying all the spent fuel, is creating future plutonium mines, or to the present reprocessing and recycling which leads to the separation of pure plutonium, the global actinide recycling allows a complete burning of all the elements which could be of interest for proliferators while avoiding any separation of pure elements in the process which involves only a very active and self protecting mixture of actinides.

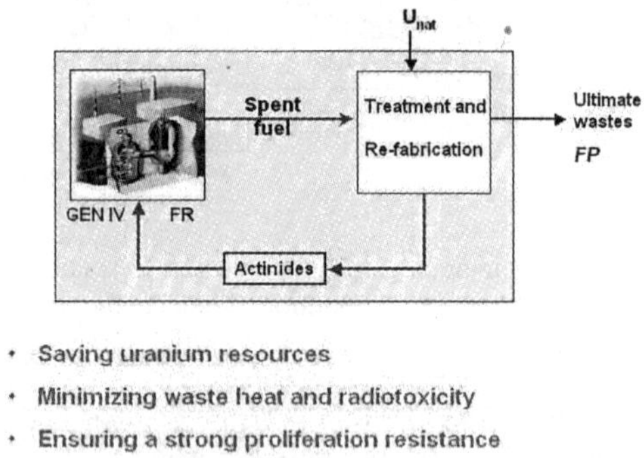


There is already a large experience in the world with fast reactors design, construction and operation. France has built the largest one, Superphenix, a 1200 MWe , in the frame of a European cooperation. Started in 1985, it was definitely shutdown in 1997 for political reasons. The country is still operating Phenix, a 250 MWe prototype, mainly used for research purposes.

While most of this experience is with sodium cooled fast reactors, other possibilities of cooling, helium, lead or lead-bismuth, supercritical water, are also considered in the Gen IV program to keep choices open for the future, each solution having advantages and drawbacks.

The industrial deployment of fast reactors will require large improvements as compared to the previous design, in order to satisfy Gen IV criteria, in particular:

- Lower capital cost;
- Easier in-service inspection;

- Better proliferation resistance.

The GIF selection of concepts includes also a Very High Temperature Reactor aimed at studying the possibility of using nuclear heat for industrial processes, in particular for hydrogen production. Several processes are being considered for producing hydrogen with nuclear power, either electrolysis at high temperature or chemical reactions. The output temperature of helium, the coolant of the VHTR should be around 1000°C. There is also a certain amount of experience with high temperature reactors. In addition to thermal efficiency, they offer some interesting safety features, the fuel being composed of microcapsules which ensure the containment of radioactive products in small volume.

CONCLUSIONS

The future of nuclear energy is foreseen in the coming decades through the industrial deployment of Generation III reactors with, in parallel, the development of Generation IV systems. This will allow a large increase of the nuclear power capacity in the world and a sustainable source for a significant part of the energy needs at reasonable costs and in a safe and secure way.

## INNOVATIVE NUCLEAR ENERGY SYSTEMS

KAZUAKI MATSUI
The Institute of Applied Energy,
Tokyo, Japan

### LONG-TERM PROSPECT OF NUCLEAR POWER GENERATION

Mid- and long-term prospect of nuclear power generation capacity in Japan is illustrated in the following figure based on the assumption of nuclear share of thirty to forty percent in power generation. In Europe, today's capacity of 120 GWe, about twice that of Japan, remains as is, but some Western countries may revert to the nuclear option and many Eastern countries including the former Soviet Union could expand it to meet fast growing demands. Contrary to the Japanese and European power scenery, the U.S. power appetite still grows steadily.

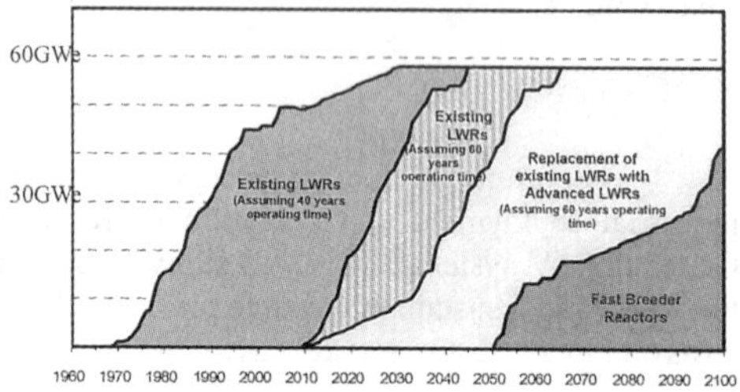


Concerning the whole world, power generation capacity has to match quickly growing power demand mostly due to the developing nations' economic needs. Based on our analysis on electricity in the 21st century under carbon emission control to maintain $CO_2$ concentration at 550 ppm and competitive nuclear cost of 40 mills per kWh, the world nuclear capacity will be about four thousand two hundreds GWe, in other word 4,200 reactors of 1 GWe by the year 2100. This means that the world needs about forty new nuclear reactors every year by the end of this century to provide energy to the world population and maintain atmospheric quality.

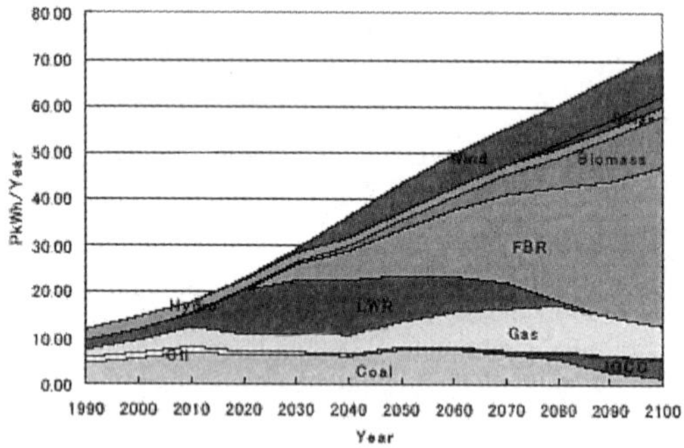


### GENERATION IV NUCLEAR ENERGY SYSTEMS

To accommodate such a large number of nuclear plants in the world, innovative nuclear energy systems and institutions are inevitable. In 2000, the world's leading nuclear countries decided to form the "Generation IV International Forum" to cooperate for R&D's.

- The Generation IV International Forum, or GIF, was chartered in May 2001:

- to lead the collaborative efforts of the world's leading nuclear technology nations,
- to develop the next generation's nuclear energy systems to meet the world's future energy needs.

- The world's first agreement aimed at the international development of advanced nuclear energy systems was signed on February 28, 2005.
- The current GIF members are Argentina, Brazil, Canada, Euratom, France, Japan, Republic of Korea, Republic of South Africa, Switzerland, the United Kingdom, and the United States.
- Russian and Chinese memberships were endorsed in July 2006.
- The Organization of Economic Co-operation and Development (OECD), Nuclear Energy Agency (NEA) as the Technical Secretariat for the GIF and the International Atomic Energy Agency (IAEA) participate in all GIF meetings.

The most promising concepts of Generation IV systems were selected based on 15 criteria of 4 goal areas of safety and reliability, economics, sustainability and proliferation resistance and physical protection. Six concepts follow with my personal observations.

- Very High Temperature Gas Cooled Reactor (VHTR) was the most popular among GIF concepts by member states because of possible Hydrogen production capability.
- Sodium Cooled Fast Reactor (SFR), once a lame duck, becomes a champion if a Gen-IV has to be demonstrated in 2020s.
- Supercritical Water Cooled Reactor (SCWR), only one light water cooled reactor, needs to clarify material challenges in supercritical water environment.
- Lead Cooled Fast Reactor (LFR) was a window for possible Russian involvement to GIF and small scale reactor development.
- Gas Cooled Fast Reactor (GFR) - why?
- Molten Salt Cooled Reactor (MSR) seems interesting but unpopular, maybe a window for Thorium?

Following the multi-national Framework Agreement of February 2005, the first SFR System Arrangement was signed in February, 2006 in Fukui, Japan, and Project Arrangements under SAs will follow. System and Project R&D plans are mostly ready to start. A new chairman with a new leadership will be in place from December 2006. Following are some additional plans:

- Strengthen collaboration with IAEA INPRO, mainly in infrastructure, safety and proliferation issues,
- Set up regular mechanisms to communicate with the senior regulators,
- Start dialogue with Senior Industrial Panel,
- Start thinking about “demo.”

## TOWARD THE END OF CENTURY, 2100

Growing world economy and population exist mainly in developing countries such as BRICS and others. Growing concerns of climate change by greenhouse gas emission and environmental destruction. "Nuclear has a mission to relieve our Globe."

Therefore, nuclear technology should be accessible, available, and affordable everywhere in the world. In this context, Safety, Economics and Proliferation Resistance should be woven in the characteristics of Innovative Reactors.

Challenges are:

- Safety is a matter of surety, but be reasonably assured from the point of comparative risk,
- Economics is a matter of business competitiveness but be on the fair ground with externalities and Life Cycle Analysis,
- Waste can be minimized, but it will be in accordance with people's preference and economics of Fuel Cycle with recycling waste management,
- No unique solution for proliferation resistance, but a real challenge for thousands of nuclear power reactor worldwide remains.

We anticipate thousands of reactors and fuel cycle facilities in the world at the end of century whether we like it or not. Most additional reactors will be needed for their survival and a better life of presently developing countries. We face the following serious questions to address:

- Do we have a clear mind or picture of thousands of reactors in the latter of the 21$^{st}$ century?
- Are they LWR or FBR or HTGR or what? (Of course Gen IV)
- Who will take responsibility for international nuclear fuel cycle which produce and reprocess fuels, manage wastes?
- Where will the waste repositories be?
- Breeders will most probably be needed, and fast spectrum reactors. This means Plutonium. How will it be managed?

There are some ideas addressing proliferation issues. One of them is "Minor Actinides containing Fuel" to prevent access to fuel due to its high radioactivity and heat emissions. Another example is "Nuclear Energy Park" where Natural Uranium or Thorium of Input and Energy/Electricity, Hydrogen, Heat as Output with Essentially No Waste outside the park. This ideal system has been

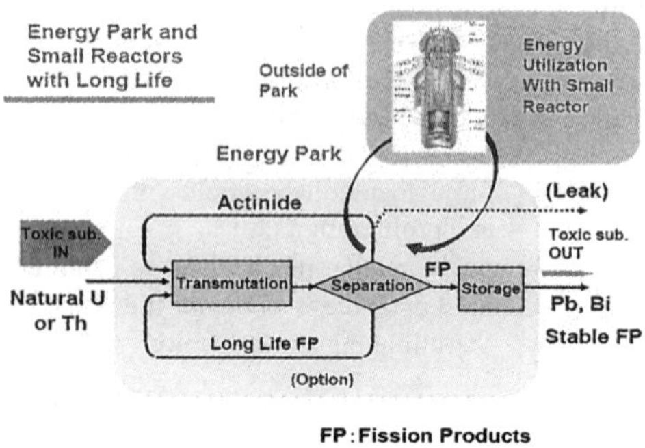

named as "Future Nuclear Equilibrium State" or "Self-Consistent Nuclear Energy System (SCNES)."

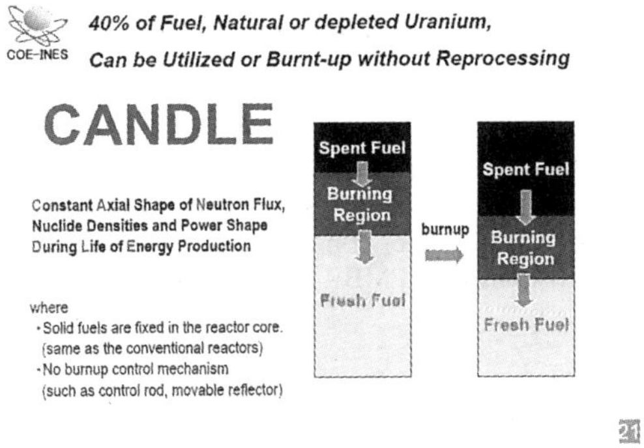


If very-very high burn-up of fuel can be achieved such as 40% of natural fissile material, enrichment and reprocessing of spent fuel would be less attractive. "CANDLE" is one of the ideas with ultra-long cycle of decades and no fuel re-load.

Many of these ideas are just at the embryo stage, but I believe we should keep ambitious R&D options open for a brighter future.

REFERENCES TO "CANDLE" CONCEPT

1. H. Sekimoto and S. Miyashita (2006). "Startup of "Candle" Burnup in Fast Reactor from Enriched Uranium Core," *Energy Conversion Management*, 47[17], 2772-2780.
2. H. Sekimoto, Y. Udagawa (2006). "Effects of Fuel and Coolant Temperatures and Neutron Fluence on CANDLE Burnup Calculation," *Journal of Nuclear Science and Technology*, 43[2], 189-197.

# THE PEBBLE BED MODULAR REACTOR PROJECT

JOHAN SLABBER
Chief Technology Officer, PBMR (Pty) Ltd, Centurion, South Africa

THE PROJECT

The Project PMBR (Pty) Ltd intends to build a demonstration module at Koeberg near Cape Town and an associated fuel plant at Pelindaba near Pretoria. The goal is to commercialize and market 165Mwe modules in single- or multi-module configuration for the local and export markets. This will transform PMBR (Pty) Ltd into a world-class company.

THE PBMR COMPANY

- Current staff of 650 people; growing to 900 in 2009
- Worlds largest technology nuclear development team
- 2000 people in total involved in project worldwide
- More than 50 PhDs and with 2.5 degrees per person on average
- Shareholders – SA Gov, Industrial Development Corporation, Eskom and Westinghouse
- Worlds Best suppliers
- SA proud of PBMR team: we know what we are doing

Supportive South African Government
- PBMR announced as National Strategic Project
- Committed to nuclear power, in the form of PBMR, as a significant portion (20-25%) of future electric supply
- Approved funding for developing Demonstration plant for commissioning in 2010
- Affirmed program to develop human capital to help enhance sustainable development in Africa

Growth in SA Electricity Demand
- Compound annual demand growth of 3.4% per year since 1992 (2004 peak 34,210MW compared to 22,640 in 1992)
- National Energy Regulator's Integrated Resource Plan shows:
  - Projected growth of ~2.8%/annum to 2022
  - New build capacity of over 20,000MW required by 2022
  - Growth at 4% would require ~40,000MW
- Eskom predicts growth in demand of 1200 MW p/a over next 20 years

Diversity of Energy Sources

- The expansion of generating capacity in South Africa should include a diversity of energy sources, including coal, hydro, nuclear, wind, solar, wave, tidal, etc.
- To meet energy development challenges, South Africa needs to optimally use all energy sources available and vigorously pursue energy efficient programmes

World Electricity Market

- World capacity in January 2002 was 3,465GW (~100 x Eskom)
- World average growth of 3% per annum since 1980 (equates to 600 PBMRs per year)
- MIT forecasts world demand to triple by 2050
- Current world spending is about $100bn per year on new power stations
- Fossil fuel costs have risen dramatically
- Environmental pressure is increasing

Resurgence of Nuclear Energy

- Thirty nuclear plants are being built today in 12 countries around the world
- Green guru James Lovelock and Greenpeace co-founder Patrick Moore calls for "massive expansion" of nuclear to combat global warming (May 2004)
- George Bush signs energy bill and describes nuclear as one of the nation's most important sources of energy (Aug 2005)
- U.S. Energy Secretary Samuel Bodman predicts nuclear will "thrive as a future emission-free energy source" (April 2005)
- Tony Blair proposes new generation of nuclear plants to combat climate change (July 2004)
- China plans to build 27x1000MW nuclear reactors over the next 15 years
- India plans a ten-fold nuclear power increase
- France to replace its 58 nuclear reactors with EPR units from 2020, at the rate of about one 1600 MWe unit per year.
- IAEA predicts at least 60 new reactors will become operational within 15 years

Views on Nuclear

*"How are we going to satisfy the extraordinary need for energy in really rapidly developing countries? I don't think solar and wind are going to do it. We are going to have to find a way to harness all energy supplies that includes civilian nuclear power."*

Condoleezza Rice, U.S. Secretary of State, Sept 2005

Views on PBMR

*"The long-term future of power reactors belongs to very high temperature reactors such as the PBMR."*

Nils Diaz, Chairman of the U.S. Nuclear Regulatory Commission, July 2004

*"I feel we made a mistake in halting the HTR programme." Klaus Töpfer, Germany's former Minister of Nuclear Power and Environment.*

Davos, January 2003

*"The PBMR technology could revolutionize how atomic energy is generated over the next several decades. It is one of the near-term technologies that could change the energy market."*

Prof. Andrew Kadak, Massachusetts Institute of Technology, January 2002

*"Little old South Africa is kicking our butt with its development of the PBMR. This should be a wake-up call for the U.S."*

Syd Ball, senior researcher at Oak Ridge National Laboratory, 11 June 2004

PBMR Uniquely Positioned

- Non-$CO_2$ emitting option in climate change debate
- Inherent safety reducing regulation burden
- Small unit flexibility with short construction periods
- Accepted as very low nuclear proliferation risk
- Close enough to commercial deployment to achieve "first to market" dominance
- Eskom build program of at least 20 000 MW over the next 15 years

Advantages to South Africa

- Ability to site on coast, away from coal fields
- RSA based "turnkey" supplier allows localisation of manufacture on sub-contractors
- Locally controlled technology limiting foreign exchange exposure
- About 56,000 local jobs created during full commercial phase
- R23 billion net positive impact on Balance of Payments

Salient Features

- Can be placed near point of demand
- Small safety zone
- On-line refuelling
- Load-following characteristics
- Process heat applications
- Well suited for desalination purposes
- Synergy with hydrogen economy

WHY PBMR COULD BE THE FIRST SUCCESSFUL COMMERCIAL GENERATION IV REACTOR

PBMR Design Objectives

(Source:A Technology Roadmap for Generation IV Nuclear Energy Systems U.S. DOE Dec 2002)

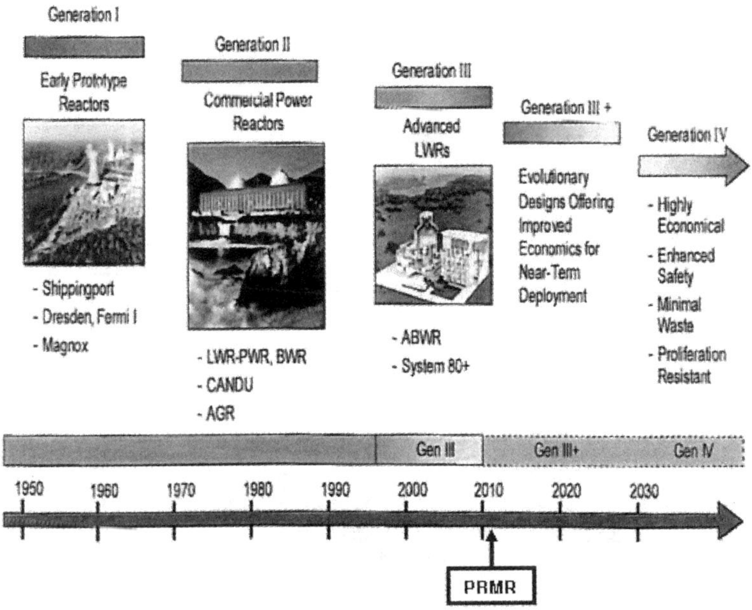


- PBMR design objectives are in line with the requirements of the Generation IV nuclear energy systems
- According to the current planning should come on line in 2010

The Technology

The PBMR is a small-scale, helium-cooled, direct-cycle, graphite-moderated, high-temperature reactor (HTR). Although it is not the only gas-cooled HTR currently being developed in the world, the South African project is internationally regarded as the leader in the power generation field.

Schematic Diagram of Power Conversion Unit

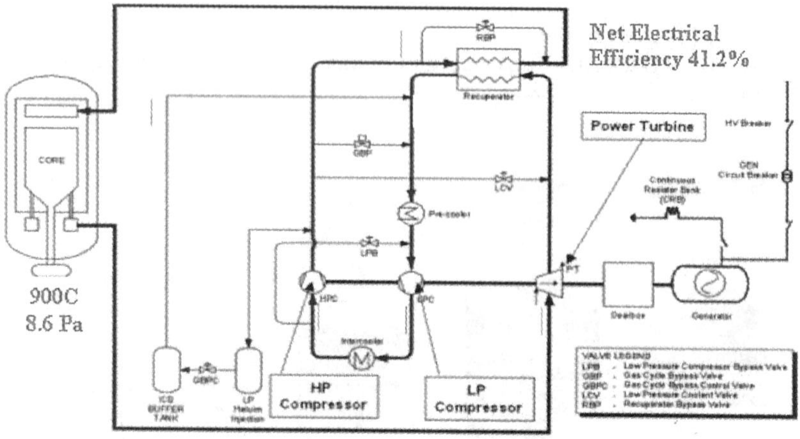


Physical Layout of PBMR Main Power System

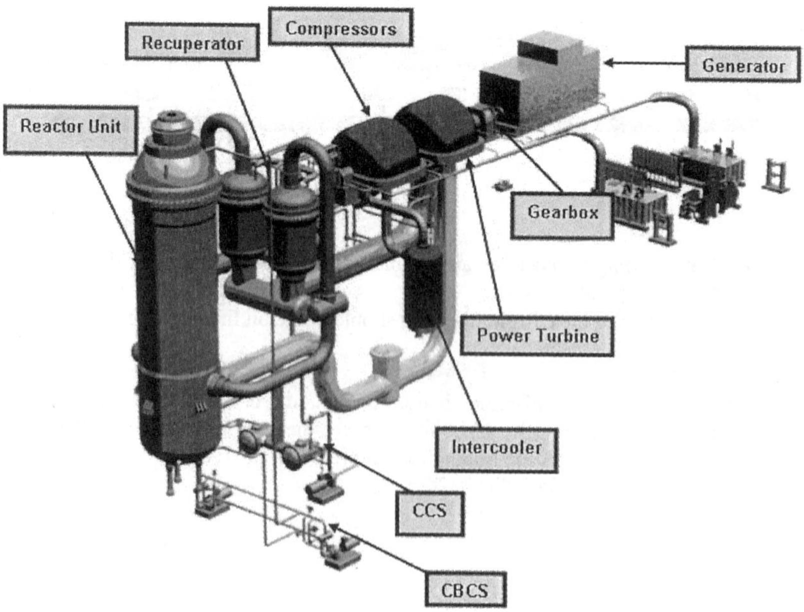

PBMR Fuel Design

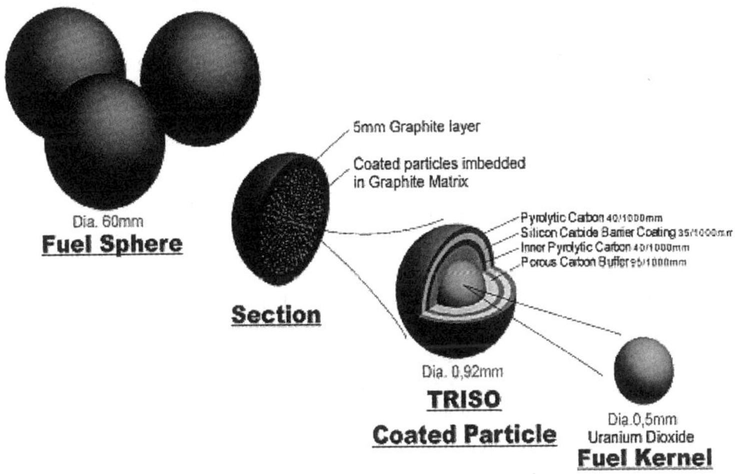

Fuel Handling System

- PBMR spent fuel to be kept on site
- A 165 MWe module will generate 32 tons of spent fuel pebbles per year, about one ton of which is uranium
- Fuel spheres are “pre-packaged” for final disposal purposes
- Draft nuclear waste management policy issued for public comment in 2003
- 

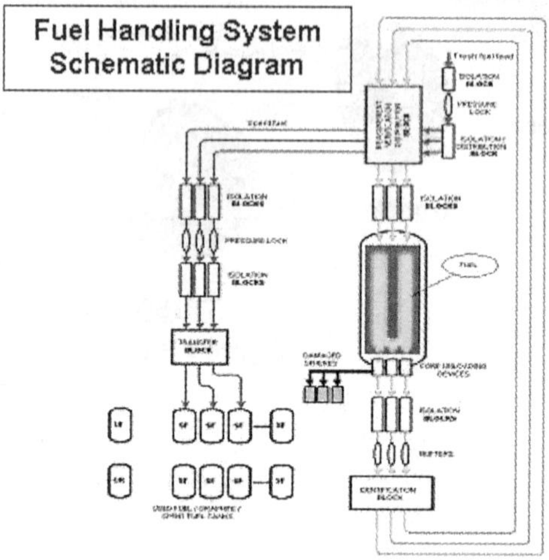


Reactor Safety Fundamentals

- Main safety objective is to preserve the integrity of the fuel under all postulated events
- To reach this objective it is therefore necessary to ensure that the fuel does not heat up or is degraded by some other means to a point where the activity retention capability is lost
- The ultimate fuel temperature and the fuel element structural characteristics determine the activity retention capability during operation and following an event.
- Three factors determine the ultimate fuel temperature during operation and following an event:
  - Production of heat in the core
  - Removal of heat from the core
  - The heat capacity of the core

PBMR Core Size and Shape

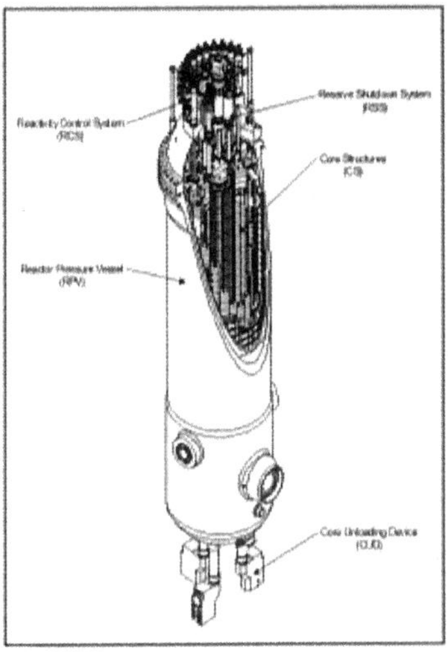


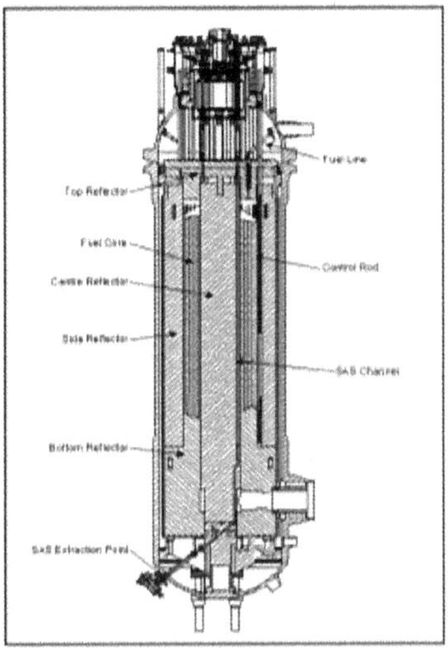


- All ceramic
- Low power density
- Large heat capacity

PBMR Passive Decay Heat Removal Mechanisms

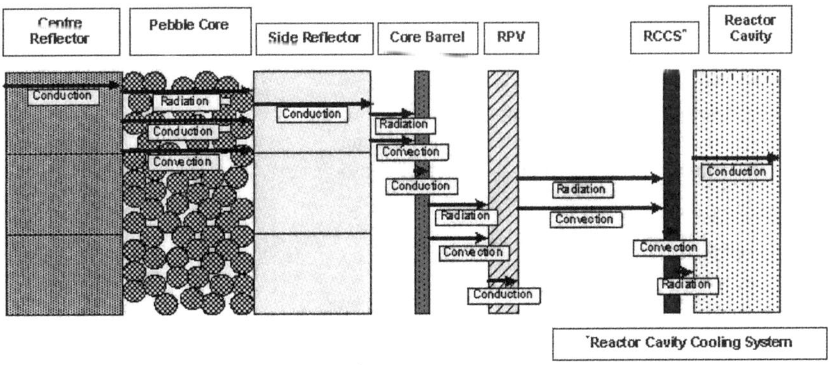

Fuel Temperature Rise Following a Depressurized Loss of Forced Cooling

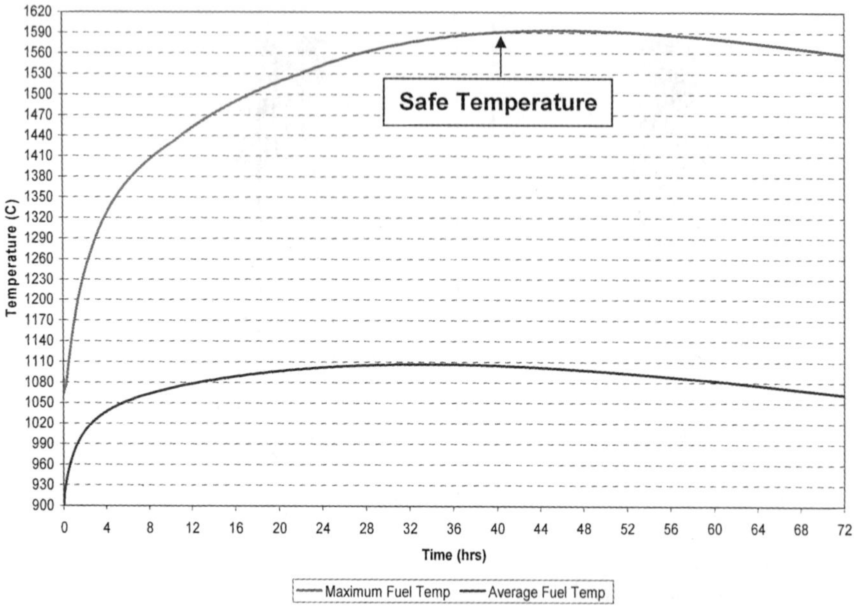


Safety Features Summary

- Simple design base
- If fault occurs, system shuts itself down
- The transfer medium (helium) is chemically inert
- Coated particle provides excellent containment for the fission product activity
- No need for safety grade backup systems
- No need for off-site emergency plans
- License application for small safety zone
- Inherent safety proven during public tests

Demonstration Plant Building

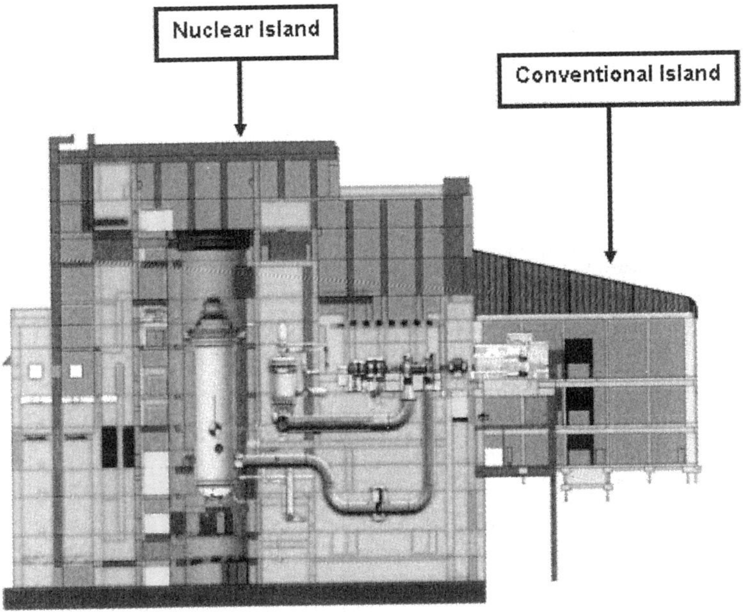


Demonstration Plant Building

Multi-Module Concept

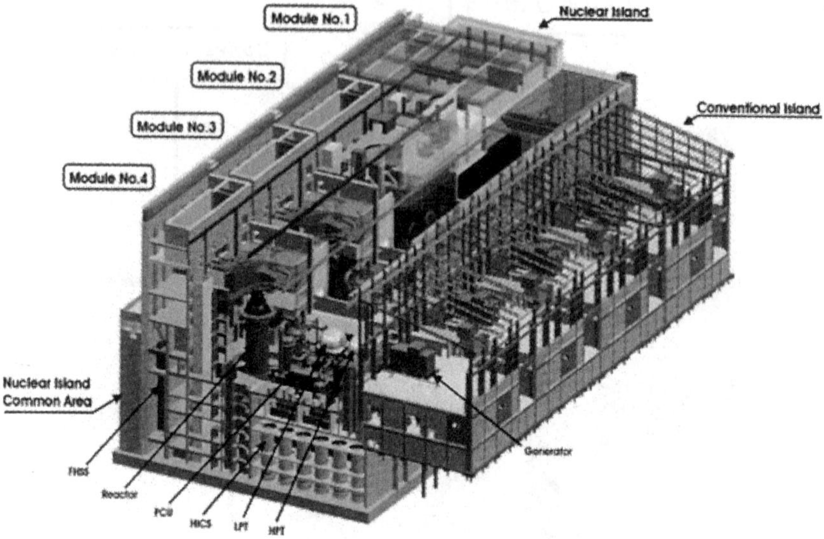


What Happens Next?

- EIA reopened and public meetings started
- Construction to start in 2007 (subject to positive conclusion of regulatory processes)
- Demonstration module and fuel plant to be completed by 2011
- First commercial modules to be completed by 2013

# THE CLIMATE CHANGE IMPERATIVE AND THE FUTURE OF NUCLEAR ENERGY

STEVE FETTER
School of Public Policy, University of Maryland, USA

Much of the recently renewed interest in nuclear energy is driven by the desire to reduce emissions of carbon dioxide and thereby mitigate global climate change. The climate-change imperative is described well in Article II of the Framework Convention on Climate Change:

> *"The ultimate objective of this Convention...is to achieve...stabilization of greenhouse gas concentrations in the atmosphere at a level that would prevent dangerous anthropogenic interference with the climate system. Such a level should be achieved within a time frame sufficient to allow ecosystems to adapt naturally to climate change, to ensure that food production is not threatened and to enable economic development to proceed in a sustainable manner."*

"Dangerous anthropogenic interference" now seems inevitable, inasmuch as there is growing evidence of changes in climate that are already well underway, which are likely to damage or destroy vulnerable ecosystems. The main challenge is not to avoid "dangerous" climate change, but to avoid "catastrophic" change.

There is, at yet, no consensus on the level at which carbon dioxide and other greenhouse gases should be stabilized. Most climate-change impact studies focus on the effect of a doubling of the carbon dioxide concentration, from the preindustrial level of 275 parts per million by volume (ppmv) to 550 ppmv. The resulting change in global-average surface temperature is expected to be about 3°C. For comparison, the change in average surface temperature from the last ice age to the current interglacial period was roughly 5°C, which demonstrates that seemingly modest changes in average temperature can correspond to dramatic changes in environmental conditions.

More significant than the change in global-average temperature—but more difficult to quantify—are the temporal patterns of regional changes in temperature, precipitation, cloudiness, soil moisture, winds and ocean currents, and changes in variability as well as the mean. There is good reason to believe that a doubling of carbon dioxide would produce a significant increase in sea level and in the intensity of storms, significant changes in the availability of fresh water, and significant decreases in biodiversity.

But, according to today's best estimates, it seems that a doubling of carbon dioxide is unlikely to trigger truly catastrophic events, such as a rapid collapse of the West Antarctic ice sheet (which would raise sea level by 6 meters), a collapse of the global thermohaline circulation (which would cause temperatures in Europe to plummet), or a run-away positive feedback in which the melting of tundra and forest die-back release of huge amounts of carbon dioxide and methane (which would trigger much larger increases in temperature).

Thus, it would seem that a not-unreasonable stabilization target would be a doubling of the carbon dioxide concentration. But one must bear in mind that carbon dioxide is not the only greenhouse gas. Other gases—notably methane, nitrous oxide, and chlorofluorocarbons and their non-ozone-depleting substitutes—also contribute to greenhouse warming. Even in a world of tight controls, it is unlikely that these could be stabilized at levels equivalent to less than 100 ppmv of carbon dioxide. Thus, carbon dioxide must be limited at about 450 ppmv to stabilize greenhouse gas concentrations at an "equivalent doubling." One should also bear in mind that although today aerosol emissions are believed to have a cooling effect, efforts to control air pollution and acid deposition, as well as carbon emissions, will result in steep reductions of global aerosol emissions—just as we have witnessed such reductions in developed country aerosol emissions in the last 25 or so years.

Figure 1 shows past fossil-fuel emissions of carbon and allowable future emissions for stabilizing carbon dioxide concentrations at 450 and 550 ppmv. Also shown are reference scenarios of carbon emissions developed by the Intergovernmental Panel on Climate Change (IPCC), based on various assumptions about population, economic growth, and technological change. The concentration of carbon dioxide in the atmosphere is determined primarily by cumulative emissions (i.e., the area under the curve). In the median reference scenario, 550 billion tons of carbon (GtC) are emitted due to fossil-fuel burning from 2000 to 2050, and 1200 GtC from 2000 to 2100. For comparison, stabilization at 450 ppmv would permit emissions of only 300 and 550 GtC, respectively; for stabilization at 550 ppmv, the corresponding emissions are 450 and 900 GtC.

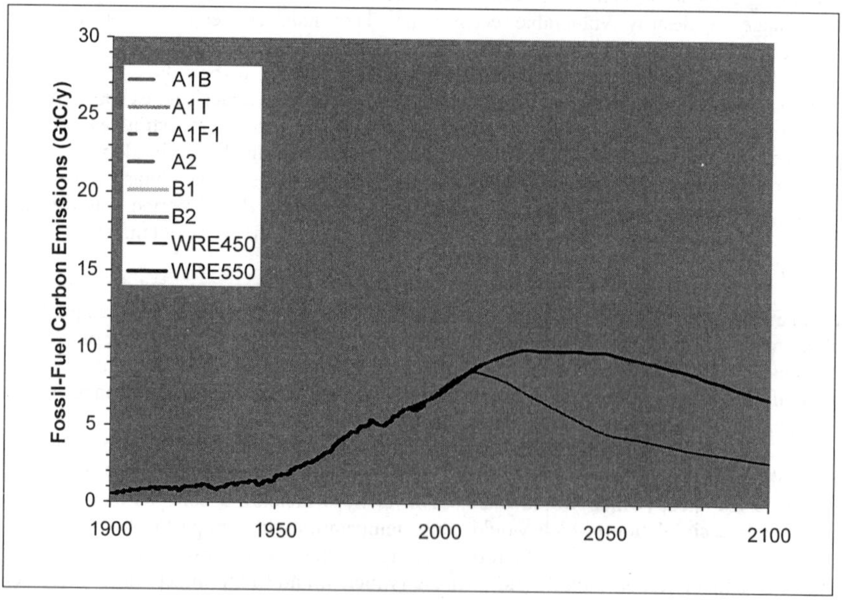


*Figure 1. Fossil-fuel carbon emissions, 1900-2000 and IPCC reference emission scenarios (A1, A2, B1, B2) and stabilization scenarios (WRE450 and WRE550), 1990-2100.*

To put this in perspective, a large coal-fired power plant with an electrical output of 1 gigawatt ($GW_e$) emits about 1 million tons of carbon per year. Thus, stabilization at 450 ppmv would require displacing roughly 200,000 gigawatt-years ($GW_ey$) of coal-fired plant operation by 2050, and 700,000 $GW_ey$ by 2100. Stabilization at 550 ppmv would decrease the required reductions by roughly a factor of two.

The following four options are available to achieve these reductions:

- Demand reductions resulting from increased efficiency and fossil-fuel prices
- Capture and storage of carbon from fossil fuels
- Renewable energy (wind, solar, biomass)
- Nuclear energy

Modeling exercises typically show that all of these options would play a significant role in stabilizing carbon dioxide concentrations at 450 to 550 ppmv. The first option—increased efficiency and prices for fossil-fuel prices—is essential. At least two of the three energy supply options are needed. Of these carbon-free supply options, only one—nuclear energy—is deployed on a large commercial scale today.

Today nuclear power supplies about 16 percent of global electricity, and about 7 percent of total world energy demand. In order to make a substantial contribution to mitigating carbon emissions, nuclear energy would have to grow significantly over the next 50 years and beyond. Consider, for example, growth from the current level of about 360 $GW_e$ to 1500 $GW_e$ in 2050 and 4500 $GW_e$ in 2100.[1] Reactors would come on line at a rate of about 35 $GW_e/y$ in 2025 (roughly equal to the historical peak in 1984), increasing to 70 $GW_e/y$ in 2050. Nuclear generation would total about 170,000 $GW_ey$ from 2000 to 2100. Assuming that the nuclear reactors displace traditional coal-fired plants, they would supply about one-quarter of the reductions needed to stabilize carbon dioxide concentrations at 450 ppmv, and about half of the required reduction for stabilization at 550 ppmv.

In order for nuclear power to expand substantially, four issues must be addressed. First, nuclear electricity must be economically competitive with alternative carbon-free sources. Second, nuclear must be judged adequately safe, both by electricity producers and consumers. Investors will not build a reactor if they believe their investment might be lost in an accident, and a single serious accident anywhere could be sufficient to halt the growth of nuclear everywhere. Third, there must be sufficient long-term nuclear waste storage to prevent continuing uncertainties about ultimate disposal from preventing the building of large numbers of new reactors. Finally, an expansion of nuclear energy must not produce corresponding increases in the spread of nuclear weapons, risks of sabotage, or the theft of nuclear materials.

Over the next 50 years, these concerns are best addressed by advanced light-water reactors (LWRs) operating on a once-through fuel cycle with long-term storage of spent fuel. The LWR is the most mature reactor technology, with licensed designs and readily available production infrastructure. The LWR is the only nuclear technology that can be deployed on the required scale in the 2010-2030 time frame. Advanced or so-called "third generation" light water reactors (LWRs) are evolutionary improvements of the reactors most commonly deployed today. Construction and operation and maintenance costs are relatively well-known, and there is reasonable confidence that advanced LWRs can produce electricity at costs that will be competitive with carbon-free alternatives—about $70 per megawatt-hour (MWh), and possibly as little as $50/MWh with modest reductions in capital cost, construction time, and operations and maintenance costs.

Regarding safety, advanced LWRs are estimated to have accident probabilities 10 to 100 times smaller than the current generation of LWRs. If these probabilistic risk assessments are correct, this would mean a probability of core damage of less than one per million reactor-years, and a probability of a large release of radioactivity (sufficient to produce one or more off-site deaths) of less than one per ten million reactor-years. In the nuclear growth scenario outlined above, there would be 0.3 to 3 percent risk of core damage at one of the more than 1000 reactors operating by 2050, and 0.03 to 0.3 percent risk of a large release. In my view, these are acceptable risks for both investors and the public, given the associated benefits of climate change mitigation.

The once-through fuel cycle will continue to be less costly than reprocessing and recycle for at least the next 50 years, and probably through the remainder of the century. All fuel cycles require a geologic repository for the disposal of either spent fuel or high-level reprocessing wastes. There is now no doubt that geologic disposal can be both safe and economical. Although repositories are being built in several countries, none have received spent fuel or high-level wastes from commercial nuclear reactors. The most pressing need, therefore, is centralized national or international dry storage for spent fuel. Spent fuel can be stored safely, securely, and inexpensively in such facilities for 50 to 100 years at low cost—less than one percent of the cost of nuclear-generated electricity. In 50 years, the future of nuclear energy will be much clearer, and we will know whether the stored spent fuel should be reprocessed or placed in a repository. In the meantime, the possibility of international geologic disposal should be explored, to avoid the need for every state with nuclear power to develop its own repository.

Advanced LWRs operating on a once-through fuel cycle are relatively easy to safeguard and are highly resistant to diversion and theft. The spent fuel assemblies are large and can be monitored easily using video cameras; the diversion of an assembly would be easy to detect. Although the spent fuel contains plutonium, very high levels of heat and radiation protect against theft for at least 150 years.

The main proliferation issue lies at the front end of the fuel cycle: the uranium enrichment required to produce the low-enriched uranium (LEU) fuel. In the MIT scenario, 16 countries would have at least 10 $GW_e$ of nuclear capacity, each requiring at least 1.5 million separative work units (SWU) per year of enrichment capacity. For comparison, only 5000 SWU are needed to produce enough high-enriched uranium (HEU) for one nuclear weapon. Enrichment plant safeguards could be improved to ensure timely detection of a significant diversion of material, or of *any* production of HEU.

The problem is that any country capable of building a large commercial LEU enrichment facility could build a much smaller facility for the production of HEU. Small clandestine centrifuge enrichment facilities would be extremely difficult to detect. For this reason, it is essential to limit the spread of enrichment technology and commercial enrichment facilities. This might be accomplished by providing guaranteed fuel supply to countries that forswear uranium enrichment (and spent-fuel reprocessing). Still more attractive would be an agreement by the fuel supplier state to take back the spent fuel and assume all associated waste disposal burdens.

Although I believe that advanced LWRs offer the best nuclear-powered prospect for significant reductions in carbon emission for the next 50 years, it is worthwhile to invest in research to develop alternatives. Two concepts that may have significant advantages over advanced LWRs are (1) gas-cooled graphite-moderated reactors, and (2) small, long-lifetime, sealed-core reactors. Several variants of the former have been developed in previous decades. This concept promises a much higher degree of safety than the LWR, perhaps eliminating the possibility of a large release of radioactivity (assuming that graphite fires can be excluded). Gas-cooled reactors also can operate at much higher temperatures, allowing higher thermal efficiencies and correspondingly lower cost of electricity.

The small sealed-core reactor concept—sometimes referred to as the "nuclear battery"—is far less developed. Several U.S. national laboratories, as well as groups in Russia and Japan, are developing lead-cooled fast reactors with generating capacities ranging from 20 to 200 $MW_e$, and core lifetimes from 15 to 30 years. The reactor would be delivered to a prepared site as a sealed unit, ready to be installed and brought on-line. At the end of the core life, the entire reactor would be returned to the manufacturer for replacement. This concept also promises a high degree of safety. The host country would require a lower degree of nuclear expertise and no fuel-cycle services, making the concept more suitable for smaller developing countries and potentially far more proliferation resistant. Although the small size and very-high-burnup fuel tend to make this concept more expensive than the LWR, substantial economies of scale might be found in the mass production of such units in factories—much like airplanes are mass produced.

Looking beyond 2050, if nuclear power continued to grow (e.g., to 4500 $GW_e$ by 2100), the price of uranium would likely increase to the point that alternatives to thermal reactors operating on a once-through fuel cycle would become attractive. The most obvious alternative is the fast breeder reactor, which would breed plutonium from natural uranium. Another possibility is the molten-salt reactor operating with a thorium fuel cycle. Although it is prudent to begin research on lowering the costs and increasing the proliferation resistance of these alternatives, there is no need to press for early commercialization of either fast reactors or reprocessing and recycle.

## REFERENCES

1. This is consistent with scenarios developed in MIT, *The Future of Nuclear Power* (2003), and Son Kim and Jae Edmonds, "Nuclear Energy in a Carbon-constrained World" (University of Maryland, November 2005).

## THE NUCLEAR FUEL CYCLE: PATHWAY TO SUSTAINABILITY

PHILLIP FINCK
Applied Science and Technology, Argonne National Laboratory
Argonne, USA

### NUCLEAR POWER: SUSTAINABILITY ISSUES

To ensure a sustainable future for nuclear energy, several requirements must be met. These include spent nuclear fuel (waste) management, uranium supply and economics and nonproliferation.

### NUCLEAR FUEL MANAGEMENT OPTIONS

Three options are being considered for disposing of spent nuclear fuel: the once-through cycle is the U.S. reference; limited recycle has been implemented in France and elsewhere and is being deployed in Japan; and full recycle (also known as the closed fuel cycle) is being researched in the U.S., France, Japan and elsewhere.

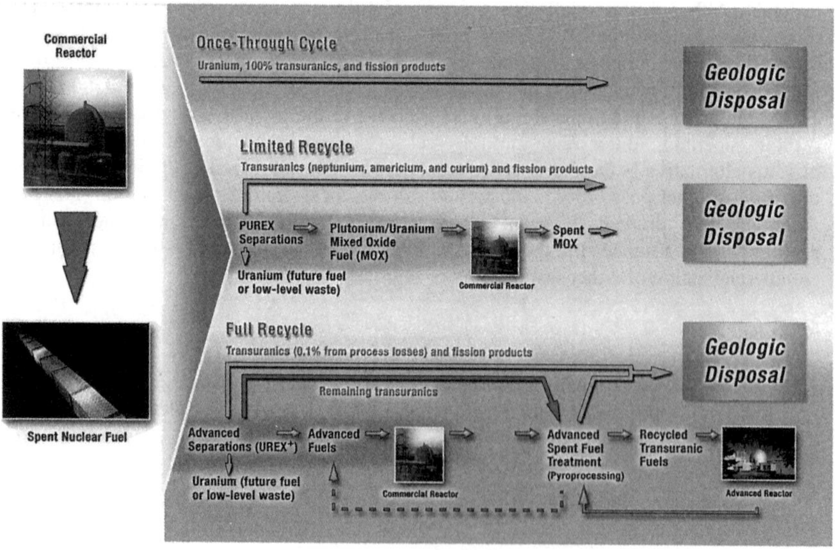


### Issues and Solutions

The issues and solutions concerning sustainability of the nuclear fuel cycle include energy utilization, proliferation, waste minimization, and the balance of fast and thermal reactors.

- Energy utilization
  - Efficiency
  - Recycle for fissioning of key isotopes and (eventually) breeding
- Proliferation
  - Destruction/stabilization of key isotopes
  - Better controls in the fuel cycle
- Waste minimization
  - Efficiency
  - Recycle for destroying key isotopes
- Fast and thermal reactors can work in symbiosis to achieve these goals
  - Thermal reactors preferentially destroy fissile isotopes
  - Fast reactors destroy all isotopes
  - Balance of thermal/fast reactors is decided on economic and policy considerations – at this point U.S. is not considering thermal recycle

TRANSMUTATION IMPACT OF ENERGY SPECTRUM

The balance of thermal and fast reactors affects the transmutation of fissile isotopes. Fissile isotopes are likely to fission in both thermal and fast spectra. The fission fraction is higher in the fast spectrum. Significant (up to 50%) fission of fertile isotopes occurs in the fast spectrum. One of the key factors is the behavior of Pu-240. The net result is less higher actinide generation in a fast reactor.

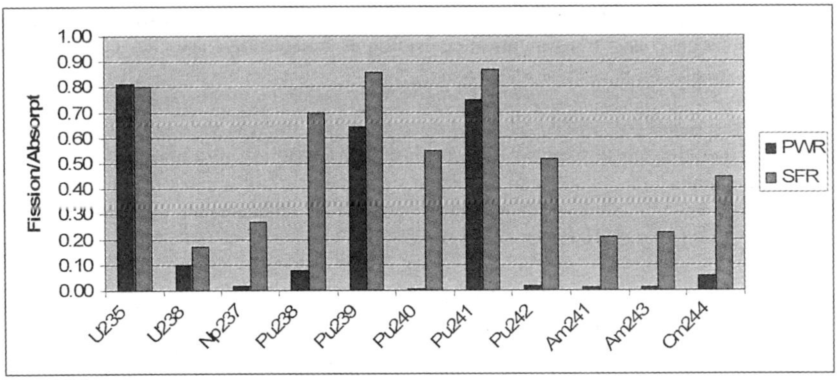

Fast Reactor with Closed Fuel Cycle Was Key to Conception of Nuclear Power

| | |
|---|---|
| Fermi: | The vision to "close" the fuel cycle |
| 1950s: | First electricity-generating reactor: EBR-I with a vision to "close" the fuel cycle for resource extension |
| 1960s-1970s: | Expected uranium scarcity—significant fast reactor programs |
| 1980s: | Decline of nuclear – uranium plentiful – two paths:<br>*– USA (and others): once through cycle and repository*<br>*– France, Japan (and others): limited recycle to mitigate and delay waste disposal. Continue closed cycle R&D* |
| Late 1990s: | Rebirth of closed cycle research and development for improved waste management (USA) |
| Now: | Long-term energy security and the role of nuclear energy |

## POTENTIAL BENEFITS OF CLOSED FUEL CYCLE: WASTE MANAGEMENT

Only three options are available for the disposal of accumulating spent nuclear fuel:

- Build more ultimate disposal sites like Yucca Mountain.
- Use interim storage technologies as a temporary solution.
- Develop and implement advanced fuel cycles, consisting of separations technologies that separate the constituents of spent nuclear fuel into elemental streams, and transmutation technologies that destroy selected elements and greatly reduce repository needs.

A responsible approach to using nuclear power must always consider its whole life cycle, including final disposal. We consider that temporary solutions, while useful as a stockpile management tool, can never be considered as ultimate solutions. It seems prudent that the U.S. always have at least one set of technologies available to avoid expanding geologic disposal sites.

- Geologic repositories and a closed fuel cycle provide interdependent benefits.
  - A geologic repository is essential for all spent nuclear fuel management options (once-through, limited, full) for long-term isolation of nuclear waste.
  - The quantity of waste that must be stored is much smaller with a closed fuel cycle (full recycle).

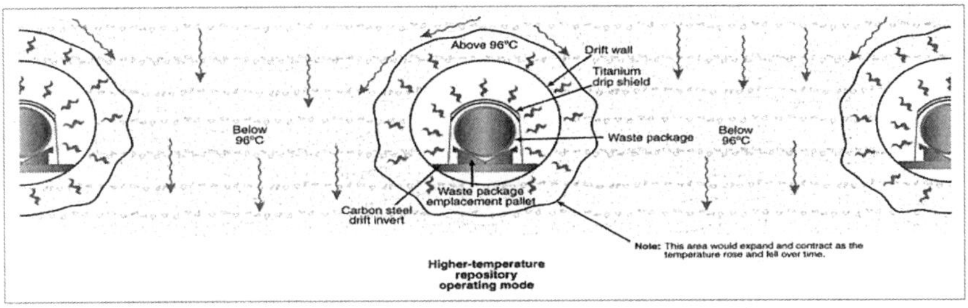


- Certain elements (plutonium, americium, cesium, strontium, and curium) are primarily responsible for the decay heat that causes repository temperature limits to be reached
- Large gains in repository space are possible by processing spent nuclear fuel to remove those elements

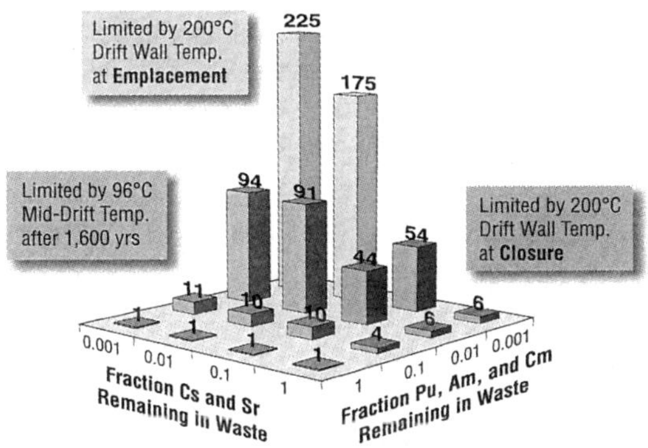


  - The recovered elements must be treated
  - Cesium and strontium must be stored separately for 200-300 years
  - Plutonium, americium, and curium can be recycled for transmutation and/or fission
    - Irradiation in reactors

Estimated Dose Rates

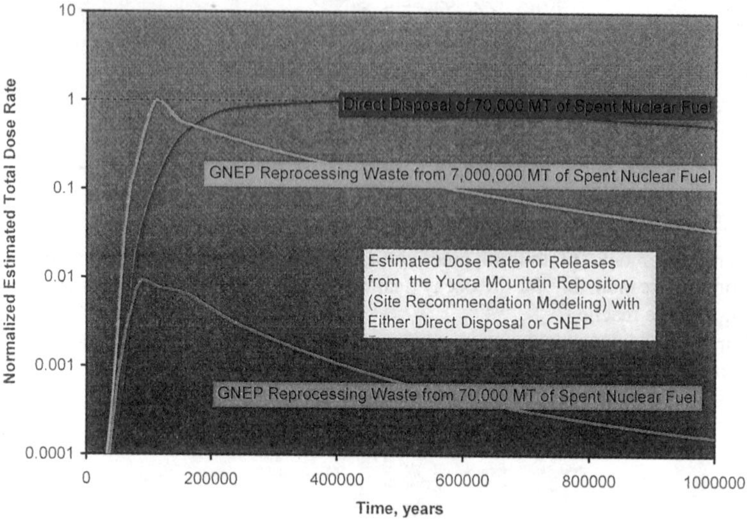


POTENTIAL BENEFITS OF CLOSED FUEL CYCLE: NONPROLIFERATION

- One of the overarching elements of recent proposals is to establish reliable fuel services
  - Establish a consortium of nations with advanced technologies to enable developing nations to acquire nuclear energy economically while minimizing proliferation risk
- This will be enabled by implementing the following technologies:
  - Demonstrate more proliferation-resistant recycling
  - Develop advanced burner reactors
  - Develop enhanced nuclear safeguards
  - Demonstrate small-scale reactors
- An unresolved issue is waste management from the spent fuel that is taken back

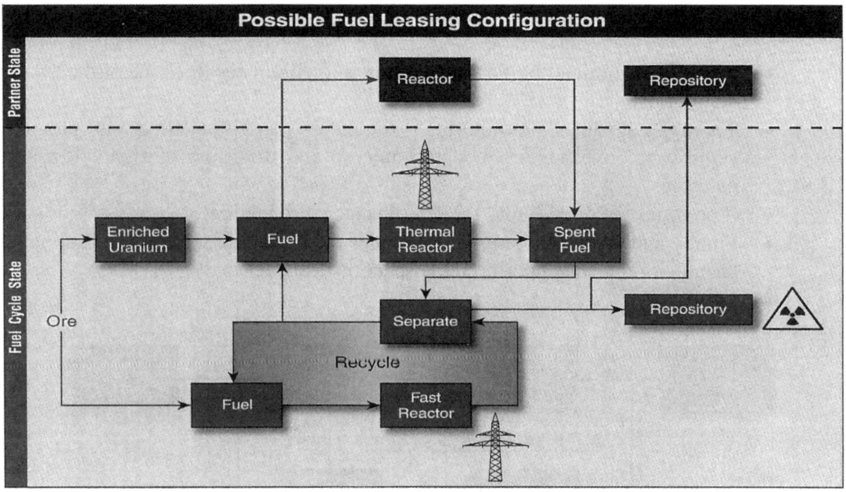


ADVANCED SEPARATIONS: AQUEOUS SPENT FUEL TREATMENT UREX+) FOR WASTE MANAGEMENT AND PROLIFERATION RESISTANCE

Full recycle approaches are being researched in France, Japan, and the United States. This approach typically comprises three successive steps: an advanced separations step based on the UREX+ technology that mitigates the perceived disadvantages of PUREX, partial recycle in conventional reactors, and closure of the fuel cycle in fast reactors.

UREX+ – Is an advanced liquid-liquid extraction process for treatment of light water reactor spent fuel. Similar to PUREX, the irradiated fuel is dissolved in nitric acid. The UREX+ process consists of a series of solvent-extraction steps for the recovery of Pu/Np, Tc, U, Cs/Sr, Am and Cm. Advantages—meets current separations requirements for continuous recycle. Builds on engineering experience derived from current aqueous reprocessing facilities such as La Hague. Disadvantage—cannot directly process short-cooled and some specialty fuels being designed for advanced reactors.

The first step, UREX+ technology, allows for the separations and subsequent management of highly pure product streams. These streams are:

- Uranium, which can be stored for future use or disposed of as low-level waste.
- A mixture of plutonium and neptunium, which is intended for partial recycle in conventional reactors followed by recycle in fast reactors.
- Separated fission products intended for short-term storage, possibly for transmutation, and for long-term storage in specialized waste forms.
- The minor actinides (americium and curium) for transmutation in fast reactors.

The UREX+ approach has several advantages:

- It produces minimal liquid waste forms, and eliminates the issue of the "waste tank farms."
- Through advanced monitoring, simulation and modeling, it provides significant opportunities to detect misuse and diversion of weapons-usable materials.
- It provides the opportunity for significant cost reduction.
- Finally and most importantly, it provides the critical first step in managing all hazardous elements present in the spent nuclear fuel.
-

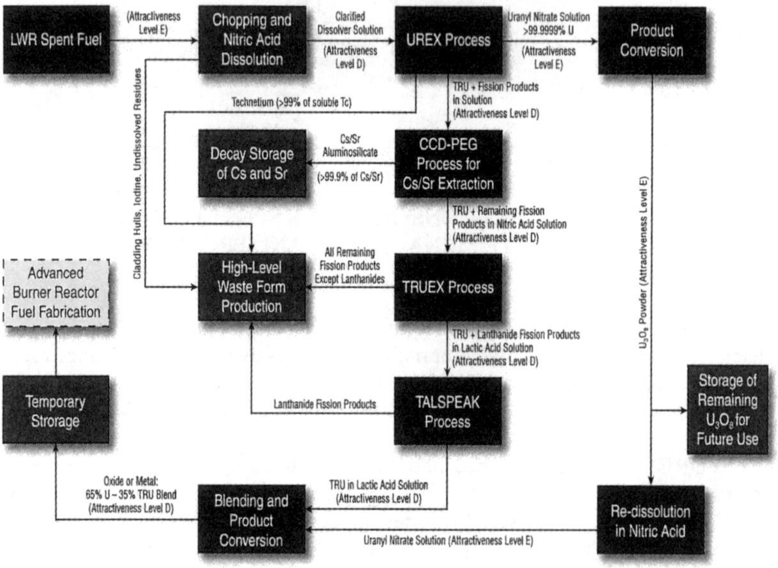

LIGHT WATER REACTOR: ONCE-THROUGH FUEL CYCLE
ELECTRICITY GENERATION COST DISTRIBUTION

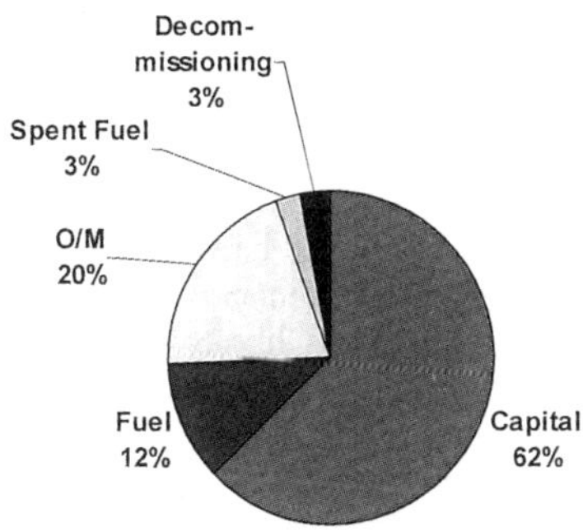


COMBINED LIGHT WATER REACTORS WITH ADVANCED BURNER
REACTOR: CLOSED FUEL CYCLE: ELECTRICITY COST DISTRIBUTION

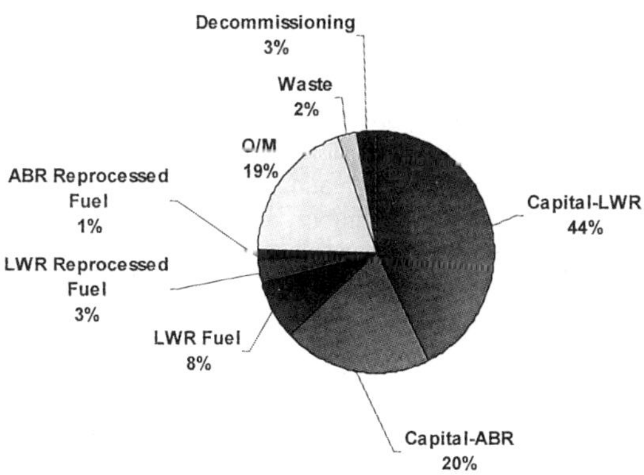

# ROLE AND STATUS OF GEOLOGICAL DISPOSAL

CHARLES MCCOMBIE
Arius Association, Baden, Switzerland

## INTRODUCTION

For many years, nuclear supporters have been talking of a possible nuclear power renaissance. Today there are definite signs that this is finally beginning to happen. New plants are being built or planned in China, Japan, Korea, Finland, France and even the USA. Phase-out policies are being rethought in countries like Sweden, Belgium and Germany. Countries like Vietnam, Indonesia, the Baltic States and even Australia are choosing or debating initiating a nuclear programme.

Support for these nuclear power developments will be strongly influenced by the progress of waste management programmes, especially final disposal. Conversely, the growing realisation of the potential global benefits of nuclear power may well lead to increased support, effort and funding for initiatives to ensure that all nations have access to safe and secure waste management facilities. This implies that large nuclear programmes must make progress with implementation of treatment, storage and disposal facilities for all of their radioactive wastes. For small nuclear programmes (and for countries with nuclear applications other than power generation) such facilities are also necessary. However, for economic and other reasons, these small programmes may not be able to implement all of the required national facilities. Multinational cooperation is needed. This can be realised by large countries providing back-end services such as reprocessing and disposal, or by small countries forming regional or international partnerships to implement shared facilities for storage and/or disposal.

This paper gives a brief summary of the status of how national waste management programmes are progressing (or not progressing) and of how the credibility of multinational concepts is being enhanced by a number of current initiatives. These include Russian proposals for international facilities, the recent Global Nuclear Energy Partnership (GNEP) initiative of the USA, studies on regional repositories in the SAPIERR project (Support Action: Pilot Initiative for European Regional Repositories) and IAEA and EC support for both types of initiative.

Prime conclusions to be drawn are that there is no long-term alternative to geological disposal and that continuing efforts are needed to ensure that the feasibility of implementing safe deep repositories is accepted by scientists, the public and the politicians. However, no deep geological repositories for high radioactive level wastes (HLW) or spent nuclear fuel (SNF) will be operating on the short timescales of the next few years, when key decisions are needed on how to expand nuclear power programmes whilst reducing nuclear threats. The important challenges in this period are therefore storing all sensitive materials safely and securely, and simultaneously working to ensure that safe disposal facilities will be available to all such programmes when they are needed.

# A BRIEF LOOK BACK AT GEOLOGICAL DISPOSAL

## A Well Founded Idea

Geological disposal was not (despite the assertions of some of its opponents) chosen as a "cheap and dirty" option to get the radioactive waste "out of sight and out of mind." The concept of geological disposal is a logical consequence of the easily observable decay of radioactivity with time, which leads to a continuous reduction in toxicity of these wastes. Finite hazardous lifetimes (and low volumes of wastes) led to:

- Development of concepts where environmental protection could be aimed at by isolating wastes from man's surroundings for long enough to allow such decay to occur, and
- A search for environments which showed sufficient stability for the time periods involved: namely thousands or even hundreds of thousands of years.

There are not many environments for which we have evidence of their evolution and their stability over hundreds of thousands of years. Old, deep geological formations are the most obvious candidate environments that can be accessed with today's technology. Other options have, in fact, been considered. A comprehensive document on all these options was published already in 1974.[1] Concepts that have been examined (more than once) include disposal in space, under ice caps, in subduction zones, etc., but all have been judged infeasible or unsafe. Transmutation is still being studied in various countries. In the view of most experts, it may eventually change the nature or quantity of radioactive wastes to be disposed, but transmutation will not remove the need for geological disposal.

Consequently, concepts for geological disposal under the continental earth's crust have been developed over many years and the concept of disposal in deep geological formations was recognised by the U.S. National Academy of Sciences as early as 1957[2] to be the most promising form of confinement for long-lived wastes from the nuclear fuel cycle.

Despite the above historical facts, accusations that nuclear power was started without any consideration having been given to the management of its wastes have often been made by anti-nuclear groups. These opponents have likened the construction of the first nuclear power plants to "building a house with no toilet." The experts in the nuclear community see this differently. They point out also that for many years, or even decades, there was no technical need for disposal. The quantities of high level waste or spent fuel were too small to justify implementing repositories and, in any case, a cooling time of around 40 years was the sensible technical choice.

## Mixed Beginnings

In retrospect, however, there was indeed too little effort invested into organising long-term management and disposal; most attention was devoted to implementing practical measures for handling and storing radioactive wastes safely. This is now recognized as a mistake. Even the famous nuclear pioneer Alvin Weinberg has been quoted as saying *"During my years at ORNL, I paid too little attention to the waste problem. Designing and building reactors, not nuclear waste, was what turned me on . . . [A]s I think about what I would do differently had I to do it over again, it would be to elevate waste disposal to the very top of ORNL's agenda."*

With time, however, things changed: dynamic waste disposal initiatives were started and, paradoxically, the nuclear opponents were in large measure to thank for this. Because they asserted that lack of demonstrated safe technologies for disposal should preclude the use of nuclear power, governments were pressured to demand specific projects that could provide this demonstration.

The first example was in Sweden, where the Stipulation Act of 1977 made credible disposal concepts a pre-condition for the start up of new power stations. This led directly to the establishment of the pioneering KBS project which developed technical disposal concepts that are valid still today. A similar situation resulted in Switzerland when the new Atomic Energy Act of 1975 and associated regulatory requirements demanded demonstration projects before the year 1985 if new nuclear plants were to be introduced to the country, or even if the existing stations could continue operation. These are clear examples of cases where nuclear skeptics or opponents have given a positive impulse to the planning of geological disposal.

There are also striking counter-examples, i.e., cases where nuclear opponents have slowed or stopped any progress in disposal. In the UK, the Government abandoned a HLW disposal in the 1980s in order to avoid public conflicts over drilling sites; in Spain a specific repository siting programme was scrapped for the same reason; in the Netherlands, the Government blocked a highly interesting programme on disposal in salt domes and ruled that storage for at least 100 years was the option to be chosen. The reasons for opposition to progressing repository programmes are diverse. Some people genuinely believe that the safety of deep geological disposals has not been demonstrated sufficiently and that allowing years or decades for further work will produce some as yet undefined better solution—a "magic bullet." Others object for tactical reasons—an accepted waste disposal solution would remove one of their last anti-nuclear arguments, now that operational safety and economics are both clearly favourable.

A real danger resulting from those tactical manoeuvres of opponents is that an "unholy alliance" could result in all efforts on preparing for geological disposal grinding to a halt. By this, I mean that indefinite storage could become the common solution that satisfies both the nuclear opponents (who wish to block a real final solution) and extremists in the nuclear industry (who know very well that the storage option is much less costly than implementing geological repositories). The losers, in this case, are our children and grandchildren, the future generations who then inherit an unsolved problem passed on to them by us because we did too little to clear up our own mess.

## AN IMPORTANT NEW DRIVER

Where are we today on all of the issues influencing efforts made towards implementing deep geological disposal? Unfortunately for the world in general, but productively for waste management, a new and frightening aspect has leapt to the forefront. This is the growing concern about the misuse of nuclear materials by nations that are intent on gaining nuclear weapons capabilities, or even more worrying, the possibility of nuclear terrorist acts. In the recent past—in particular since the terrorist attacks on the USA in 2001—the security issues associated with management of nuclear materials, including wastes, have assumed high profile.

The concerns about the spread of sensitive technologies such as enrichment and reprocessing have correctly taken front place. These concerns have led directly to the Russian and American fuel cycle initiatives described later. However, the back-end of the nuclear fuel cycle cannot be neglected when we are trying to minimise security concerns. Spent nuclear fuel and HLW must be kept away from persons, organisations or governments that might misuse it. A very effective way to make these materials inaccessible is to emplace them in a limited number of highly controlled national or multinational underground facilities. The latter of these options is discussed in more detail later in this paper. First, however, a brief overview is given of how national programmes are progressing with implementation of safe, secure and environmentally friendly final repositories for spent nuclear fuel and HLW.

## STATUS OF GEOLOGICAL DISPOSAL PROGRAMMES

For at least 25 years after the original 1950's publications on the concept of geological disposal, the validity of this approach was not questioned. It was formally adopted as a final goal, through policy or legal decisions, in many countries, including the USA, Canada, Sweden, Finland, Belgium, Switzerland, France, Spain, South Korea, and Japan. As mentioned above, several of these countries initiated active scientific and technical programmes aiming at implementing disposal, usually some 20 years or so into the future. International organisations such as the OECD/NEA, the IAEA, and the EC established working groups and networks of the organisations involved. Special journals started up. Innumerable conferences were organised around the world; for example the major annual International Waste Management Conference in Tucson, Arizona, USA was held in 2006 for the 32$^{nd}$ time.

### Slow Progress...

However, virtually every geological waste disposal programme in the world ran into difficulties in keeping to originally proposed schedules. For example, in the U.S. programme, in 1982,[3] a target date for repository operation of 1998 was set. In stages afterwards, the target for a U.S. repository at Yucca Mountain was moved back to 2010 because of unresolved technical, licensing and legal issues. Today, 2010 is also recognised as unachievable and the official target date specified by U.S. DOE is still later. Other programmes have also been compelled to move target dates back. Through to the present, the only active programme that met its early deadlines has been Finland.

Slippages in deadlines, however, are common in large projects: disposal programmes are not unusual in this respect. Less common are decisions of the type taken in some countries–namely to indefinitely postpone implementation of geological repositories. This has happened several times, in each case due to public opposition leading to governmental decisions to halt siting processes. Examples are the Netherlands, Spain, the United Kingdom, Argentina and the Czech Republic.

In a few countries, there has been a still more radical political reaction to problems encountered by geological disposal programmes. This began in France, where intense opposition to siting efforts in crystalline rock areas, together with growing opposition to disposal per se, led in 1990 to a new law in which the geological disposal option was treated as one of three lines to be followed. The other two, transmutation and long-term storage, were to be studied with equal intensity at least up to a decision date set for 2006. A key result of the major project that resulted from this French programme is the decision taken in by the French Parliament this year that a geologic repository for HLW should be implemented by the year 2025.

Backing off from the choice of geological disposal as the preferred national strategy has taken place in two further countries, namely the UK and Canada. The UK government decided to re-open all alternatives and to have a very wide public debate before choosing a preferred future course. This decision followed on the loss of the proposed Sellafield site as a result of a public hearing that severely criticised the scientific, engineering and societal aspects of work by UK Nirex. In Canada, the Government also decided to re-open discussion on all conceivable long-term spent fuel management options following the review by the Seaborn Committee[4] of the major study submitted by AECL. In the Canadian case, the science and technology was not faulted; the proposed repository concept was judged technically capable of providing safety. However, it was also judged that the public confidence in the safety was insufficient to allow an implementer to proceed to specific repository siting.

For carrying out these re-evaluations, the governments of the UK and Canada set up special bodies, respectively the Committee of Radioactive Waste Management (CoRWM) and the Nuclear Waste Management Organisation (NWMO). After extensive consultation exercises, both have recently produced recommendations that geological programmes should move ahead, although in an extended staged process. (See www.corwm.org.uk and www.nwmo.ca)

But Some Progress…

The above, rather sobering, look at the slow progress of geological repositories in some countries contrasts with the advances made in some other parts of the world. In the USA, the WIPP deep repository for transuranic wastes has been operating successfully for some years and has recently been recertified to continue doing so. Furthermore, since the U.S. congress has decided that a licensing application should be prepared for the Yucca Mountain Project in Nevada, a deep repository for used nuclear fuel may well be constructed and operated in the United States in the foreseeable future, despite the significant hurdles still faced. In the Northern European countries, Finland and Sweden, the deep repository programmes are very advanced and steering towards definitive dates for implementation. More influential, perhaps, than the technical developments that have been initiated in these countries, are the societal processes that have been invoked to try and ensure that the repository has a sufficient level of acceptance. In most other countries of the world, the combined technical and societal approaches employed in the Scandinavian countries are looked upon as role models for how things might be arranged also in other programmes.

In the European Union, a 2002 draft directive instructed all European Union member States that specific deadlines for siting repositories and for implementing these facilities must be set. Although the over-ambitious deadlines proposed in the initial draft were dropped, the thrust of the initiative will likely remain. This thrust confirms, at least for the European Union, that deep geological disposal is indeed the preferred waste management strategy for used nuclear fuel and high-level wastes.

Achievements To Date

A broad look at the actual situation around the world today reveals the following. The present position is that technologies for implementing deep geological disposal have been developed and extensively tested in a number of countries, although fully implemented in only very few cases. These technologies are based on different conceptual designs for a deep repository, including the choice of the engineered barriers that enclose the used nuclear fuel and also the geological medium in which the repository will be sited. In all of these different programmes, the safety of the deep geological system—as assessed by the range of methodologies developed for this purpose—is invariably shown to be very high. The development of the safety assessment methodology itself has involved many man-years of intellectual effort and also extensive collaboration between researchers in different countries around the globe. Assessing the safety is based upon analysing how the entire repository system will behave far into the future. This estimation in turn is based upon a sound scientific understanding of how the materials will evolve in the deep geological environment, and of how any radionuclides released might be transported through the deep underground, back towards the environment of humans. The safety assessment is not a purely theoretical desk exercise. The models are based upon experimentation in the laboratory and in the field. The understanding that is built up is checked by observing how natural systems with similar properties behave over the very long time-scales considered. Although there are still dissenters to be found, in the scientific community there is general acceptance of the feasibility of safe disposal. Unfortunately, this general consensus does not yet extend to the majority of members of the public.

As a complement to these overarching comments on the status of geological disposal, many publications include good overviews of programmes world-wide. A recent example is the review by Witherspoon and Bodvarrson.[5] In addition, the IAEA maintains a web site that documents current general trends and also developments in individual countries. Finally, most national waste disposal organisations have their own web sites.

The current status of national geological disposal programmes is thus well documented and it illustrates that progress is being made in many countries, but that this is a slow process. For some countries, national repositories may be difficult or unfeasible because of the lack of favourable geological formations, shortage of technical resources, or unacceptably high costs. For these, multinational repositories are a potential solution and, in recent years, there has been a rapid increase in interest in this possibility as described in the following section.

## MULTINATIONAL INITIATIVES

In the early years of nuclear development, the concept of nuclear fuel cycle centres, including international repositories, was topical. The IAEA charter itself allowed the Agency to be involved in centralized plutonium storage and management. Various studies were performed on regional nuclear fuel cycle centres and on international spent fuel management. These are documented in.[6]

The past five years have seen a continual growth in the interest of many national waste management programmes—especially those of small countries—in the concept of multinational or regional disposal facilities. The prime drivers were originally the economic and political problems that might be lessened by being shared between countries facing the same challenges. The potential safety and safeguards benefits were also recognised at this early stage. Increasingly—in particular after the terrorist attacks in the USA in 2001 and in connection with nuclear proliferation concerns—attention focused on the security advantages that could result. The most recent manifestation of this is the Global Nuclear Energy Partnership (GNEP) promoted currently by the U.S. Government. The IAEA, honoured in 2005 with the Nobel Prize for its efforts to reduce nuclear risks, has been careful to point out that these risks can also be important at the "back-end of the back-end" of the nuclear fuel cycle, i.e., not only in enrichment and reprocessing but also in storage and disposal, in particular of spent fuel.

In its publications in this area and in recent statements of representatives of the IAEA, two potential routes to achieving international disposal have been described. One of these, the "add-on approach," is the inclusion of disposal within a broader scheme of internationalised fuel-cycle services provision. The other, which does not require global strategic developments and agreements, is the "partnering scenario," in which a number of countries agree to look for a common disposal solution involving one or two shared repositories. These should be sited in locations to be decided by the multinational participants in the same democratic, consensual approach that has been used by potential siting communities in the more successful national programmes.

In both potential disposal approaches to multinational disposal, significant progress is being made. Below, we describe the add-on approach, using the topical example of Russia and then examine the partnering scenario, using experience gained in the SAPIERR project of the EC.

## THE ADD-ON OPTION

A single country, or a network of countries with appropriate facilities working together, by providing extended fuel-cycle services to countries adhering to the NPT and wishing to use nuclear power, could limit the spread of those sensitive technologies that are allowed under the Treaty, namely enrichment, reprocessing and storage/disposal of fuel. Crucial pre-requisites would be security of supply of services to all co-operating users (as emphasised by the Multilateral Approaches Group established by the IAEA[7]) and close international monitoring by the IAEA. The whole concept has been raised again recently by IAEA Director General, Mohammed El Baradei.[8,9] It is very topical because of the concerns with nations such as Iran expanding their nuclear capabilities to include fuel enrichment.

Although emphasis is on the front end of the fuel cycle, where most security concerns arise, back-end services would also be offered as part of this suite of provisions, either by countries establishing new, dedicated multinational storage and disposal facilities to fit into the scheme or by countries with existing facilities that could be extended for international use.

Within this international fuel cycle scheme, the fuel leasing component is certainly the closest to being an accepted practice. This is almost the practice followed by the former USSR with its satellite States. More recent global concerns about security have led to it being the universally preferred solution, if nuclear power plants are to be operated in countries such as Iran and North Korea. Recent proposals from the U.S. Government have indicated its support for such a scheme. Should it come to pass, the gate will be opened for other large nuclear fuel suppliers to improve the attractiveness of their fuel services, while at the same time enhancing global security. Potential network partners in internationalising the fuel-cycle would all have to be NPT signatories and could clearly include the major suppliers of uranium or of fuel cycle services or of power reactors, i.e., the list includes countries such as Argentina, Australia, Canada, France, Japan, Russia, the UK and the USA.

The most likely country to offer to act as host in this add-on scenario is recognised to be the Russian Federation. Support has been expressed at Government level. The law currently allows import of spent fuel for storage or for reprocessing with return of residues. However, there is solid support for expanding this service to include final acceptance of fuel or even high level radioactive wastes (and, it is acknowledged, also strong opposition). Moreover, once a first move is made, it is not impossible that competition could even arise. Supporters of hosting an international repository have spoken up in Kazakhstan and China in the past and recently again in Australia. Acting as a host is economically attractive for Russia since it would provide either income from provision of services or fuel for the future, or both. However, as has been recently pointed out,[10] the law would have to be changed and a number of other conditions would have to be fulfilled if a range of important international stakeholders are to be comfortable with what is offered and the conditions attached.

The recent GNEP proposal from the USA is primarily aimed at making the nuclear fuel cycle more secure. It asserts that this could be achieved by restricting the processes of enrichment and reprocessing to a limited number of trustworthy countries (or existing weapon States) that should then provide services to other countries wishing to use nuclear power for peaceful purposes. For this to be attractive to these customer countries there must be sufficient incentives and the supply of services must be guaranteed. One incentive would be to have no HLW or spent fuel to be managed long-term and intimately disposed. This requires the fuel suppliers to take back the spent fuel—probably under a leasing arrangement—or for a third party, trustworthy country to offer storage and disposal services. Proposals to host an "international nuclear waste dump" have, not unexpectedly, led to public and political opposition. However, offering a global service that enhances world security, and is for the host country both safe and profitable, maybe more acceptable.[11] From a waste management perspective, GNEP does not add much to the existing Russian proposals. In fact, the additional elements in GNEP, in particular the very ambitious or even unrealistic intentions to develop wholly new fuel cycles, may be counterproductive. They may lead to the more pragmatic proposals, such as fuel leasing, being postponed for the long times needed for such fuel cycle developments.

A fundamental point is that purely unilateral initiatives (whether they are in Russia, the USA or elsewhere) will very probably not succeed: a proper multinational approach is absolutely essential. The time is now ripe for initiating such an approach by bringing the key players together in a free and open discussion to develop plans for how a specific project can be established.

### THE PARTNERING APPROACH: SAPIERR

The second option for implementing multinational repositories—partnering by smaller countries—has been particularly supported by the European Union through its promotion of the potential benefits of regional solution, i.e., facilities shared by contiguous or close Member States. For the "partnering" scenario, in which a group of usually smaller countries cooperate to move towards shared disposal facilities, exploratory studies have been performed most recently by the Arius Association, which also co-manages the European Commission SAPIERR project on regional repositories.[12]

The Support Action: Pilot Initiative for European Regional Repositories (SAPIERR) project finished at the end of 2005 after 2 years of work involving organisations from 14 different countries. This should be succeeded by a follow-on SAPIERR-2 project (Strategic Action Plan for Implementation of European Regional Repositories – Stage 2). This would establish a dedicated multinational organisation that would develop the shared repository option in a staged process similar to that favoured by national programmes. The SAPIERR-2 project looks in more detail at the following topics: multinational legal and business structures; legal liabilities; economics (costs, benefits); safety and security; public and political attitudes.

## WE DEFINITELY NEED GEOLOGICAL REPOSITORIES—BUT WHEN?

As argued above, geological disposal is a necessary final step in the fuel cycle if nuclear power is to be sustainable in the sense that unnecessary burdens are not passed on to future generations. Continuing with nuclear power is therefore justified only if there is a sufficient consensus that safe geological repositories can be implemented. It is often argued that the public confidence needed to achieve this consensus can be achieved only by having operating repositories. This is a dangerous argument for several reasons:

- Even the most advanced programmes will not have operating HLW/SNF repositories for 10-15 years. Decisions on expanding nuclear power are, however, needed much more urgently. In many countries the lifetime of first generation reactors is ending and replacements are needed. In developing countries with rapidly expanding energy needs, use of fossil fuels should be restricted.

- Given the timescales for which repositories must guarantee the safe containment of radioactive wastes, a facility that has been operating, even for years, still does not "prove" that long-term containment will function as expected.
- Technical developments in the nuclear fuel cycle may make the direct disposal of spent nuclear fuel a less attractive option than recovering the fissile material for further use. Retaining the possibility to access the fuel is a prudent approach—provided of course that adequate security measures are taken.
- Small countries will for a long time have too little waste and too little funds available to make implementing a national repository feasible. They should not be forced into this route prematurely. Multinational repositories can and will solve this problem, but not in a short timeframe.

What then is the long-term solution that can justify continuing with and expanding nuclear power production? Disposal solutions must be demonstrated to be feasible. This is not accomplished by simply building a facility. The following requirements are both necessary and sufficient:

- A technical concept involving engineered and natural safety barriers must be developed and its expected performance analysed using appropriate scientific modelling, backed up by comprehensive data collection. The safety level that the facility offers must be recognised by scientists—and by the public.
- The engineering skills needed to implement such facilities must also be recognised as being available today. This can be best done by ensuring that construction of the facilities requires only geotechnical and engineering skills that have been applied already in comparable projects.
- The funding needed to implement repositories must be conservatively estimated and the required funds should be accumulated in dedicated funds that can not be diverted to other uses.
- Finally, given the considerable societal and technical challenges involved in selecting a suitable site, this step should ideally also be accomplished. This means, in the best case, that a specific site has been identified and, in all cases, that the feasibility of doing so is accepted.

When all of these conditions have been satisfied, then the repository implementer can, with a good conscience, sit back and leave the decisions on when to move to implementation to be taken in a broad societal context. Today, the conditions are not satisfied for most countries. For the few with chosen repository sites (such as Finland) they are; for some more with acceptable siting areas identified (including for example Switzerland) they almost are; for many, the funding condition is not met; and for some there is not yet consensus on the achievable safety.

## STORAGE IS NOT A MAJOR PROBLEM

If storage of spent fuel is to continue for some long time—as is inevitable—then certain requirements have to be fulfilled. There must be adequate storage capacity available and the facilities must be safe and secure. The storage options available are in the pools that all reactors have for initial cooling of unloaded fuel, or in away-from-reactor dedicated storage facilities. The basic techniques employed are wet storage with the fuel under water that cools and shields it, or dry storage in casks or vaults.

In many of the countries with nuclear power, storage of spent nuclear fuel is not, or is no longer, a major problem. This positive situation often results, ironically, from the unsuccessful attempts of national waste management programmes to move ahead with disposal projects. Delays on the repository front have compelled some countries to increase their storage capacities, either by re-racking pools at reactors or by constructing new storage facilities. In any case, many programmes have planned for long periods of interim storage (up to 40 or 50 years) to allow the fuel to cool sufficiently before moving to geological disposal, or else simply to postpone the expensive task of implementing disposal, and thus allow time for funds to accumulate. Examples of the former include Sweden, Switzerland, Finland, and Japan; the latter approach is illustrated by the Netherlands and Slovenia.

There are, however, some prominent exceptions: in those countries that urgently need expanded storage capacities the reasons are usually political or societal rather than technical. The USA has manoeuvred itself into a corner by trying to implement an aggressive disposal strategy at Yucca Mountain, while centralised storage schemes have been blocked by law (at Yucca Mountain) or by opponents (in Utah: the private spent fuel storage initiative). In Japan, there have been problems in gaining public acceptance at potential centralised storage sites. This problem is even greater in Taiwan and South Korea.

The IAEA keeps track of the storage capacities world wide. Table 1 below shows that there are many facilities operating and many in construction or planned. The new facilities being implemented are mostly of the dry storage type since this can be done in a more cost-effective modular way, since the storage may be very long—and thus suited for low maintenance and also because reservations have been expressed about the security aspects of wet stores.

**Table 1: Capacities of Nuclear Fuel Storage Facilities**

| | WET | DRY | | |
|---|---|---|---|---|
| **Country & Unit** | **In Operation** | **In Operation** | **Construction** | **Planned** |
| Argentina (t HM) | 986 | 1000 | 0 | 0 |
| Armenia (t HM) | 0 | 74 | 0 | 0 |
| Belgium (t HM) | 1760 | 2100 | 0 | 0 |
| Bulgaria (t HM) | 600 | 0 | 0 | 0 |
| Canada (t HM) | 0 | 75 | 0 | 0 |
| Czech Republic (t HM) | 0 | 600 | 1340 | 0 |
| China (t HM) | 500 | 0 | 0 | 0 |
| Finland (t HM) | 1694 | 0 | 0 | 0 |
| France (t HM) | 18000 | 0 | 0 | 0 |
| Germany (t HM) | 560 | 10123 | 8220 | 0 |
| Germany (Cask-Bund.) | 0 | 0 | 440 | 0 |
| Hungary (t HM) | 0 | 580 | 0 | 0 |
| India (t HM) | 275 | 590 | 0 | 0 |
| Japan (Cask-Bund.) | 7755 | 408 | 0 | 0 |
| Japan (t HM) | 3140 | 0 | 0 | 0 |
| Korea, Republic of (t HM) | 0 | 1212 | 0 | 2000 |
| Lithuania (t HM) | 0 | 360 | 0 | 0 |
| Romania (Cask-Bund.) | 0 | (300000) | 0 | 0 |
| Russian Federation (t HM) | 14960 | 0 | 0 | 38000 |
| Slovakia (t HM) | 1690 | 0 | 0 | 780 |
| Spain (Cask-Bund.) | 0 | 1680 | 0 | 0 |
| Sweden (t HM) | 8000 | 0 | 0 | 0 |
| Switzerland (t HM) | 0 | 2500 | 600 | 0 |
| Ukraine (t HM) | 2518 | 0 | 0 | 0 |
| Ukraine (Cask-Bund.) | 0 | 9120 | 0 | 21350 |
| United Kingdom (t HM) | 10300 | 700 | 0 | 0 |
| USA (t HM) | 2127 | 5838.4 | 2729 | 82393 |
| **Total** | **74865** | **36960.4** | **13329** | **144523** |

The capacities shown as '0' here might mean the capacity is really small or unknown!
Table adapted from IAEA website http://www-nfcis.iaea.org/

The table shows that a global value for reactor storage capacity of around 270,000 t HM will be available in the near future. How does this compare with the existing stock levels and the annual production of spent fuel?

At the beginning of 2003, the IAEA recorded that, of the spent fuel generated to date, about 171,000 t HM of spent fuel were stored in storage facilities of various types (see Table II from Reference 13). Worldwide the spent fuel generation rate, now at about

10,500 t HM/year, is expected to increase to about 11,500 t HM/year by 2010. As less than one third of the fuel inventory is reprocessed, about 8,000 t HM/year on average will need to be placed into interim storage facilities. The IAEA assumes that current plans are maintained, resulting in the following regional trends:

- West Europe will have slightly decreasing quantities of spent fuel to be stored, due to reprocessing of spent fuel;
- East Europe will double the amount of spent fuel to be stored in the coming ten years;
- America will store all discharged fuel, thus the amount of spent fuel is constantly increasing;
- Asia & Africa like East Europe, will double the amount of spent fuel to be stored in the coming ten years.

The conclusions to be drawn from the above statistics are that of the 270,000 t HM storage capacity that will be available soon, only around 200,000 t HM are already in store. Expansion is also relatively easy—which implies that storage could continue for a few decades at the present rate of generation. Lack of available storage is, therefore, not a strong driver leading to repository implementation. This does not allow for local shortages of capacity of course, but also does not allow for the relative ease of increasing storage capacities.

**Table 2: Status of spent fuel stored in world regions t HM**

| Region | Amount |
|---|---|
| | |
| West Europe | 36,100 |
| East Europe | 27,700 |
| America | 83,300 |
| Asia & Africa | 23,900 |
| World | 171,000 |

(Status 1 January 2003)

## CONCLUSIONS

The conclusions that can be drawn from this review of the past history and present status of geological disposal can be summarised as follows:

- Nuclear power did not live up to early expectations: it was technically more complex than assumed, economically less attractive than expected and socially became progressively less supported after its promising start. Opposition stopped or slowed growth: the nuclear industry did not make sufficient efforts to inform and consult with the public, leaving the field open for intensive and effective lobbying by anti-nuclear forces. The list of counter-arguments (often recycled) focused on reactor safety, economics, security and waste management.

- Despite some severe setbacks, nuclear power over some decades proved itself increasingly to be reliable, safe and economic: many of the objections were thus countered. In addition, the positive environmental aspects of nuclear power are becoming increasingly recognized by a public that is becoming ever more aware of the catastrophic consequences that can result from unabated consumption of fossil fuels.
- Despite the widening acceptance or support for nuclear, serious reservations continue to be expressed on two issues: nuclear security and long-term waste management. These issues are linked and are both being addressed today by intensifying efforts to ensure that all hazardous radioactive materials (and, in particular, fissile materials) are being moved into well safeguarded storage facilities.
- Deep geological repositories are an essential component of the long-term management of radioactive wastes. These do not need technically to be implemented on a short timescale. They cannot be implemented on timescales affecting urgent decisions on expansion of nuclear power. However, enough must, be done to establish technical and political confidence in the feasibility of safe disposal. Many nations are trying to progress plans and projects for implementing deep the geological repositories that will be needed to provide long-term safety and security in any credible waste management system.
- For some countries, it will be infeasible or impossible to implement the costly deep repositories that will be needed to safely store their relatively small quantities of hazardous long lived wastes and/or spent fuel. Therefore national efforts must be complemented by multinational cooperative initiatives that will make appropriate storage and disposal facilities available to all countries that make use of nuclear technologies. Implementation projects that arise from such cooperation could bring huge and mutual benefits to both host countries and user countries of shared multinational repositories.
- The most effective ways forward to ensure security and long-term safety are that immediate efforts are made to ensure secure storage of all hazardous radioactive materials; that advanced disposal programmes continue towards realisation of repositories; and that active steps are taken towards the realisation of shared multinational facilities for both storage and disposal of HLW and SNF.

REFERENCES

1. BNWL, "High-Level Radioactive Waste Management Alternatives", 4 Vol. BNWL-1900, Richland, Washington, Pacific Northwest Laboratories, May 1974
2. NRC (1957), "The Disposal of Radioactive Waste on Land" Washington, D.C., National Academies Press.
3. EnPA (1982), "Energy Policy Act of 1982: Section 801: Nuclear Waste Disposal"
4. Canadian Environmental Assessment Agency (CEAA), "Nuclear Fuel Waste Management and Disposal Concept (Seaborn Report)". Report of the Nuclear Fuel

Waste Management and Disposal Concept Environmental Assessment Panel. B. Seaborn (chairman), 1998

5. Witherspoon, P.A. and Bodvarsson, G.S. "Geological Challenges in Radioactive Waste Isolation" Third and Fourth Worldwide Reviews. Berkeley National Laboratory, University of California, 2002 and 2006.
6. IAEA, "Developing and implementing multinational repositories: Infrastructural framework and scenarios of co-operation", TECDOC 1413, 2004.
7. IAEA, "Multilateral Approaches to the Nuclear Fuel Cycle", Expert Group Report submitted to the Director General of the IAEA, 22nd February 2005.
8. El Baradei M. (2003): "Statement to the Forty-seventh Regular Session of the IAEA General Conference 2003", www.iaea.org
9. El Baradei M. (2003): "Towards a Safer World," The Economist, 16th October 2003.
10. Chapman N., McCombie C., "What will it take to develop an international repository in Russia?", Safety Barrier No 3-4, 2005, Radon Press Moscow.
11. Choi J.S., Isaacs T.H, "Toward a New Nuclear Regime", Proceedings of ICAPP 2003, May 2003.
12. Chapman N., McCombie C., Stefula V., "Possible Options and Scenarios of Regional Disposal and Future RTD Recommendations conclusions" www.sapierr.net
13. Fukuda K., Danker W., Lee J.S., Bonne A. and Crijns M.J. "IAEA Overview of global spent fuel storage", IAEA, Vienna, www.pub.iaea.org/MTCD/publications/PubDetails.asp?pubId=6924

# PROLIFERATION RESISTANCE AND PHYSICAL PROTECTION FOR INNOVATIVE NUCLEAR REACTORS AND FUEL CYCLES

RICHARD HOSKINS
Office of Nuclear Security, Department of Nuclear Safety and Security
Vienna, Austria

Let's start out by defining the two main themes of this discussion: proliferation resistance and physical protection.

## PROLIFERATION RESISTANCE

The characteristics of a nuclear energy system that impede the diversion or undeclared production of nuclear material, or misuse of technology, by States intent on acquiring nuclear weapons or other nuclear explosive devices.

## PHYSICAL PROTECTION

That combination of features necessary for the prevention of unauthorized removal of nuclear material, or sabotage, or other malicious acts by non-State actors involving nuclear materials in use, storage or transport.

## STRATEGIC OVERVIEW

Proliferation resistance

- PR measures are working against the State on behalf of the international community
- Can deter by increasing likelihood of detection, through effective verification-and raise costs of covert activities
- Technical measures can prevent or inhibit diversion or covert production at declared facilities
- Not much can be done, technically, to prevent or inhibit covert facilities

Physical protection

- PP measures work with the State in the interests of the State and the international community
- Higher chance of success

## INTERNATIONAL PROJECT ON INNOVATIVE NUCLEAR REACTORS AND FUEL CYCLES (INPRO)

Initiated in 2000

*"all interested Member States to combine their efforts under the aegis of the IAEA in considering the issues of the nuclear fuel cycle, in particular by examining innovative and proliferation resistant technology"*

Objective: to develop methodology for the assessment of innovative nuclear reactors and fuel cycles (INS) with respect to the following performance characteristics:

- Safety
- Proliferation Resistance
- Physical Protection
- Waste
- Infrastructure
- Economics

<u>Outputs:</u>

- *"Guidance for the Evaluation of Innovative Nuclear Reactors and Fuel Cycles"* (IAEA TECDOC-1362 June 2003)
- *"Methodology for the Assessment of Innovative Nuclear Reactors and Fuel Cycles"* (IAEA TECDOC-1434 December 2004) Available on: <u>www.iaea.org</u>
- INPRO Manuals (draft) on the six performance characteristics including:
  - 'Proliferation Resistance: Assessment for Innovative Nuclear System'
  - 'INPRO Manual on Physical Protection'

<u>INPRO Methodology</u>

- Extrinsic measures: States decisions and undertakings. Related to international legal framework.
- Intrinsic measures: technical measures taken by the State.
- Recognize synergies between safety, physical protection (security measures) and proliferation resistance (safeguards) measures

PROLIFERATION RESISTANCE

<u>Proliferation Resistance International Legal Framework</u>
Obligations are placed on the State by:

- Nuclear Non-Proliferation Treaty (NPT)
- IAEA INFCIRC/153 (Corr.) Full scope safeguards
- IAEA INFCIRC/540 (Corr.) Additional Protocol
- IAEA INFCIRC/66 Rev.2 In non-NPT States
- UNSCR 1540

<u>Proliferation Resistance INFCIRC/153 and 540</u>
Safeguards:

- Designed to detect diversion of material or undeclared activities
- Provide timely notification of breaches
- Provide credible assurances that there has been no diversion and that there are no undeclared activities

Note: INFCIRC/66 pre-dates full scope safeguards and used for specific material or facilities in non-NPT States.

Proliferation Resistance Extrinsic features

- National legislation and regulations to implement obligations under NPT and INFCIRCs
- Establish national system for accounting for and control of nuclear material (SSAC). Requirement of INFCIRC/153
- *"States shall take and enforce measures to establish domestic controls to prevent the proliferation of nuclear weapons....including controls over related materials" SCR 1540*
- Declare all nuclear material and provide accounts for inspection
- To facilitate implementation of safeguards inspections
  - Provide design information and agree Facility Attachments
  - Agree to Subsidiary Arrangements
- Accept inspections including swipe sampling and short-notice inspections

Proliferation Resistance Intrinsic (technical) features

A. Preventing or inhibiting: Diversion

- Reduce attractiveness of material for nuclear weapon programmes:
  - Material characteristics, isotopic content, chemical form, bulk, mass, radiation properties, heat etc.
- Confine material to areas with limited access/egress
- Make material difficult to move without being detected:
  - Through size, weight, radiation, detectability, seals
- Design to reduce MUF

B. Preventing or inhibiting: Covert production

- Design to prevent undeclared target materials being irradiated near the core
- Use reactor cores with small reactivity margins to prevent operations with undeclared targets
- Design fuel cycle facilities so that they are difficult to modify
- Design facilities so that they can be effectively monitored by safeguards
- Note: aside from technical features which prevent or inhibit diversion of nuclear materials, not much can be done, in technical terms to prevent or inhibit the development and operation of covert facilities

PHYSICAL PROTECTION

International Legal Framework

*Convention on Physical Protection of Nuclear Material (CPPNM):*

- Covers requirements for the protection of nuclear material in international transport

*Amended CPPNM:*

- Covers material in use, storage or transport
- Requires 'appropriate physical protection regime' to protect against theft, to ensure rapid recovery of stolen material, to protect against sabotage and to mitigate against its consequences
- Requires a State to establish a legal an regulatory framework to govern establishment of PP system
- Requires that a competent authority is established by State to implement PP system
- Establishes Fundamental Principles of Physical Protection
- Criminalizes specified actions

*Physical Protection Fundamental Principles:*

- Responsibility for PP lies entirely with State
- State is responsible for adequate protection of NM in international transport
- State is responsible for establishing legislative and regulatory framework to govern PP including requirements, evaluation, licensing, and inspection
- State should establish competent authority
- Prime responsibility for implementation of PP lies with license holders
- Must give priority to establishing a security culture
- PP should be based on State's evaluation of the threat
- PP should be based on graded approach taking account of threat, attractiveness of material and potential consequences of loss or sabotage
- PP system should adopt defence in depth involving technical, personnel and organizational aspects
- Should establish quality assurance system
- Should establish contingency plans to respond to loss of material or sabotage
- Information on PP system should be confidential

*UN Security Council Resolution 1540:*

- Chapter VII – Binding - Requires States:
    - To criminalize the manufacture, acquisition, possession, development, transport, transfer, or use of nuclear weapons by non-State actors
    - To account for and secure nuclear material in production, use, storage, and transport
    - To develop and maintain appropriate physical protection measures
    - To develop and maintain effective border controls etc. to detect, deter, prevent and combat illicit trafficking
    - To establish domestic controls over nuclear weapons related materials to prevent proliferation
    - To establish export controls
    - To cooperate multilaterally in the framework of the IAEA

*Convention on the Suppression of Acts of Nuclear Terrorism:*

- Criminalizes:
  - Possession of material or device with intent
  - Use of material or device, or sabotage, with intent
  - Threats

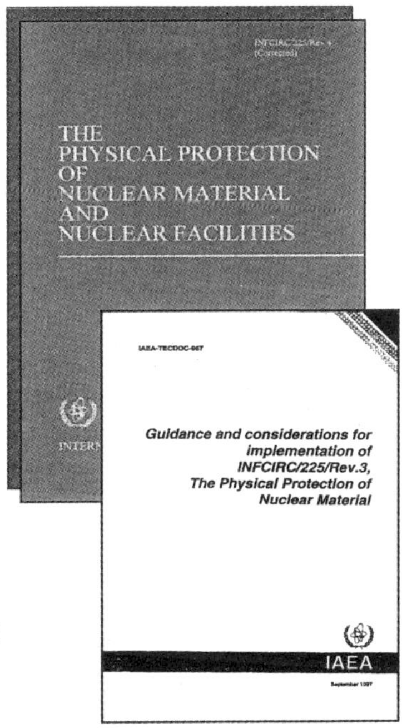


International Guidelines and Recommendations

INFCIRC/225/Rev.4:

- Defines elements of a PP system: evaluation of threat, technical hardware measures, legislative and regulatory measures, establishment of competent authority
- Sets requirements for PP systems against theft: protected and inner areas, intrusion detection, guarding and patrolling, response forces
- Sets requirements for PP against sabotage: mix security devices, procedures and design, establish 'vital areas', intrusion detection, access, synergies with safety measures
- Categorizes nuclear material
- Sets requirements for PP in international transport

New unified IAEA *Nuclear Security Series*

- Those in pipeline include:
  - Development and maintenance of Design Basis Threat
  - Physical protection of nuclear materials against sabotage
  - Identification and PP of vital areas at nuclear facilities
  - Guidelines on self-assessment of engineering safety aspects of the protection of NPPs against sabotage
  - Nuclear security culture
  - Security during transport
  - Security of radioactive waste
  - Security of computer systems at nuclear facilities
  - Preventive measures against insider threats

Intrinsic Measures in INS:

- Apply PP Fundamental Principles
  - Threat based approach
  - Graded approach
  - Defence in depth
  - Security culture
- Apply INFCIRC/225
  - PP system
- Minimize potential for sabotage
  - Engineering design
  - Protect vital areas
  - Insider threat
  - Safety measures
  - Emergency response
- Material accountancy and control

Observations:

- Both PR and PP are best achieved through a combination of international legal measures, national legislative and regulatory infrastructures, and technical measures. The latter alone are not enough.
- Proliferation resistance addresses an 'insider threat' which it cannot prevent by technical means alone. The safeguards regime acts as a deterrent, an inhibitor and as an early warning system. But in the end, political decisions start or stop proliferation.
- Technical features which will inhibit proliferation activities can be identified for reactor design. Few have been yet identified for the design other fuel cycle facilities (except perhaps for efforts to reduce MUF).
- Lateral approaches such as leasing fuel and central repositories for spent fuel are an alternative strategy.
- But the increase in transports brings its own risks of theft or sabotage
- PP measures address an adversarial threat external to the operator. The operator and international community share the same objectives of prevention and mitigation
- For PP an extensive legal framework is emerging which will be supported by internationally developed guidelines and recommendations related to implementation. The Amended CPPNM is a major step forwards.
- PP of future nuclear facilities, Gen III and IV, will benefit from this new framework. The design of new facilities can take PP requirements into account from the beginning – designing for security.
- PR already has international legal framework; the NPT/Full scope safeguards/Additional protocol. Improvements will come from changes in technical features.
- PP can be improved now. PR will have to wait for new reactors.

Evaluation and Advisory Services

- International Physical Protection Advisory Service (IPPAS)
- International Team of Experts (ITE)
- Integrated Regulatory Review Service (IRRS)
- IAEA State System of Accounting and Control Advisory Service (ISSAS)
- International Nuclear Security Service (INNServ)

Strategic Overview

- Tendency to see solutions to PR and PP requirements solely in technical terms.
- Solutions are:
  - international obligations
  - international standards, recommendations and guidelines
  - national legislative and regulatory systems
  - technical features

# COAL WITH CARBON CAPTURE AND STORAGE: THE MAIN COMPETITOR

CARMEN DIFIGLIO[1]
Deputy Assistant Secretary for Policy Analysis
U.S. Department of Energy, Washington, DC, USA

ABSTRACT

This paper outlines the most promising technological approaches to achieve $CO_2$ capture and storage (CCS). The relative costs of CCS and nuclear power are presented for currently available and advanced technologies. The paper also discusses the risks of $CO_2$ storage, ways to remediate those risks and U.S. Department of Energy programs that can lead to safe $CO_2$ storage. Lastly, estimates from the International Energy Agency (IEA) are used to show that CCS holds the greatest potential, next to energy efficiency, to reduce global $CO_2$ emissions. In addition, it is shown that if CCS is not developed, the marginal cost of $CO_2$ emission reduction will be significantly higher, especially in coal-rich countries such as China.

## INTRODUCTION

$CO_2$ capture and storage (CCS) could be one of the most important methods to reduce emissions of $CO_2$. It has the potential to be of comparable importance to energy efficiency, renewable energy or nuclear power. However, CCS is a relatively undeveloped technology. Without CCS, world-wide coal use would have to be substantially constrained to, for example, reduce 2050 $CO_2$ emissions to current levels. Deployment of CCS would allow continued use of this valuable energy resource. This is not only of economic importance but would support energy security policies in many countries (e.g., the U.S.). With nuclear power and renewable energy, CCS completes a strategy to achieve secure supplies of electric power while producing very low $CO_2$ emissions.

CCS can be applied to other processes and fuels than coal-fired electric power plants. Natural gas power plants, cement manufacturing, steel production, fuel processing and ammonia production are all candidates for cost-effective use of CCS. CCS is basically a very simple concept. Instead of allowing $CO_2$ to be exhausted into the atmosphere, $CO_2$ is captured and sent via pipeline to an injection well, where it is pumped underground to depths generally greater than 800 meters to maintain critical pressures and temperatures. Candidate sites for geologic storage include deep saline formations, depleted oil and gas reserves, and unminable coal beds. Suitable sites have a caprock, or an overlaying impermeable layer, that prevents $CO_2$ from escaping back towards the surface.

---

[1] The author supervised the development of the International Energy Agency's Energy Technology Perspectives model used to develop the quantitative estimates cited in this paper (IEA, 2004 and IEA, 2006) during the time he was Head of IEA's Energy Technology Policy Division (1998-2004).

## $CO_2$ CAPTURE TECHNOLOGIES

Three promising approaches for $CO_2$ capture are being developed. In the first approach, $CO_2$ is absorbed from an integrated gasification combined cycle plant (IGCC). In this process, coal is gasified resulting in a synthetic gas stream (CO and $H_2$). A shift reaction converts the syn-gas to $CO_2$ and $H_2$ allowing the removal of up to 90% of the $CO_2$ that would normally be released to the atmosphere. $CO_2$ removal occurs before combustion. Oxygen-blown gasifiers would improve the efficiency and reduce the cost of $CO_2$ capture in IGCC plants (see discussion of air separation technologies below). Turbines that can operate on high concentrations of $H_2$ are also needed.

The second approach involves removing the $CO_2$ from the flue gas of a steam-coal plant using amine scrubbing or membranes. Since the percentage of $CO_2$ and pressure in the flue gas is relatively low, this process tends to add more cost to a steam plant than the first approach adds to an IGCC plant and may not lead to as high a percentage of $CO_2$ removal.

The third approach would eliminate nitrogen from the combustion of coal by substituting oxygen for air in a steam-coal plant. The flue gas would then have a very high percentage of $CO_2$ resulting in cost-effective removal. The primary disadvantage of this approach is the high energy requirements and cost required to separate oxygen from air. Improved air separation technologies will help reduce the cost of oxy-fueling. An alternative oxy-fueling method would be to use metal oxides in fluidized-bed plants. In one reactor a metal reacts with air to form a metal oxide. In another reactor the metal oxide reacts with the fuel to produce syngas and metal. However, this technology is relatively undeveloped (see IEA, 2004, pp. 51-54).

Table 1 presents the estimated costs of CCS plants using technology that could be developed today. These costs assume that there has been some experience and would not necessarily be the costs the first plants built. Nonetheless, no technologies are assumed that have not been used in other industrial applications. A super-critical coal plant that would other wise cost $1,535/kW would cost $2,655/kW. The steam coal with CCS plant would produce electricity for 8.0 cents/kWh (instead of 4.8 cents/kWh without CCS). An IGCC plant that would otherwise cost $1,658/kW would cost $2,438/kW with CCS. The IGCC plant with CCS would produce electricity for 7.1 cents/kWh (instead of 5.2 cents/kWh without CCS). For cost comparison purposes it is estimated that a Generation III nuclear reactor would cost $2,014/kW and produce electricity at a cost of 6.0 cents/kWh. Consequently, while we estimate that nuclear power would be less expensive than CCS, the estimated cost difference (1.1 cents/kWh) between nuclear and an IGCC plant with CCS is not very large.

**Table 1: Cost Comparisons of CCS with Nuclear Power: Current Technologies.**

| **Technology** | **Available when** | **Investment Cost ($/kW)** | **Capture Cost ($/T $CO_2$)** | **Electricity Cost (c/kWh)** | **Capture Cost (c/kWh)** |
|---|---|---|---|---|---|
| Super-Critical Coal | NOW | 2655 (1535) | 39 | 8.0 (4.8) | 3.2 |
| IGCC | NOW | 2438 (1658) | 26 | 7.1 (5.2) | 1.9 |
| GEN III+ | NOW | 2014 | NA | 6.0 | NA |

Source: Department of Energy PI-ETP Model

Table 2 presents the estimated costs of CCS plants technologies that could be if CCS took hold in the market place and there was an opportunity to employ less certain technologies. A steam coal plant that would other wise cost $1,025/kW would cost $1,400/kW with oxy-fueling and CCS. The oxy-fueled CCS plant would produce electricity for 3.8 cents/kW instead of 2.9 cents/kWh without CCS. An IGCC plant that would otherwise cost $1,260/kWh would cost $1,635 with CCS to produce electricity at 4.1 cents/kWh (instead of 3.3 cents/kWh without CCS).

In the current-technology scenarios, CCS adds 2-3 cents/kWh to electricity cost and costs between $26 to $39 per ton of $CO_2$ removed. With advanced technologies these costs could drop to less than 1 cents/kWh and achieve $CO_2$ control for $11 to $14 per ton of $CO_2$. Significantly, the cost of power with CCS is estimated to be similar to the cost from Generation IV nuclear power plants.

**Table 2: Cost Comparisons of CCS with Nuclear Power: Advanced Technology Scenario.**

| Technology | Available when | Investment Cost ($/kW) | Capture Cost ($/T $CO_2$) | Electricity Cost (c/kWh) | Capture Cost (c/kWh) |
|---|---|---|---|---|---|
| Steam Oxy Fuel | 2020 | 1400 (1025) | 14 | 3.8 (2.9) | 0.9 |
| IGCC | 2020 | 1635 (1260) | 11 | 4.1 (3.3) | 0.8 |
| GEN IV | 2020-2030 | 1000 to 1400 | NA | 3.0 to 4.5 | NA |

Source: Department of Energy PI-ETP Model

## $CO_2$ STORAGE

The technologies required for geologic sequestration are well developed. There is a considerable amount of commercial experience in injecting $CO_2$ in oil reservoirs as a means to increase the yield from partially-depleted oil fields. More recently, $CO_2$ has been stored in deep saline reservoirs. The main issue is not how to transport and store $CO_2$, but the risk of doing so. Safe disposal of $CO_2$ will require proper siting, operation and maintenance, and long-term monitoring. Capture costs and concerns with long-term liability for storage sites are major considerations still being addressed by ongoing R&D and policy development. In addition to technical and economic hurdles to commercial deployment, public awareness and acceptance of projects to store very large volumes of $CO_2$ need to be achieved. Also, while there is experience with regulations and permits for smaller amounts of materials, i.e., hazardous waste and waste injection wells, there is no set of regulations for $CO_2$ storage, and in addition to environmental issues, questions remain about ownership and liability for the $CO_2$ and for ownership of the storage space.

In the U.S., large point sources of $CO_2$ (each emitting more than 100,000 tons of $CO_2$ per year) originate from various industrial sectors including coal-fired power plants, ammonia production, and cement manufacture among others. There are approximately 1,700 of these sources in the U.S. that collectively emit more than 3 gigatons of $CO_2$ (Gt $CO_2$) per year. Initial assessments show there is an abundance of geologic storage capacity, well distributed throughout the U.S. Although capacity estimates vary, recent estimates show a U.S. capacity as high as 3,900 Gt $CO_2$. Worldwide, estimates range as high as 10,000 GT (IEA, 2004). Since these capacities are an order of magnitude higher than the cumulative global emissions that would occur over the next 100 years, $CO_2$ storage can be viewed as a relatively long-term option, not just a temporary stop gap. While some authors suggest that CCS is a transition strategy (NAE, 2004), since this transition can last over 100 years, this characterization should not be interpreted to diminish the significance of CCS as a $CO_2$ control strategy for the foreseeable future.

The geological formations of primary interest to sequestration include:

- Existing oil and gas fields and potential enhanced oil/gas recovery opportunities
- Depleted oil and gas fields
- Deep saline formations
- Deep unminable coal seams, possibly with coal bed methane recovery

Other possibilities include storage in mafic/basalt rock formations and above ground conversion of $CO_2$ to solid carbonate materials.

Enhanced recovery with $CO_2$ floods is used commercially in North America. There were some 70 $CO_2$ floods in the United States in 2000 that resulted in almost 200,000 bbl of oil per day, approximately 5 percent of total U.S. oil production during the same period. The majority of the $CO_2$ for enhanced oil recovery (EOR) operations comes from natural sources, because $CO_2$ captured from most anthropogenic sources is currently too expensive to compete with other $CO_2$ sources.

Nevertheless, sequestration as part of an EOR operation has the attraction of being a revenue producing process, and is very likely to be some of the first sequestration opportunities to be implemented at large scale. For example, the British Petroleum (BP) Carson Hydrogen Power project will convert the carbon in petroleum coke, a by-product of the refining process, and recycled waste water into hydrogen, a clean-burning gas, and $CO_2$. The hydrogen gas will be used to fuel a power station capable of providing the California power grid with 500 MW of electricity. At the same time, about 4 million tons of $CO_2$ per year will be captured, transported and stored in deep underground oil reservoirs where it will enhance existing oil production.

If EOR projects are to include a $CO_2$ sequestration component, changes may be needed to the facility and/or operations. For example, different project goals may necessitate additional site characterization, the use of multiple geologic formations, or temporary $CO_2$ storage. A critical component will be monitoring and verifying the volume of $CO_2$ stored and additional site closure practices to ensure $CO_2$ is sequestered for the long time frames required.

Injection of $CO_2$ into depleted oil and gas fields would be similar to commercial EOR experience. While one of the main attractions for using the fields is that large amounts of geological data will be available, the existing fields will also have numerous old wells that may no longer be sealed and could leak the $CO_2$ back to the atmosphere. Before sequestration, the existing field would have to be closely examined and issues such as old wells would have to be addressed.

Deep saline deposits could hold the most $CO_2$ (apart from ocean storage – see below). Commercial-scale projects dedicated to geologic $CO_2$ storage in saline formations are at the Sleipner West field in the North Sea and the In Salah gas field in Algeria. Sleipner West is a natural gas/condensate field operated by Statoil and is located about 500 miles off the coast of Norway. The natural gas has a $CO_2$ content of about 9 percent which, to meet commercial specifications, must be reduced to 2.5 percent. At Sleipner, the $CO_2$ is compressed and injected via a single well into the Utsira Formation, a 500 foot thick, brine saturated formation located at a depth of about 2,000 feet below the seabed. The operation is commercially driven by a carbon tax imposed by Norway.

In 2004, BP launched a $CO_2$ capture and storage project at the In Salah gas field, in the Algeria desert. In Salah is a joint venture between Sonatrach, the Algeria national energy company, BP and Statoil. Approximately 10% of the gas in the reservoir is made up of $CO_2$. Rather than venting the $CO_2$, which is the established practice on other projects of this type, the project is compressing it and injecting it in wells 1,800 meters deep into a lower level of the gas reservoir where the reservoir is filled with water. Around one million tons of $CO_2$ will be injected into the reservoir every year.

The most important trapping mechanism to contain $CO_2$ in deep saline reservoirs is hydrodynamic trapping, where a caprock prevents upward movement of $CO_2$. Saline and other types of reservoirs also have two additional trapping mechanisms that help contain the $CO_2$: solubility and mineral trapping. Solubility trapping is the dissolution of $CO_2$ into the reservoir fluids; mineral trapping is the reaction of $CO_2$ with minerals in the host formation to form carbonates. As the $CO_2$ moves through the deposit, it comes into contact with uncarbonated formation water and reactive minerals. A portion of the $CO_2$ dissolves in the formation water and becomes permanently fixed by reactions with minerals in the host rock. Over long periods of time, the $CO_2$ might all dissolve and be fixed by mineral reactions, essentially becoming permanently sequestered.

Sequestration into deep coal seams has been proposed as a means to safely store $CO_2$ because the $CO_2$ will both react with the coal materials, and displace methane from the coal. Some tests have been performed for the purpose of enhancing coal-bed methane recovery, but little has been done to examine the sequestration issues. As with the other EOR technologies there is the potential benefit of increased energy production that could offset some of the $CO_2$ capture and storage costs. Below is a list of major operation $CO_2$ capture and geologic storage projects worldwide.

Ocean storage of $CO_2$ (in the water column) is also possible but, because of the environmental risks, this remains one of the more uncertain approaches to $CO_2$ storage. Nonetheless, there are ongoing studies to develop instrumentation and test the behavior of $CO_2$ stored at various depths. Still, the impact of injected $CO_2$ on adjacent marine life is not well understood. It is fair to say that, apart from a few large countries (e.g., Japan), ocean storage is not needed to implement a CCS program.

## RISKS OF $CO_2$ STORAGE

Risks to human health and safety arise (almost) exclusively from elevated $CO_2$ concentrations in ambient air either in unconfined outdoor environments or in buildings. At concentrations above ~2%, $CO_2$ has strong effects on respiratory physiology, and at concentrations above 10% it can cause unconsciousness and death. Exposure studies have not revealed any effect of chronic exposure to concentrations below 1%.

Increases in dissolved $CO_2$ concentration that might occur as $CO_2$ migrates from a storage reservoir to the surface can alter groundwater chemistry, potentially affecting drinking water aquifers. The direct effects of dissolved $CO_2$ are likely minor, as carbogaseous water is currently used to provide bottled mineral water and gas. Dissolved $CO_2$ forms carbonic acid, altering the pH of the solution, and potentially causing indirect effects including mobilization of toxic metals, sulfate, or chloride; and possibly giving the water an odd odor, color, or taste. In the worst case contamination could reach dangerous levels, excluding the use of groundwater for drinking or irrigation.

Carbon dioxide storage may have an impact on flora and fauna that come in contact with the injected, or subsequently leaked, $CO_2$ and any accompanying substances. Impacts might be expected among microbes in the deep subsurface and among plants and animals in shallower soils and at the surface. The probability of impact and its severity will likely be inversely related: microbes in the injection site are certain to be affected, but the impacts are likely to be of little consequence; whereas surface releases into ecologically sensitive areas require a leakage pathway to the surface for prolonged periods, and while the impacts would be of more concern, the likelihood of exposure to leaked $CO_2$ is relatively low. Impact due to leakage is mediated by several factors: type and density of vegetation; exposure to other environmental stresses; prevailing environmental conditions like wind speed and rainfall; low-lying areas, and the density of nearby animal populations.

It is possible, that under some circumstances, $H_2S$, $SO_2$, $NO_2$ and other traces gases may be stored along with $CO_2$. In this case, the risks may be different than the risks described above. For example, $H_2S$ is considerably more toxic than $CO_2$, so well blow-outs containing $H_2S$ may present greater risks than from wells blow-outs from storage sites that only contain $CO_2$. Similarly, dissolution of $SO_2$ in groundwater creates a far stronger acid than does dissolution of $CO_2$. Thus, the mobilization of metals in groundwater and soils may be higher, leading to greater risk of exposure to hazardous levels of trace metals. A systematic and comprehensive assessment of how these additional constituents would impact the risks associated with $CO_2$ storage has not been conducted yet. In addition, it is not yet clear whether or not it is advantageous to consider co-storage of these gases. When and if it is determined that co-storage of $CO_2$, $H_2S$, $SO_2$ and $NO_2$ is desirable, composition specific risk assessments will be needed.

At a properly designed and well-managed $CO_2$ storage site, the chance of $CO_2$ leakage should be small. Properly-designed sites will have one or more injection zones that can accept and store large quantities of $CO_2$, overlain by suitable caprocks, and will not be located in areas that have a high incidence of seismic activity. Fortunately, within the United States there are relatively few areas where seismicity would be a significant concern, allowing for carbon capture and storage (CCS) deployment across a wide range of locales. $CO_2$ storage sites can be designed against sudden large releases by avoiding areas with significant risk of seismicity and by mitigating leakage pathways such as faults and abandoned wells.

Seismic surveys can be undertaken at candidate sites to assess whether there are any faults that might allow injected $CO_2$ to migrate out of the target injection zone. Seismic surveys, however, are just one aspect of a comprehensive pre-injection site evaluation that would need to be performed at each prospective $CO_2$ storage site. This pre-injection site evaluation would also need to identify the extent and condition of any abandoned wells (e.g., decades-old oil and gas production wells). Adequate sealing of abandoned wells that penetrate the storage zone would need to be assured to prevent these man-made structures from becoming pathways for $CO_2$ to migrate back to the surface.

Measuring, monitoring, and verification (MMV) systems will be needed to ensure that injected $CO_2$ remains in the target formation. Some technologies needed to monitor certain aspects of $CO_2$ storage are commercially available. However, the large-scale deployment of CCS technologies will depend in part on developing a much more robust and accurate suite of MMV technologies. While the issue of leakage from $CO_2$ storage in deep geologic formations remains a subject of debate and intense research, several points are worth stressing:

Because the majority of any potential large-scale CCS deployment is still likely decades away, we can use the next decade's worth of planned field experiments and potential early commercial CCS deployments to fundamentally improve our knowledge base about this key issue. There is a pressing need to amass field data to better bound likely leakage rates.

Sudden releases of $CO_2$ are unlikely. To the extent that leakage does occur, the most likely pathways are transmissive faults and unsecured abandoned wells. In order to migrate back to the surface, a molecule of $CO_2$ would have to find its way through many layers of low-permeability rock, through which it might move only centimeters per century. Finding its way to the surface by moving upward through thousands of meters of solid rock could take millennia.

The likelihood and extent of any potential $CO_2$ leakage should slowly decrease as a function of time after injection stops. This is because the formation pressure will begin to drop to pre-injection levels, as more of the injected $CO_2$ dissolves into the pore fluids and begins the long-term process of forming chemically stable carbonate precipitates.

Remediation options are available for most of the leakage scenarios that have been identified, namely (1) leaks within the storage reservoirs; (2) leakage from active or abandoned wells. In order to develop these remediation options and to advance the technical knowledge of sequestration, the U.S. Department of Energy (DOE) has established Regional Sequestration Partnerships. The seven partnerships include 40 States and 4 Canadian Provinces. More than 200 industry and government organizations are participating with the primary contractors. DOE will deploy a geographic information system (GIS) database that will be available to partnership members and the public. DOE will use the regional data to develop a National/North American sequestration GIS.

## GLOBAL POTENTIAL OF CCS TO REDUCE $CO_2$ EMISSIONS

A recent report by the International Energy Agency, *Energy Technology Perspectives*, provides a comprehensive assessment of all major technologies that could be deployed to reduce world-wide $CO_2$ emissions by 2050. It establishes a Baseline Scenario in which current policies continue. The technology deployment under this scenario provides the basis for estimating future $CO_2$ emissions. This technology deployment is altered in several alternative technology scenarios:

- The ACT Map scenario estimates the technologies that would result from substantially increased government R&D and deployment policies.
- The ACT Map scenario without carbon capture and storage.
- The ACT Map scenario with lower-energy efficiency uptake.
- The ACT Map scenario with lower-nuclear power uptake.
- Lastly, a scenario is defined that assumes high success in all technology areas.

In the analysis used to develop these scenarios, a \$25/ton value of $CO_2$ is used to represent the collective effect of all of the deployment policies. Under the Baseline Scenario, worldwide $CO_2$ emissions would increase over current levels by almost 140% (from 25 gigatons to almost 60 gigatons. The ACT Map scenario results in sufficient low-emission technology deployment to almost return emissions to 2003 levels by 2050 (+6 percent). The optimistic Tech Plus scenario would achieve significantly reduced emissions (-16 percent). These show the important potential of advanced technology development and deployment as these reductions are achieved without reducing energy service demand, an unreasonable expectation, especially in the developing world.

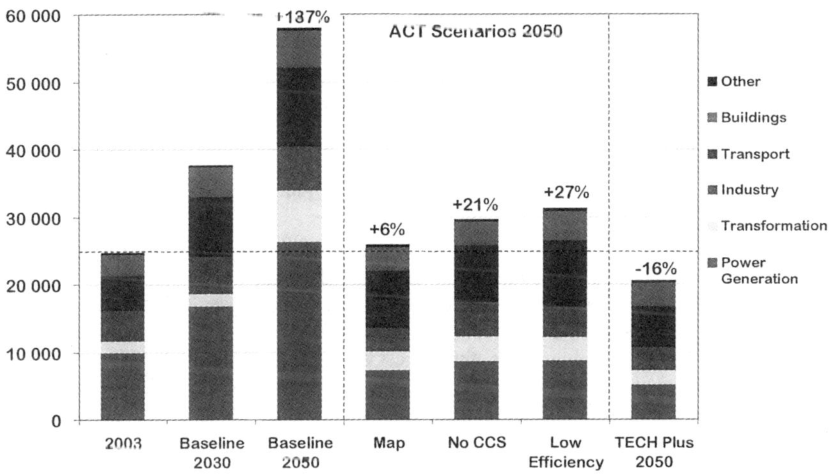


*Figure 1: Global $CO_2$ Emissions 2003-2050.*

Figure 1 also shows the consequences of not achieving technology goals in two areas: CCS and energy efficiency. Without CCS technologies, emissions would be 21 percent higher in 2050 than 2003. Similarly, a low-efficiency scenario would increase emissions 27 percent above 2003 levels. These results show the importance of both CCS and energy efficiency in reducing world-wide emissions. No other technologies, including nuclear, are estimated to be as important.

In Figure 2 we can see that CCS is responsible for 20 percent of the estimated emission reductions in the ACT Map scenario. Without CCS, other sources of emission reductions would partly make up for the loss of CCS but total emission reductions remain significantly lower.

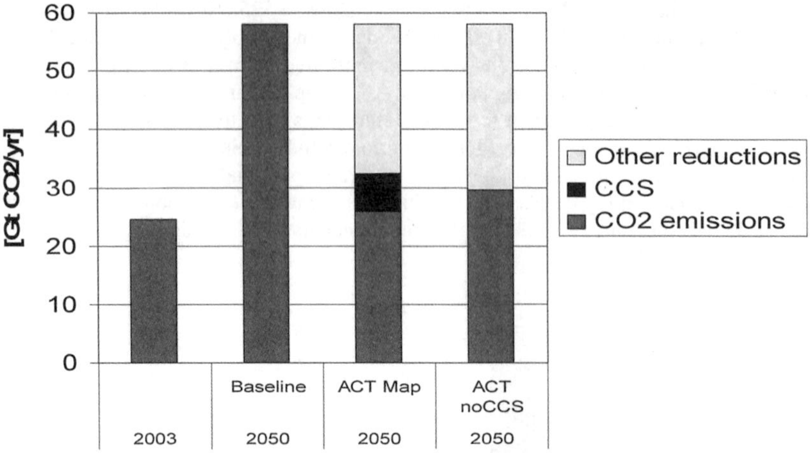


*Figure 2: CCS Accounts for 20% of Global Emission Reductions.*

Another illustration of the importance of CCS is shown in Figure 3. It shows the sources of all estimated emission reductions in the ACT Map scenario for each technology. The two most important areas are energy efficiency (most important) and reductions in power sector emissions (almost as important). Within the power sector, CCS accounts for, by a wide margin, the largest emission reductions; larger than nuclear power, renewables or fuel switching. In addition, CCS achieves significant reductions in the industry and fuel-processing sectors. Combined, CCS is only second to energy efficiency as a source of emission reduction.

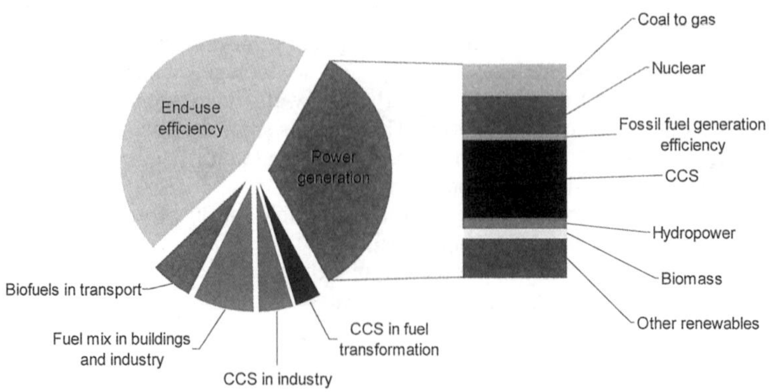


*Figure 3: Emission Reduction by Technology.*

Marginal cost analysis can show how important CCS would be by comparing the estimated marginal cost of emission reduction with and without CCS. Marginal cost is important as it establishes the price of $CO_2$ control. Also, with higher marginal costs, it will be more difficult to establish the policy measures needed to bring about reduced emissions. In Figure 4 we see the marginal cost of the ACT Map, low nuclear, low renewable and no CCS scenarios for two world regions: Europe and China. In each region, the no-CCS scenario results in the largest marginal costs of emissions (40 US$/ton of $CO_2$ or higher), up from around $25 US$/ton of $CO_2$. This figure is particularly interesting as it shows how important CCS is to China as a way to keep the cost of emission reductions close to $25. In Europe, the low nuclear and low renewable scenarios also cause fairly large increases of marginal cost (above $35), but these scenarios do not greatly affect the marginal cost in China. It is only the no-CCS scenario that significantly raises the marginal cost of achieving lower emissions.

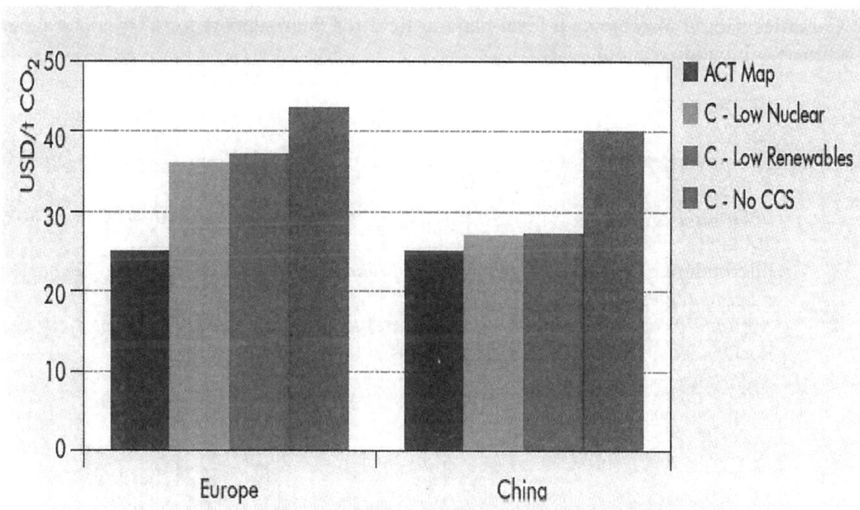


*Figure 4: Marginal Costs of Low-Technology Development.*

In conclusion, IEA's *Energy Technology Perspectives* study shows how important it will be to achieve success in overcoming the current barriers to widespread deployment of CCS. These are:

- Establishing whether widespread CCS can be achieved with an acceptable environmental risk.
- Identifying protocols for implementing CCS in such a way as to assure low risk.
- Establishing the legal and policy frameworks necessary for the private sector to engage in CCS.

- Reducing the cost of $CO_2$ capture through RD&D, emphasizing a range of technologies applicable to both IGCC and steam coal plants.

## CONCLUSION

This paper has summarized why CCS is needed to reduce power-sector emissions to near-zero levels and achieve important reductions in other industrial sectors. Next to energy efficiency, CCS may be the most cost-effective strategy to achieve very large emission reductions. Since, CCS is almost always an added cost, incentives are needed to stimulate commercial development. Legal and policy frameworks are needed as well as commercial-scale projects to demonstrate the needed technologies. Governmental cooperation (such as the IEA Greenhouse Gas R&D Programme and the Carbon Sequestration Leadership Forum) is essential. Capture technology development should remain a top energy R&D priority as well as projects to test and map repositories. Countries should also create a level-playing field for their support for CCS along other climate-mitigation technologies.

## REFERENCES

1. Intergovernmental Panel on Climate Change Special Report, *Carbon Dioxide Capture and Storage*, ISBN 92-9169-119-4, 2004.
2. International Energy Agency, *Prospects for $CO_2$ Capture and Storage*, Paris, 2004.
3. International Energy Agency, *Energy Technology Perspectives 2006, Scenarios and Strategies to* 2050, Paris, 2006.
4. National Academy of Engineering, "The Hydrogen Economy, Costs, Barriers and R&D Needs," National Academies Press 2004.

# 3. GLOBAL MONITORING OF THE PLANET

## PROLIFERATION
## FOCUS: NUCLEAR WEAPONS

# PROLIFERATION OF NUCLEAR WEAPONS: THE 2006 OUTLOOK

RICHARD WILSON
Department of Physics, Harvard University, Cambridge, USA

As I introduce this session I want to make several general observations.

Firstly, the World Federation of Scientists, meeting at Erice, make their best contributions when they look at long-term rather than short-term issues. I suggest that this will be the case when we consider the problems of nuclear weapons; the possible proliferation thereof both in number and in numbers of countries possessing them.

Secondly, I note that it is now 61 years since the world's scientists understood that mankind now knew how to destroy itself. A few of them understood when they saw the test at Alamagordo in July; the rest when the bomb dropped on Hiroshima. In 1945, I was running a Boy Scout camp in southern England when a boy came from the nearby farm with the news: an atomic bomb had been dropped on Hiroshima. I knew it was the end of a 6-year war that had killed 80 million people. I was overjoyed. In 1940, I had seen a house destroyed by a 50-pound bomb and in 1944 I had seen a U1, carrying a 1 or 2 ton warhead, head on, 150 yards away just before it crashed into a house killing the family and knocking out half the windows in our house. I had studied mathematics one year and physics another in Oxford, and was expecting to enter the Royal Air Force as a radar officer, and although I did not know the strength of the explosion I knew it was large (now known to be 15,000 tons TNT equivalent). The Nuovo Zembla blast was 5 million tons. A billion times the strength of the bomb that destroyed a house in 1940. We still have 10,000 of them though only a few that large.

Thirdly, I note that the Hiroshima and Nagasaki bombs forced the end of World War II in which America emerged as the most powerful nation. British imperialism was replaced by a reluctant American imperialism. But it was an America that projected hope, optimism and generosity. Domestically this was displayed in the GI bill under which returning veterans from all walks of life were paid to acquire an education. In Europe it was expressed in the Marshall plan in which victors France and England, and vanquished Germany, were rebuilt at U.S. expense. This contrasted with the short-sighted demands of reparations by the victors of World War I. The Japanese, expecting to be treated as slaves, as had often occurred with vanquished nations, were treated well and the Japanese showed their gratitude by taking the U.S. national game, baseball as their own. A big compliment most of the U.S. ignored. In nuclear matters the generosity was first shown in the Baruch plan, to share all secrets, and their control, with the fledgling United Nations but this plan did not get accepted. But Eisenhower in his famous Atoms for Peace speech of 1952 displayed this generous trend. Nuclear information was to be shared with all nations if it had a peaceful intent. This was later embodied in the Nuclear Non-Proliferation Treaty, (NPT) which we will discuss today. Unfortunately the generosity was not always respected. Canada supplied a heavy water reactor to India but carelessly did not insist on guarantees that it not be used for military purposes. My Canadian friends were appalled that India used it to help make a bomb. I suspect that the French, who supplied a heavy water reactor to Israel were less surprised and not appalled. But neither country, Canada or France, is likely to repeat this mistake.

Fourthly, I note that in the last 50 years all this has changed. America is no longer seen as a land of hope, but a land of fear, pessimism and greed. Fear seems to be the emotion that President Bush, the younger, uses to keep power. The greed was all too obvious in the behavior of Ken Lay and others of Enron. Many of us also feel that way about Halliburton. Why did we enter Iraq and not Sudan, if it were not to control most of the world's oil? Many have noted that in the last 60 years the USA has steadily increased the military option and has with increasing frequency used it as a substitute for diplomacy in Grenada and Panama. The CIA, established as a civilian agency to analyze data, is now reduced to its covert activities and the analysis is "safely" in the defence department. Many friends of mine say: "we don't hate Americans we hate American policies." Don't let us off the hook so easily. 18 months ago the American people re-elected George W. Bush, knowing full well what the policies were. Don't let me off the hook either. Although I did not vote for President Bush, I did not campaign for his rival. I did not go and explain to Ohio voters why the world depended upon them.

In nuclear matters it is true also. Most outside observers feel that the USA is the biggest violator of the Non-Proliferation Treaty. Article IV calls upon the nuclear weapons states to disarm. While that seemed inappropriate during the cold war, it seems stupid to have so many bombs now. No doubt Dick Garwin will correct me if I have the details wrong, but I believe the USA has 10,000 nuclear weapons of which some 1,500 are on trigger alert or close to it. The number of weapons required to instil caution in an opponent is probably no more than a hundred or two; explosion of these in major cities would inflict unacceptable damage on any opponent. Of course an opponent who is ignorant of the destructive power of these weapons might think that his country could survive such an attack. Indeed in Mutually Assured Destruction, it is the perception of the enemy politician that matters. I was scared at the thought of a hundred or so bombs in the Cuban missile crisis. But some politicians may be harder to scare. Let us hasten their learning. I remember Marshal Yazov telling me in May 1991 that Chernobyl had persuaded reluctant Russian generals that a nuclear war could not be won.

Worse still, the USA, both under President Clinton and under President Bush has declined to ratify the nuclear test ban treaty even though committees of the nations leading scientists have agreed that under any scenario they can think of, the USA would be more secure after signing the treaty than before. And the USA is making more and different types of weapons. The USA is also in violation of title VI which encourages weapons states to share their knowledge for peaceful purposes. This was done with enthusiasm 50 years ago, as both the USSR and the USA provided research reactors to their client states. This enthusiastic generosity has lessened and many nations feel the stick of obtrusive inspections more than the carrot of help in nuclear matters. Indeed, I have argued that for the last 60 years the biggest single incentive for a nation to develop nuclear weapons has been the attitude of the United States. This started when, in September 1945, British scientists, returning from Los Alamos, found that scientific communication was cut off. They did not even have access to the papers they themselves had written! Britain needed no atomic bomb of its own for its defence but the British cabinet decided that one was necessary to be taken seriously by the USA and maintain the "special relationship." There was an implicit promise in the Atoms for Peace speech that nations that did not make nuclear bombs were to be treated by the weapons states more generously, particularly in nuclear matters, than those that made them. This implicit promise was bent 30 years ago when Israel was treated more generously than Egypt or

Syria. The United States is bending it again with the proposal to provide assistance on nuclear power to India.

To get special treatment should not any sensible nation develop its own bomb?

In the forthcoming talks Ahmad Kamal will no doubt tell us, as he briefly did 2 years ago, that NPT was dead on arrival. Others will devise band-aids to fix the weak points. But I suggest that we must go beyond these two extreme approaches.

What do we as scientists think is a reasonable way of controlling the genie that escaped from the bottle 61 years ago? Is NPT, with all its faults, a good starting point?

And more generally, how can we get back the hope, optimism and generosity that was the prevailing feeling in 1950?

We have with us Indian scientists who will tell us their perspective on the latest proposal for the U.S. to provide nuclear technology for nuclear power. And we have, of course, Pakistani scientists also. But I think that the most important will be the talks from Japan, Germany and Switzerland—three countries that have the capacity and infrastructure to make bombs within months if not weeks. Why have they decided not to do so? Do they, as a Japanese said in Singapore about 5 years ago, depend on the nuclear umbrella of the USA? Is that enough?

We had hoped to have scientists from Iran and Uzbekistan to add their perspective but, alas, they are unable to be here.

In 1978, just after the High Energy Physics Conference, I lectured on this general subject at Keidanren in Tokyo. First I had to apologize. In 1945 I should not have been overjoyed. Glad that the war was over, of course. But in retrospect I am ashamed of my joy; I should immediately have had intense sorrow and sympathy for the 200,000 people who lost their lives in those two cities. Then I reminded my listeners that if all the atomic bombs then possessed by the USA and USSR were exploded, it might be the end of civilization as we know it. "Please do not be lazy" I pleaded with the audience. "It is your world as well as America's. If we foul up, we will drag you into the mire with us. Help us think through this problem and establish procedures, and guidelines so that the world will stay at peace for ever." I make this same plea now not merely to the Japanese but to the delegates to this conference and through you to the whole world.

Unless the hope, optimism and generosity return, the imperial USA will continue to choose the short-term path that was espoused by General Groves and Edward Teller. To ensure that no nation can equal the USA in military strength. This was the policy of ancient Rome. But when the Roman empire fell, there were civilized people in distant countries and less civilized people around in Europe to pick up the pieces. If and when the American empire falls there could be Armageddon.

Many U.S. scientists, particularly physicists, have tried to influence the congress and administration over the years with diminishing success. I hope Dick Garwin can tell us how to remedy this. Maybe the U.S. scientists should have a massive education campaign, beginning even in elementary schools, and including the southern and midwestern states. Maybe the hope will not come from America. Maybe from a united Europe. Maybe from Russia. Maybe from Japan. Maybe from China. Let us start with hope from Erice.

"We will try to remain serene and calm when Alabama gets the bomb." (T. Lehrer, circa 1970).

## THE DEMISE OF THE NON-PROLIFERATION TREATY

AHMAD KAMAL
Senior Fellow, United Nations Institute for Training and Research
New York, USA

Conventional wisdom tells us that the tentative nuclear deal between the United States and India will mark the death of the Non-Proliferation Treaty (NPT). That is only partially correct.

The NPT has already died a thousand deaths. It started life still-born at birth, based as it was on an unequal and discriminatory "contract" between signatories, most of whom were beguiled into giving up their sovereign rights in exchange for a deceitful promise which the five nuclear powers of the time had no intention whatsoever of fulfilling. Glass beads for gold all over again. Article 4 of the NPT had stated that:

> *"Each of the Parties to the Treaty undertakes to pursue negotiations in good faith on effective measures relating to cessation of the nuclear arms race at an early date and to nuclear disarmament, and on a Treaty on general and complete disarmament under strict and effective international control."*

There was never any "good faith." The NPT died when all five nuclear weapons states used the cover of the NPT, immediately after its signature, and then consistently over the next two decades, to embark in enthusiastic nuclear proliferation by building up their own nuclear weapons stockpiles hundreds of times over, and their fissile material stockpiles by the thousands of times.

The NPT died again in 1974, when India challenged it frontally by exploding its first nuclear device, calling it the Smiling Buddha, and in the process thumbing its nose publicly at the discriminatory nature of the NPT regime. We should not forget either that it was the United States, Canada, and Russia which had helped build up Indian nuclear capacity in the first instance, and were now reaping the seeds of their own actions.

It died again when Israel built up its own covert nuclear capacity with the active scientific help and collusion of its closest ally, who saw it as a strategic partner living behind a stockade in an unfriendly environment, or a sort of a forward outpost in the oil-rich territories of the Asian Wild West.

It died yet again, when India and Pakistan both conducted almost a dozen nuclear tests within three weeks, in a tit-for-tat race to prove their respective virilities to each other. Once they had done so, both declared themselves "nuclear weapon states." A frontal challenge once again to the limiting definition contained in Article 9 of the NPT, which stated unequivocally that:

> *"For the purposes of this Treaty, a nuclear-weapon State is one which has manufactured and exploded a nuclear weapon or other nuclear explosive device prior to January 1, 1967."*

It died all over again when signatories to the NPT, who had ostensibly signed on to its non-proliferation provisions, were found to be cheating under the table, and building up nuclear weapons capacity, accompanied by in one case by dangerous missile delivery capacity. Once these illegal concealments were discovered, these states were either bombed back into a *status quo ante* as in the case of Iraq, or prevailed upon to give up their ambitions as in the case of South Africa and Libya, or as in the case of North Korea, the state just withdrew from the NPT.

It received another death blow from an outsider to the NPT, when it was discovered that a prominent Pakistani scientist was peddling his nuclear wares to all who were willing to pay the price in his illicit Nuclear Wal-Mart. He was then granted a gracious pardon despite all his criminal actions.

So, the nuclear deal between the United States and India is nothing new. It is part of a pattern under which key nuclear weapons states are only too willing to subject their ostensible nuclear non-proliferation policies to their real foreign and security policy interests. All is permissible when it is a question of national security, and no international treaty is going to be allowed to stand in the way. The focus is on national security, not Rule of Law. No Article 1 of the NPT shall stand in the way, even if it states categorically that:

> *"Each nuclear-weapon State Party to the Treaty undertakes not to transfer to any recipient whatsoever nuclear weapons or other nuclear explosive devices or control over such weapons or explosive devices directly, or indirectly; and not in any way to assist, encourage, or induce any non-nuclear weapon State to manufacture or otherwise acquire nuclear weapons or other nuclear explosive devices, or control over such weapons or explosive devices."*

Once it is understood that the underlying foundation of nuclear non-proliferation is security and not Rule of Law, the whole of the current debate falls into place.

India first. The foreign policy objective is to somehow contain a China which is seen as a growing threat because it appears to be too huge to fall into line in recognizing the primacy of the sole surviving super power. So, we have to build up Indian "peaceful" nuclear capacity, Article 1 of the NPT and thirty-six years of categorical policy statements notwithstanding. It does not matter if, in the process, the Indian production of nuclear weapons goes up two-fold or three-fold. After all, that helps build up India as a counter-weight to China, and that is what we need most of all. It does not matter either if this increased Indian nuclear capacity sets off a dangerous arms race with Pakistan. If the choice is between containing China and an arms race in South Asia, the former is obviously going to remain more attractive.

Iran next. Under Article 4 of the NPT, Iran has the full legal right to build up its nuclear fuel cycle, as long as this is for the "peaceful uses of nuclear energy." Article 4 of the NPT states as follows:

1. *Nothing in this Treaty shall be interpreted as affecting the inalienable right of all the Parties to the Treaty to develop research, production and use of nuclear energy for peaceful purposes without discrimination and in conformity with articles I and II of this Treaty.*
2. *All the Parties to the Treaty undertake to facilitate, and have the right to participate in, the fullest possible exchange of equipment, materials and*

> *scientific and technological information for the peaceful uses of nuclear energy. Parties to the Treaty in a position to do so shall also cooperate in contributing alone or together with other States or international organizations to the further development of the applications of nuclear energy for peaceful purposes, especially in the territories of non-nuclear-weapon States Party to the Treaty, with due consideration for the needs of the developing areas of the world.*

Note that the NPT uses the word "inalienable." In other words, the right to the peaceful uses of nuclear energy is not conferred by the NPT, but is recognized as pre-existing and fundamental.

Iran affirms that this is indeed the case, and produces scientific evidence about the finite nature of fossil fuels, and a thirty-year old testimonial from a President of the United States encouraging it to develop a nuclear fuel cycle. The counter-arguments of today are based on suspicions and political considerations, and on the fact that mastery of the nuclear fuel cycle inevitably adds on to capacity to produce nuclear weapons. That is a problem to which there is no scientific solution. Either we have to depend on the legal terminology of the NPT, or we have to base policy on extra-legal considerations. The latter hold sway.

North Korea finally. Security is a primary concern of all states, not just of the few. All states have the sovereign right to analyse their security threats and to find durable solutions which meet their specific requirements in their own specific regions. North Korea sees a threat to its own security from a South Korea, which has the physical presence of U.S. troops, armed with nuclear weaponry, albeit unconfirmed.

In actual fact, our concerns about the North Korean nuclear weapons capacity is not so much about North Korea itself, but the consequence that any forward policy or nuclear and missile delivery tests on the part of that country may produce on Japan. The latter has tens of tons of plutonium in the country, and should it face a security threat from North Korea, it would have to change its decades old non-nuclear policy. That is nightmare scenario, for a nuclear Japan would completely change the geo-politics of the Far East, and indeed of the world as a whole.

Security is also far more comprehensive than just military security. Energy security, or freedom from energy dependence on outsiders is perhaps just as understandable in a world that hungers for energy as a way out of the low status to which they have been condemned in an unjust environment. No effort at denying them that fundamental right to sovereign life and security can be successfully implemented, at least not in any long-term time frame.

So, we clearly have a major problem on our hands. We all want a nuclear weapons-free world, but at the same time we want to retain our own nuclear weapons and to deny them to others. We all want energy self-sufficiency, and know that the recourse to nuclear energy is inevitable in a fossil-fuel shrinking world, but at the same time we do not want to allow the concomitant technological knowledge and capacity which can be used for nuclear weapons production also. We all want our own security, but at the same time want to deny the same right to others.

While these factors slowly sink in, regional bush-fires continue to burn in many parts of the world. While that happens before our eyes, scientists continue to serve their masters in building up nuclear weaponry to ever more deadly and usable proportions. There is scientific pride in building smaller nuclear weapons, of miniaturising them into battlefield proportions, and of giving them greater killing capacity.

So much then for a Non-Proliferation Treaty which is not worth the paper it is written on. There is no point lamenting the demise of a treaty that was so deeply flawed from the start. What is necessary is to agree on the way forward, of a world in which nuclear weapons can be finally outlawed, and the benefits of nuclear energy spread around the world. This will require a complete change of attitude, in which the well-endowed understand that they live in a globalised world, in which the aristocratic privileges that they have enjoyed for so long will have to give way to tolerance and compassion and Rule of Law.

The answer lies, now as always, in verifiable nuclear disarmament, and not in nuclear non-proliferation. We have less than ten nuclear weapon states today, so a negotiation towards nuclear disarmament should require an agreement between this handful of countries only. If they understand the need, we can eliminate this dangerous weapon of mass destruction, once and for all, just as we did in the case of chemical weapons. If they do not, and given the arrogant attitudes of most of them, the chances are that they will not, then we must adjust ourselves to a world in which more and more countries will move into the nuclear club. The tide may perhaps be slowed, but it cannot be stopped. In which case let us just relax and enjoy it, and stop pontificating about the advantages of nuclear virginity.

# SCIENTISTS AND (NON)PROLIFERATION OF NUCLEAR WEAPONS

RICHARD L. GARWIN
Thomas J. Watson Research Center
IBM Research Division, Yorktown Heights, USA

Scientists have played an important role in the creation of nuclear weapons and in the attempt to prevent their proliferation and indeed to greatly reduce their numbers and the possibility of use.

The Pugwash movement and others have attempted to eliminate nuclear weapons, with the judgement that they are more a threat to humanity than they are a tool for survival. To date, these efforts have met with very mixed success.

Scientists have also been an important means for proliferation, either intentional or as an unwanted or inadvertent consequence of their actions. So we are both the good guys and the bad guys, recalling the character in the Peanuts comic strip, "We have met the enemy and he is us!"

As is known to all, the first nuclear weapon was tested by the United States in New Mexico on July 16, 1945. This plutonium implosion weapon was then used to destroy the Japanese city of Nagasaki, three days after the $^{235}$U gun-type weapon destroyed Hiroshima. Since then, no nuclear weapon has been used in warfare, but more than 2000 have been exploded in test by the United States, the Soviet Union, England, France, China, India, and Pakistan. (The United States has conducted 1,149 nuclear tests, the Soviet Union some 1,100, Britain 45, France 210, and China 43. In addition, India and Pakistan have conducted a handful of nuclear tests in 1998.) I have previously written on non-proliferation, with many papers available at www.fas.org/RLG/, in particular a 1996 paper, "The Post-Cold War World and Nuclear Weapons Proliferation."[1] Dozens of countries have had nuclear weapons development programs, but essentially all except Israel, India, and Pakistan have signed the Non-Proliferation Treaty of 1970 and have committed themselves not to develop or to obtain nuclear weapons. It is perfectly legal, however, for any state belonging to the NPT as a non-nuclear weapon state (NNWS) to give three-month notice and it is then free to develop or acquire nuclear weapons. Although any NNWS has that right, it might not be prudent to exercise it. In any case, the Parties have agreed:

> *Such notice shall include a statement of the extraordinary events it regards as having jeopardized its supreme interests.*

The program to develop the U.S. nuclear weapon was greatly aided by British participation and the story is told very well in Richard Rhodes' "The Making of the Atomic Bomb."[2]

On the more technical side, the work at Los Alamos, where the bombs were actually designed and built, is described in considerable detail in a book, "Critical Assembly."[3]

The first technical information about the nuclear weapon was provided in the report by Henry DeWolf Smyth, in a document universally dubbed, "The Smyth Report."[4] The explosion of the first two weapons in Japan, of yield 13,000 and 20,000 tons of high-explosive equivalent, revealed the existential secret of the atomic bomb: that it could be done. Whatever the blind alleys of conception that had prevented full understanding by German scientists working on nuclear weapons, the mystery was largely eliminated by the fact that a nuclear weapon could be built, and in both, the gun-type and implosion-type approaches.

Scientists served also as spies, especially Klaus Fuchs, a German-born physicist who worked with the British mission to Los Alamos throughout the war. It was a great surprise to his colleagues when, in 1950, he was revealed to have been spying for the Soviet Union throughout his activity at Los Alamos.

Immediately after Hiroshima, scientists in the United States turned to the control of the nuclear weapon, in order that it not be used imprudently or accidentally by the United States. On the technical side much was done, but not until 1962 were the Permissive Action Links (PAL) implemented, first in 7,000 U.S. nuclear weapons deployed in Europe, and more recently in far more sophisticated style in essentially all U.S. nuclear weapons. In 1945, J. Robert Oppenheimer turned his attention to the international control of nuclear weapons and especially to their abolition in what has been universally called the Baruch Plan, in view of its presentation by U.S. financier Bernard Baruch at the United Nations on June 14, 1946.

It was of great interest to me to learn recently of the different views held by Hans Bethe and Oppenheimer in 1945. Oppenheimer in November 1945 in a public speech, had predicted of the future use of nuclear weapons: "not a few, but thousands, or tens of thousands in case of armed conflict in a world of nuclear weapons." In contrast, Hans Bethe, who had led the Theoretical Division at Los Alamos in designing the first nuclear weapons, and who again led the theoretical design of the first U.S. hydrogen bomb, wrote in 1995:[5]

> *"I feel the most intense relief that these weapons have not been used since World War II, mixed with the horror that tens of thousands of such weapons have been built since that time, one hundred times more than any of us as Los Alamos could have ever imagined."*

While Bethe recalls Los Alamos views in 1945 that hundreds of weapons might be built, Oppenheimer was already predicting that they could number in the tens of thousands.

The Baruch plan envisaged the elimination of national possession of nuclear weapons, but it got nowhere, in view of lack of Soviet favorable response, because the United States already had full knowledge of nuclear weaponry and the possession of some, whereas the Soviet Union had nothing. Whether the Baruch Plan would have been adopted by the United States, had the Soviets agreed, is also problematical, as is seen by the course of arms control measures over the years.

Any scientist involved in the early work on nuclear weapons for a country is contributing to "proliferation." Many such scientists do this out of patriotism and even from a view that it would be beneficial to the security of the entire world. We have a lot of information now on the early Soviet nuclear weapon pioneers, and also on Andrei Sakharov in his pioneering work on the Soviet hydrogen bomb. Much of this is available in readable form through another Richard Rhodes book.[6]

How could a country, in support of its own national security goals, in good faith have signed the Non-Proliferation Treaty? But there are benefits to the NNWS status afforded by the NPT, in that the nuclear weapon states commit themselves to providing information and other help to the NNWS in the peaceful applications of nuclear energy. It does, of course, have the escape hatch of the three-month announcement and the perfectly legal abandonment of membership in the NPT.

Benefits to the NNWS under the NPT include applications in industry in medicine, as well as in basic research, and in the use of nuclear power from chain-reacting fission systems in order to provide electricity or, ultimately process heat.

Unfortunately, the NPT has the flaw, perfectly evident from the beginning, that materials and facilities acquired during the NNWS status of the country involved, can be used legitimately in a nuclear weapon program after the country opts out of the NPT. This particular flaw has come to the fore in the behavior of North Korea, which had for years a 5 MWe (25 MWt) natural-uranium/graphite reactor for which the spent fuel (6,000 fuel rods under IAEA safeguards) was stored for many years in a cooling pond. It is believed that North Korea more than a decade ago had separated some 6-8 kg of Pu from this spent fuel, but more recently North Korea gave the required 3-month notice and resigned from membership in the NPT; it subsequently stated that it had reprocessed the spent fuel formerly under IAEA safeguards.

More recently, Iran revealed that for 18 years it had not properly reported its activities to the IAEA—the dedicated agency for inspection and reporting under the NPT—and that it had been working during that period on centrifuge enrichment of uranium, according to Iran, for peaceful uses in nuclear power. Western Europe insists that Iran was thereby in violation of the NPT and must suffer consequences. Iran responds that the United States, in particular, was not fulfilling its obligations under the NPT and had imposed sanctions against Iran for totally different reasons, that prevented the U.S. and other countries from fulfilling its obligations under the NPT. Iran goes on to say that if they had revealed their perfectly legal activities under the NPT, then further "illegal" sanctions would have prevented this work, rather than fulfilling the obligations of the NWS to support the "peaceful" work of Iran.

Not being a member of the NPT either as a NWS or an NNWS, India did not violate the NPT by building and testing nuclear weapons, first in 1974, and then in 1998. But it did violate its undertakings to Canada and the United States in doing so, because it misused heavy water and the reactors that had been built with the help of other states.

More recently, A.Q. Khan of Pakistan constituted a one-man proliferation machine, in selling packages of information and materials (including centrifuges and centrifuge plans) to other states, including Libya and North Korea. Libya has revealed all of its former activities toward the acquisition of nuclear, chemical, and biological weapons, and has cooperated in the removal from its country of the centrifuges and other materials involved.

Complicating matters, of course, is that although the use of nuclear weapons by sovereign states could in principle be deterred, that is often not the case with sub-national groups such as terrorists. And so non-proliferation to sub-national groups is really more important than non-proliferation to states.

Furthermore, there was a tendency in the arms race with the Soviet Union (and there is still in the nuclear weapon establishments) to consider that the job of security is preventing access by foreign countries to the latest nuclear weapon information, because earlier nuclear weapons are "obsolete" and less performing. There are two major flaws in this. First, U.S. nuclear weapons technology has not improved much since 1962, although weapons can more readily be packaged to fit demanding environments. The more important observation, though, is that a terrorist could be perfectly satisfied with a nuclear weapon that has the yield of the Hiroshima bomb (13 kilotons) and especially if it could be assembled surreptitiously and detonated in a city.

My 2002 paper at Erice[7] provides details.

So where in the past, non-proliferation activities concentrated on limiting information and the knowledge of how to build nuclear weapons, that information has accumulated through leaks and official publications and has been widely spread and commented via the Internet, and so for the last 20 years the more important tool has been to limit access to nuclear-usable materials. There the problem is that nuclear materials have become "dual use" because there is the important nuclear power sector that also uses enrichment of uranium to provide on a very large scale LEU or even HEU for power reactors. Each 1000 MWe power reactor that is fed about 1000 kg of $^{235}U$ per year in the form of 3-5% low-enriched uranium (LEU), requires about 150,000 "separative work units (SWU)" per year. The 50,000 centrifuges that Iran plans to deploy at its facility in Natanz probably have an annual production capacity per machine of about 3 SWU, so that the entire Natanz plant, if successful, could barely provide the enriched fuel to continue to feed a single 1000 MWe power reactor such as that being completed at Bushehr. On the other hand, a nominal 20 kg of highly enriched uranium (e.g., 95% $^{235}U$) would require about 220 SWU/kg, and thus only about 4400 SWU per weapon. At 150,000 SWU per year, the Natanz plant could thus enrich enough uranium for 150,000/4400 = 33 uranium implosion weapons per year, of the type built by China and then by Pakistan. This illustrates the threat that is for the most part kept in check by the IAEA safeguards program.

Although plutonium has been used for years in the form of MOX fuel—mixed oxide fuel—in French nuclear reactors, in anticipation of the arrival of a generation of fast neutron reactors, it is clear in retrospect that, although affordable, this has not been an economically advantageous choice. But plutonium (and even civil plutonium as opposed to "military plutonium") can be used to make high-performance nuclear weapons, as published in 1994 by nuclear-weapon expert J. Carson Mark, who led the Theoretical Division for many years at the Los Alamos National Laboratory. He emphasized that an implosion that would have produced 20,000 tons of high-explosive yield with weapon-grade Pu could under no circumstances yield less than 1000–2000 tons with reactor-grade Pu and might with considerable probability yield much more.

Although nuclear weapons and nuclear reactors alike depend on the fission chain reaction, carried out by the neutrons of which 2 to 4 are liberated in each fission, the technologies involved are so different that reactor scientists and engineers have often believed that the plutonium that is the often unsought byproduct of heat from a nuclear reactor could not readily be used for a nuclear explosive. Non-proliferation efforts are inherently hampered by an inability to describe precisely the hazards to be avoided—e.g., precisely how reactor-grade Pu can be fabricated into a reliable, high-efficiency nuclear explosive—while at the same time motivating the international community to accept the fact that such Pu is a hazard of the same magnitude as the "weapon-grade" Pu that has less of the isotope Pu-240 and thus a smaller neutron background and less self-heating.

Much of the safeguards effort of IAEA properly goes toward the care of the "back-end" of the reactor fuel cycle containing plutonium. Surely non-state actors such as terrorists would prefer HEU, and after that, weapon-grade plutonium; if we are not to abandon our responsibility for the safety of our citizens, we must and will increase the priority and resources to secure HEU and weapon-grade Pu against theft and diversion as well as to prevent access to reactor-grade Pu, also usable in simple implosion weapons.

Despite its deficiencies and its inability to prevent the acquisition of nuclear weapons by North Korea, India, Pakistan, and Israel (the last three not having signed the NPT in any case), in my opinion the NPT has been a great success. It is true that the five nuclear-weapons states have not fulfilled their obligations under the NPT to reduce their holdings of nuclear weapons and are far from complete nuclear disarmament, that is in any case stipulated only in the context of general and complete disarmament. Specifically, in the body of the NPT[8] we find:

> *"Each of the Parties to the Treaty undertakes to pursue negotiations in good faith on effective measures relating to cessation of the nuclear arms race at an early date and to nuclear disarmament, and on a Treaty on general and complete disarmament under strict and effective international control."*

It is not nuclear disarmament to which the NPT members have committed themselves, but to *negotiations* toward this end. The United States, in particular, should be criticized for resisting such negotiations, but not for its failure to disarm. On the other hand, the highly misleading "Treaty of Moscow"[9] of 2002 states in Article I:

> *"Each Party shall reduce and limit strategic nuclear warheads, as stated by the President of the United States of America on November 13, 2001 and as stated by the President of the Russian Federation on November 13, 2001 and December 13, 2001 respectively, so that by December 31, 2012 the aggregate number of such warheads does not exceed 1700-2200 for each Party. Each Party shall determine for itself the composition and structure of its strategic offensive arms, based on the established aggregate limit for the number of such warheads."*

The actual number of U.S. nuclear weapons is likely to be more like 10,000. From the George W. Bush Letter of transmittal to the U.S. Congress (at the same URL):

> *"The Treaty requires the United States and Russia to reduce and limit their strategic nuclear warheads to 1700-2200 each by December 31, 2012, a reduction of nearly two-thirds below current levels. The United States intends to implement the Treaty by reducing its operationally deployed strategic nuclear warheads to 1700-2200 through removal of warheads from missiles in their launchers and from heavy bomber bases, and by removing some missiles, launchers, and bombers from operational service."*

It seems to me that this expresses the contempt for treaties evident in the renunciation of the 1972 ABM Treaty[10] by the George W. Bush administration in December 2001, without the statement of "extraordinary circumstances" required by Article XV of the treaty:[11]

1. *This Treaty shall be of unlimited duration.*
2. *Each Party shall, in exercising its national sovereignty, have the right to withdraw from this Treaty if it decides that extraordinary events related to the subject matter of this Treaty have jeopardized its supreme interests. It shall give notice of its decision to the other Party six months prior to withdrawal from the Treaty. Such notice shall include a statement of the extraordinary events the notifying Party regards as having jeopardized its supreme interests.*

The U.S. president did announce:

*"I have concluded the ABM treaty hinders our government's ability to develop ways to protect our people from future terrorist or rogue-state missile attacks..."*

but I know of no discussion with Russia as to whether the limited defensive system envisaged could be interpreted by Russia as permissible under the ABM Treaty itself.

The nations of the world, especially my own, must return to the respect of international treaties. They will then be in a position to demand that more resources be provided by the international community and spent effectively by Russia to rapidly blend down its many tons of surplus HEU to a level below 20% $^{235}U$, where it is not usable for nuclear weapons without further enrichment. Similarly for the 100 tons or more of surplus weapon Pu in Russia, where the expenditure will not be repaid by the use of the material as power reactor fuel, as in the case of the HEU.

In general, I believe the NPT has been and remains valuable, but it must be supplemented by additional protocols and by major actions far beyond the approximately $100 million annual level at which its safeguards activities operate.

REFERENCES


1. R.L. Garwin, The Post-Cold War World and Nuclear Weapons Proliferation April 19, 1996 The 29th JAIF Annual Conference Session 5 "Nuclear Non-Proliferation and Plutonium" Nagoya, Japan.
2. R. Rhodes, "The Making of the Atomic Bomb," Simon & Schuster (New York), 1987.
3. L. Hoddeson, P.W. Henriksen, R.A. Meade, and C.L. Westfall, "Critical Assembly: A Technical History of Los Alamos during the Oppenheimer Years, 1943-1945," 1993.
4. H. DeWolf Smyth, "Atomic Energy for Military Purposes: The Official Report on the Development of the Atomic Bomb Under the Auspices of the United States Government," August 1945.

5. H.A. Bethe, Public Interest Report, Journal of the Federation of American Scientists 48, September_October 1995; also available at http://www.pugwash.org/about/bethe.htm.
6. Rhodes, "Dark Sun: The Making of the Hydrogen Bomb", Simon & Schuster, 1995.
7. R.L. Garwin, "Nuclear and Biological Megaterrorism," presented at 27th Session of the International Seminars on Planetary Emergencies, Erice, Sicily, August 19-24, 2002. (A shorter version was published in MIT's Sept. 2002 Technology Review, titled "The Technology of Megaterror" at http://www.technologyreview.com/articles/garwin0902.asp)
8. http://www.state.gov/t/np/trty/16281.htm#treaty
9. http://www.state.gov/t/ac/trt/18016.htm
10. http://www.state.gov/t/np/trty/16332.htm#treaty
11. http://www.state.gov/t/np/trty/16332.htm#treaty

# IMPLICATIONS OF THE INDO-U.S. NUCLEAR AGREEMENT

RAMAMURTI RAJARAMAN
Jawaharlal Nehru University, New Delhi, India

SUMMARY

The Indo-U.S. agreement represents a fundamental transformation of the relations between the two countries. It aims to lift existing sanctions on nuclear commerce with India which have handicapped India's progress in developing nuclear energy. The Agreement is also a natural recognition on the part of the U.S. of India's growing emergence as a major power in the world. In this talk we explain the main features of the Agreement and its possible consequences

On July 18, 2005, India and the U.S. announced a wide-ranging agreement, of which the nuclear component was the most critical. The nuclear deal calls for India to identify and separate its nuclear facilities into civilian and military categories and place the former under international safeguards. In return, the U.S. would resume full civil nuclear energy cooperation with India, work with the U.S. Congress to adjust U.S. laws to enable such cooperation and persuade allies in the Nuclear Suppliers' Group to lift their sanctions.

The civil-military separation plan was negotiated and announced in March in New Delhi. But it remained to get the U.S. Congress to pass the required legislation exempting India from nuclear sanctions and for the Nuclear Supplier Group countries to agree to do the same. As of now, the desired legislation has been passed by the U.S. House of Representatives. It is up before the U.S. Senate where chances are that it will again be passed, perhaps with some caveats, since the relevant Senate committee has already voted in favor.

On the face of it, it would seem that the Agreement will be very beneficial to India. Currently, international nuclear cooperation and transfer of technology to India is being withheld, not only by the U.S., but by the entire Nuclear Suppliers Group, which includes all the countries in the world with advanced nuclear technology. The sanctions will be lifted if the Agreement fructifies. The long isolation of India's nuclear scientists would end. India can also hope to purchase badly needed supplies of uranium, build more reactors for enlarging our nuclear energy program and reduce the reliance of our increasingly energy-hungry country on scarce fossil fuels.

But the Agreement has also raised a lot of criticism from non-proliferation activists. Briefly, their concerns are:

1. By giving special exemption to India, the deal may undermine the NPT regime and non-proliferation efforts in other countries.
2. It leaves India with considerable un-safeguarded capability for producing weapons grade fissile material, should it choose to do so.
3. India may be able to spend all its indigenous uranium ore for military purposes since the Agreement allows it to import fuel for its civilian reactors.

Pakistan, not surprisingly, is unhappy for all these reasons, plus what it feels is a preferential and "asymmetric" treatment of India.

Let us explore these concerns by examining the features of the separation plan contained in the Agreement and its implications at a technical level.

The separation plan, in brief, is:

1. Of the 22 Water moderated reactors, 8 were declared as military. The remaining 14 are offered for IAEA safeguarding in perpetuity.
2. The Dhruva (100 MWth) and Cirus (40MWth) Pu production reactors.
3. The Fast Breeder Test Reactor (13 MWe).
4. The Prototype Fast Breeder Reactor (PFBR; 500 MWe).
5. Three Plutonium Reprocessing plants.
6. Uranium Enrichment plant (~5000 SWU).
7. All the spent fuel stocks until safeguards take over.

What are the implications of this separation plan on fissile material production in India? My colleagues, Drs. Mian, Nayyar, and Ramana at the International Panel of fissile materials and I, have calculated the fissile material implications of the deal in great detail.[1] The results are summarized in the table below. The table also gives the uranium requirement associated with the different possibilities.

As one can see from the table, India would have had the capability of producing about 160 kg of weapon grade Plutonium per year even if there were to be no nuclear Agreement with the U.S. What the Agreement does, by allowing India to import uranium as fuel for its 8 newly safeguarded reactors, is use the excess domestic uranium to run an un-safeguarded power reactor at low burn-up to produce more weapons grade plutonium. This procedure can yield, depending on how much they can enhance their fuel re-loading and spent fuel re-processing capabilities, anything up to a maximum of 200 more kg/year, which is worth about 40 more warheads per year.

Whether India will actually exploit the Nuclear Deal to enlarge its arsenal so much is a matter of opinion and judgment at this stage. I personally do not believe so. In a laudable act of transparency, India publicly announced its draft nuclear doctrine in 1999. It stressed on the principle of minimal deterrence as its basis. This principle has to be translated into a concrete estimate of the number and types of nuclear weapons it calls for.

**Table 1. Unsafeguarded Plutonium production capability in India**
(The rows below the red line indicate additional capacity enables by the deal.)

| Reactors | Burn up (MWd/tU) | Uranium demand (tons/year) | Reactor-grade plutonium (kg/y) (10 kg/warhead) | Weapon-grade plutonium (kg/y) (5 kg /warhead) |
|---|---|---|---|---|
| Dhruva | 1000 | 29 | | 26 (5) |
| Breeder | | | | 135 (27) |
| Seven reactors in power mode and one 220 MWe reactor in production mode | | 528 | 1147 (114) | 200 (40) |
| Seven reactors in power mode with partially depleted uranium cores and one 220 MWe reactor in production mode | | 467 | | 200 (40) |
| All eight reactors in power mode | 7000 | 338 | 1265 (126) | — |

Minimal deterrence does not require a boundless open-ended arsenal, nor that one's weapons match in number and strength those of your adversaries. It only demands that you have enough capability, in a second strike, to inflict "unacceptable damage" to the other side. As we have repeatedly argued in detail elsewhere,[2,3] half a dozen modest Hiroshima-level weapons, if dropped on major Asian cities, can kill more than half a million people in minutes. That is more than enough to be unacceptable to even a remotely rational government anywhere, including Pakistan and China.

If the adversary is controlled by such irrational and suicidal leadership that they find this acceptable as a price for military adventure (as can conceivably happen), then no arsenal of any size can deter them anyway.

By all accounts India already possesses half a ton of weapon grade plutonium worth about a 100 nuclear weapons, of which several dozen have already been assembled. Thus, for purposes of minimal deterrence, the existing stockpile already gives a comfortable margin of redundancy, several times over, to permit a second strike with a dozen bombs after taking into account factors of survivability, reliability and interception.

Nevertheless, India has been responsible for giving the impression of going for large arsenals, by invoking national security as a reason for keeping the breeder and 8 other reactors outside safeguards.

Therefore I believe our government should make very effort, consistent with sovereignty and national security, to erase this impression and reassure its neighbors and the world that it has no plans to enlarge its arsenal by virtue of the deal.

REFERENCES


1. Z. Mian, A.H. Nayyar, R. Rajaraman and M.V. Ramana, (2006) "Fissile materials in South Asia and the implications of the U.S.-India nuclear deal" To appear in *Science and Global Security*, volume 14, nos. 2-3. See also the website www.fissilematerials.org.

2. R. Rajaraman, (2005) "Save the Indo-U.S. Agreement," *Hindustan Times*, 5 November 2005.

3. R. Rajaraman, (2005) "Cap the Nuclear Arsenal Now," *The Hindu*, 25 January 2005, R. Rajaraman, "Towards De-Nuclearisation of South Asia" (paper presented at the 2nd Pugwash Workshop on South Asian Security, Geneva, Switzerland, 16–18 May 2003).

## PROLIFERATION AND THE NUCLEAR FUEL CYCLE ISSUES IN JAPAN

KAZUAKI MATSUI
The Institute of Applied Energy, Tokyo, Japan

### EXISTING JAPANESE NUCLEAR POWER GENERATION AND NEW NATIONAL ENERGY STRATEGY

Fifty-five nuclear power plants in operation, produce about one-third of the total electricity demand in Japan, contributing to:

- increase of energy self-sufficiency from 4% without nuclear to 19%.
- stability of electricity price in spite of the extreme rise in the price of fossil fuels.
- significant reduction in $CO_2$ emission/kWh, contributing to fulfillment of Kyoto Protocol.

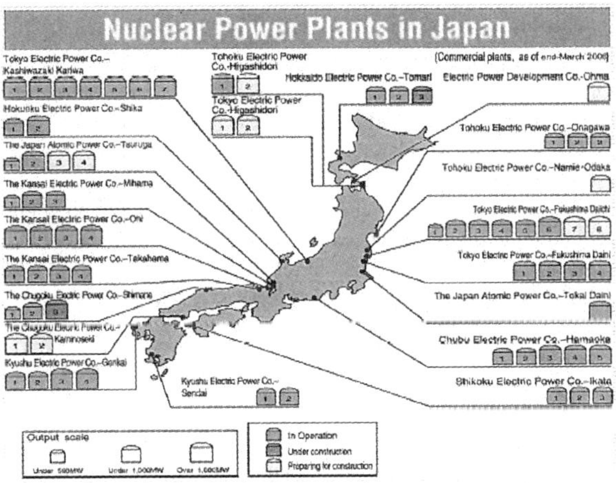


Another two plants are under construction, three more are under safety review by the Government and eight plants are in preparation for the application of construction permission within 5–10 years.

New National Energy Strategy,[1] May 2006, defines the following three objectives:

- To establish Energy Security relied on by the public.
- To establish Sustainable Infrastructure by the unified solution of energy and environmental problems.
- To contribute actively to solve Asian/World energy problems.

One of the five numerical targets to formulate the above strategy specifies the nuclear role as follows. "The ratio of nuclear power to all power production will be maintained or increased at the level of 30 to 40% or more up to 2030 or later."

EXISTING AND PREPARING NUCLEAR FUEL CYCLE RELATED FACILITIES

Japan has or plans to construct the nuclear fuel cycle facilities as commercial activities as listed below:

| | |
|---|---|
| Uranium enrichment | |
| Rokkasho uranium enrichment plant | 1,500 tSWU/y |
| Spent fuel reprocessing | |
| Rokkasho reprocessing plant | 800 t/y |
| Interim storage of spent fuel | |
| Recycling fuel storage center (Mutsu/Aomori) to be constructed by 2010 | 5,000 t |
| MOX fuel fabrication | |
| Rokkasho MOX fuel fabrication plant to be constructed by 2012 | 130 t HM/y |

MOX fuel utilization is mainly planned in 16–18 LWRs (Pu-thermal) for the moment. But above-capacity for nuclear fuel cycle does not meet the domestic demand and further measures would be needed.

Following the above line up of the fuel cycle facilities and further comprehensive related R&DD's, fast breeder reactors would be ready to be installed on a commercial scale around the middle of the century to close the fuel cycle for better resources utilization and waste management.

Base Scenario Image

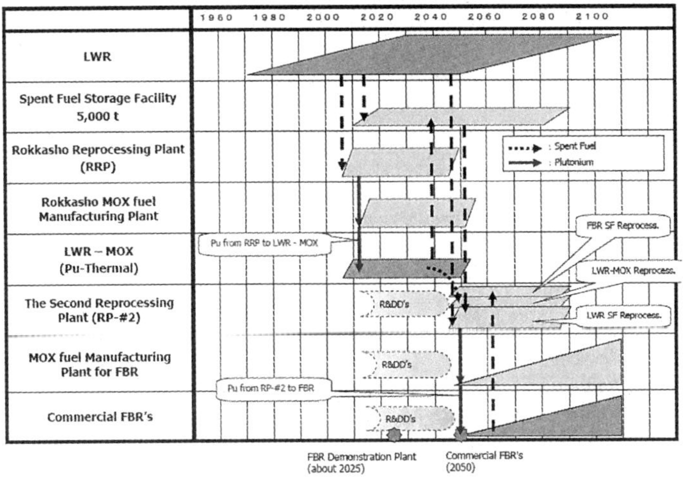


## TRANSPARENCY OF PLUTONIUM UTILIZATION IN JAPAN

In order to promote Plutonium utilization even for a peaceful power production purpose, the related transparency and accountability are musts. Following are domestic and international institutional measures and policies to secure transparency and accountability.

Law and Declaration

- Atomic Energy Basic Law, December 1955.
- Three Principles for Peaceful Use in the second article of the above law; "Nuclear development and application should be limited for peaceful objectives with full safety, and promoted *under the democratic management, on the own motive to disclose the accomplishments* for International cooperation." (author's translation)

International Schemes

- NPT; Japan as one of the most faithful states.
- IAEA Safeguards; First nuclear country to qualify "Integrated Safeguards" in 2004.

Report Plutonium Inventory to IAEA

- " Guidelines for Plutonium Management" (INFCIRC/549, since 1997).

Plutonium Guideline by Japanese Government

- Basic Position on Japan's Use of Plutonium (August, 2003); Not to possess plutonium without any peaceful utilization purpose.

*Reserves (as of December 31, 2004)*

- In Japan: 4.0 tons of Pu fissile—5.7 tons of Pu total
- Overseas: 25.3 tons of Pu fissile—37.4 tons of Pu total

*Production*

- Rokkasho RP: tons of Pu fissile/year (at full operation)
- Tokai RP: 0.2 tons of Pu fissile/year (in 2005)

*Utilization plans*

- 16-18 LWRs (including Oma, ABWR) will use MOX fuel which accounts for 5.5 –6.5 tons Pu fissile/year
- Prototype FBR, Monju and experimental FBR, Joyo consume about 0.6 tons Pu fissile/year

## REMARKS

Comprehensive fuel recycling is important and indispensable to sustainable development of nuclear energy in order to effectively use the limited uranium resource as much as possible and minimize high-level radioactive waste.

In order to accomplish comprehensive fuel recycling, the development and establishment of the Fast Breeder Reactor (FBR) fuel cycle system which is safe, reliable, economical, proliferation resistant, etc., is, we believe, a key issue for the world.

Japan has been devoted to feasibility study of advanced FBR cycle for the last several years and is going to proceed to the next step of collaborating with the international framework such as Generation IV International Forum (GIF), and Global Nuclear Energy Partnership (GNEP). Meanwhile, LWR fuel cycle is progressing in Japan.

These works must be strictly limited to peaceful purposes and transparency of plutonium utilization and non-proliferation are indispensable, as Japan has shown ever since the beginning.

Japan wishes to be a model of a non-nuclear weapon state with nuclear capabilities for peaceful use by practice, and we anticipate to be ready to be one of the world nuclear fuel cycle centers to serve and support a peaceful world and states. For the ultimate peaceful world without nuclear weapon, nuclear weapon states have an obligation to reduce world nuclear fear in general by promoting nuclear disarmament

August for the Japanese and Japan is a special month because of the Atomic Bombs which caused more than 200,000 deaths in 1945: on August $6^{th}$ for Hiroshima and August $9^{th}$ for Nagasaki. The Emperor ordered surrender on August $15^{th}$ after more than three million sacrifices. We wish "No more war, no more A-Bomb, anymore, nowhere" and to be a World model of a no nuclear weapon state with a peaceful use of it.

## TO SURVIVE TOWARD 2100

It seems a new scheme to overcome today's NPT deficiencies is needed. Multilateral Approaches to the Nuclear Fuel Cycle (MNA) prepared by IAEA, Mr. Putin's proposal and GNEP to supply nuclear fuel and receive waste are good starting points to discuss. But a fundamental question is whether Fuel Cycle capabilities are right or obligation still remains, leaving discrimination issue between having and not having, similarly to NPT itself.

There are still strong needs and rooms to continue due efforts for innovation in science and technology together with wisdom for institutions for human to survive in the $21^{st}$ century.

(The material of this paper is mainly based on the presentation by Dr. S. Saito, vice chair of Japanese Atomic Energy Commission to ICAPP 2006, Reno, USA, June 2006 but all the responsibility is with the author.)

## REFERENCES

1. http://www.meti.go.jp/press/20060531004/senryaku-houkokusho-set.pdf.

# NUCLEAR NON-PROLIFERATION: CURRENT STATE AND PROSPECTS

ROLAND TIMERBAEV
Center for Policy Studies in Russia, Moscow, Russia

*"Physicists felt a peculiarly intimate responsibility for suggesting, for supporting, and in the end, in large measure, for achieving the realization of atomic weapons...In some sort of crude sense which no vulgarity, no humor, no over-statement can quite extinguish, the physicists have known sin; and this is a knowledge they cannot lose."*

Robert Oppenheimer, 1947

Nuclear non-proliferation is, perhaps, the most complex and thorny of international issues facing mankind today and for years to come—if not forever—since the use of nuclear energy cannot be abandoned and nuclear weapons uninvented.

The Nuclear Non-Proliferation Treaty has almost 190 parties, but also a number of holdouts or states party with concealed nuclear-weapon ambitions. Since the conclusion of the NPT in 1968, intensive attempts have been and are being made to put an end to nuclear proliferation (and much has been achieved in this respect), but this appears to be a never-ending process.

For sixty years, nuclear weapons have not been used, but they continue to be seen by many, if not by a majority, of countries as an omnipotent instrument of power, prestige and status. And the very fact that the NPT has divided the world into two categories of states—nuclear-weapon (NWS) and non-nuclear weapon (NNWS)—has created a situation of never-ending inequality among sovereign states, which is deemed unacceptable by many. This, as well as concerns about their insecurity and vulnerability, feeds nuclear ambitions of at least some NNWS.

## DEVELOPMENT OF NPT REGIME

Soon after the conclusion of the NPT, it became evident that the NPT regime needed continuous upgrading. First, it was necessary to reconsider and redraft the International Atomic Energy Agency's system of safeguards to make it *comprehensive*, since the treaty, unlike the existing system, required safeguarding *all* nuclear activities of the NPT states party. This was achieved in 1971. Next steps included:

- Establishment of the Zangger Committee to elaborate a "trigger list" of nuclear materials and equipment governing the export of those items to NNWS (1971);
- Creation of the Nuclear Suppliers Group (NSG) to work out and to constantly develop Guidelines for nuclear and nuclear related exports (1975);
- Establishment of zones free of nuclear weapons in various regions of the world: in Latin America and the Caribbean—Treaty of Tlatelolco (1967); South Pacific—Treaty of Rarotonga (1985); Southeast Asia—Bangkok Treaty (1995), and Africa—Pelindaba Treaty (1996);

- Convention on the Physical Protection of Nuclear Material (1980) as amended in 2005;
- Comprehensive Nuclear Test Ban Treaty (1996);
- Additional Protocol to the agreements between states and the IAEA for the application of safeguards (1997);
- Global Partnership against the Proliferation of Weapons and Materials of Mass Destruction (2002);
- Proliferation Security Initiative (PSI) launched by U.S. in 2003 which gathered a coalition of states, including Russia, that agreed to use their national resources, including force if necessary, to interdict and seize international shipments of goods believed to be illegally destined for use in WMD programs;
- UN Security Council Resolution 1540 (2004) directed against providing any support to non-state actors (such as the network of Abdul Qadeer Khan) that attempt to develop, acquire, manufacture, possess, transport or use WMD or their means of delivery. The resolution established the Committee to promote the implementation of the said resolution, and in 2006 the UNSC extended the mandate of the Committee for two years;
- Conversion of civilian research reactors operating with weapons-grade highly enriched uranium (HEU) to using low enriched uranium (LEU). However, up to 100 such civilian facilities in the world still use some amounts of HEU;
- On the part of some NWS, primarily U.S. and Russia, several agreements were accomplished for implementing Art. VI of the NPT, which calls for negotiations in good faith on "effective measures relating to cessation of the nuclear arms race at an early date and to nuclear disarmament," including the conclusion of the START-1 and SORT Treaties in 1991 and 2002, respectively.

It can be seen from this extensive but still not exhaustive list that, over the years, numerous and time-consuming efforts have been exerted to enhance the NPT regime and its verification capability. However, the regime is still by no means faultless, facing serious challenges and threats.

Let us consider some of the major problems that the NPT regime is encountering at the present time.

## UNIVERSALIZING ADDITIONAL PROTOCOL TO SAFEGUARDS AGREEMENTS

The model Additional Protocol approved by the IAEA Board of Governors in 1997 was aimed at closing some of the loopholes in the comprehensive system of safeguards adopted in 1971. That system did not provide sufficient powers of inspection in case of clandestine nuclear activities in states party to the NPT. As a result of the 1991 Gulf War, such activities were uncovered in Iraq which prompted the IAEA to approve a strengthened set of safeguards rules making it more likely to detect *undeclared* nuclear activities in NNWS.

Although thus improved safeguards can hardly give 100% confidence about compliance, they marked a significant leap forward. Nonetheless, one has to admit that the strengthened safeguards system endorsed by the IAEA in 1997 probably marks the maximum that states are ready to accept today. It might not catch a non-compliant state red-handed, but it could sound an alarm.

In terms of real achievement, as of July 2006, the Additional Protocol was in force in no more than 76 countries, and the Agency was able to conclude—having found no indication of the diversion of declared nuclear material and no indication of undeclared nuclear material or activities—that all nuclear material remained in peaceful activities in only 24 of these countries.

Iran signed the Protocol and previously was abiding by its rules without ratifying it, but early this year withdrew its agreement to comply with the Protocol commitments.

The strengthened safeguards system adopted by the IAEA through the Additional Protocol should become a *minimum standard* for the NPT states, and supplier states should make acceptance of this standard by recipient countries as a *condition* for contracts involving nuclear items.

## PREVENTING NUCLEAR TERRORISM

The most crucial way to prevent nuclear terrorism is to keep potential terrorists from acquiring access to plutonium and, especially HEU, since would-be terrorists probably would choose HEU as fissile material that is simpler to be made into gun-type weapons than plutonium which needs a more sophisticated weapon design. Nuclear terrorists may seek to make not only nuclear weapons, but also dirty bombs, which in fact constitute radiological weapons.

Prevention of nuclear terrorism requires strict implementation of physical protection measures, material control and accounting and other security routines in dealing with such materials. Much has been done through international efforts to improve the protection of nuclear materials during recent years after the break-up of the Soviet Union, which increased the awareness of the international community to this challenge to the NPT regime. However, one should not rest on laurels and should continue to further upgrade existing systems of nuclear material protection.

In 2005, the UN General Assembly approved the International Convention for the Suppression of Acts of Nuclear Terrorism, which requires domestic criminalization of acts of nuclear terrorism and commits its parties to international cooperation in the prevention, investigation and prosecution of acts of nuclear terrorism. The Convention has been signed by over one hundred countries. States should now proceed to early ratification and implementation of the Convention. In June 2006, the President of the Russian Federation submitted the Convention to the State Duma for ratification.

In July 2006, on the eve of the Saint-Petersburg G-8 Summit, Presidents Bush and Putin announced their decision to launch the Global Initiative to Combat Nuclear Terrorism in order to pursue the necessary steps with all those who share their views to prevent the acquisition, transport, or use by terrorists of nuclear materials and radioactive substances or improvised explosive devices using such materials, as well as hostile actions against nuclear facilities.

THREE-STATE PROBLEM (INDIA, PAKISTAN, ISRAEL) AND U.S.-INDIA NUCLEAR DEAL

The existence of three countries with advanced nuclear capability which are outside the NPT—India, Pakistan, and Israel—has always been a source of grave concern and a striking blow to the NPT regime, as well as to regional and international security. For a long time, India and Pakistan have been manufacturing nuclear weapons and in 1998 tested them. As to Israel, it is generally considered to possess nuclear weapons, but is believed not to have tested nuclear explosives. Neither of these states is expected to renounce their nuclear-weapon capability and join the NPT as NNWS or form or participate in zones free of WMD.

The international community so far has not been able to find a reasonable and realistic solution to this intricate problem that would help to create a more universal NPT regime. One possibility, suggested by some experts, could be to stop requiring that India, Pakistan, and Israel immediately give up their nuclear weapons and join the NPT as NNWS. Instead, these countries are to be persuaded to commit themselves to:

- Join the CTBT and until its entry into force observe a moratorium on testing nuclear explosive devices;
- Provide public assurances that they will no longer produce fissile materials for nuclear weapons and other nuclear explosive devices and fulfill those assurances, as well as do their best to help negotiate fissile material cut-off treaty;
- Strictly observe non-proliferation rules, including rigorously implementing national export control systems that meet the highest international standards;
- Provide assurances that they will not export enrichment or reprocessing equipment or technologies;
- Apply physical protection, control, and accountancy measures to all nuclear devices, equipment, materials and installations, both military and civilian;
- Place under IAEA safeguards their civilian nuclear facilities, including all types of nuclear power reactors;
- Accept safeguards commitments under the IAEA Additional Protocol.

Although I do not believe that such an arrangement could be acceptable to many NPT states, this or some other similar ideas leading to non-proliferation objectives should be carefully explored and promoted. Any arrangement should certainly take into account the views of interested parties and requirements for strengthening international non-proliferation regime.

In July 2005, the U.S and India issued a Joint Statement under which India agreed to take several steps to demonstrate its commitment to being a responsible nuclear power and a supporter of non-proliferation goals. In exchange, the U.S. Administration agreed to seek changes in the U.S. law and multilateral commitments to permit exports of nuclear equipment and technology to India. It was reported that India would place under IAEA safeguards 14 out of 22 of its existing reactors, with the exception of military and fast-breeder facilities, as well as future thermal nuclear reactors. It was also reported that India is planning to build a new military plutonium production reactor.

In March 2006, the U.S. submitted a proposal for making changes in the NSG Guidelines that would treat India as a "special case," permitting to put aside the NSG requirement for full-scope safeguards in supplying it with nuclear equipment and technology. The U.S. proposal was supported by three NWS (France, Britain and Russia), but opposed by such countries as Sweden, Norway, Austria, Ireland, New Zealand and some others. China did not support the American proposal for a special treatment of India, and has been hinting that the same rule should be applied to Pakistan as well. The U.S. did not press for adoption of its proposal until after the U.S. Congress has changed its national legislation. Decisions in the NSG are taken by consensus.

In my view, the U.S.-India nuclear deal would fail to accomplish the goal of furthering the non-proliferation regime and universalizing international safeguards. It will secure no meaningful constraint on the growth of India's nuclear weapons stockpile and will not require India to accept the equivalent of the non-proliferation obligations under Articles I and VI of the NPT. The limited amount of additional safeguards that India has pledged to accept does not limit India's nuclear weapons program. The proposed arrangement could also trigger erosion of the NSG Guidelines.

Moreover, the U.S.-India nuclear deal risks fueling a nuclear arms race in the subcontinent. Pakistan might respond to any perceived increase in India's nuclear weapons capability by ramping up its nuclear weapons program. Pakistan has recently been reported to have embarked on a dramatic expansion of its nuclear arsenal with the construction of a new heavy water reactor capable of producing enough plutonium for up to 50 warheads a year. When completed, it would be 20 times the size of the existing plutonium production reactor at Khushab, which began its operation in 1998 and has already accumulated a certain amount of Pu.

At the end of July 2006, the U.S. House of Representatives voted overwhelmingly to approve the nuclear deal with India, and the Senate, according to *New York Times,* is expected to endorse the deal later this year, but before it goes into effect both houses will have to approve the specifics of a nuclear cooperation accord with India. Similarly, India will have to reach relevant agreements with the IAEA and the NSG.

## MULTILATERAL APPROACHES TO NUCLEAR FUEL CYCLE

Articles XII.A.5 and IX of the IAEA Statute contain provisions dedicated to the strong support for international cooperation in storing and supplying member-states with nuclear materials under the Agency safeguards. The potential modalities of such possible measures have been considered by the Agency on numerous occasions ever since 1970s.

Recently, this proposition has acquired a new dimension and significance as a way to strengthen the international nuclear non-proliferation regime, in particular since uranium enrichment technology and reprocessing of spent fuel have become more accessible to a larger group of states, some of them not in very good standing as far as the NPT is concerned.

So far neither the IAEA nor states possessing enrichment technology and capable of providing enrichment services to recipient states have been able to agree on and offer attractive proposals, based on multilateral approaches, to those who need nuclear materials for their existing and developing nuclear power industry.

In May 2006, six states producing about 95% of world enrichment supplies (Russia, U.S., France, UK, Germany and the Netherlands) agreed and announced a concept of a multilateral mechanism of assured access to nuclear fuel. They stated that the existing commercial market of nuclear fuel functions in a satisfactory way, but that it would be useful to create a mechanism for solving the problem of assurances of supply that may arise in the future. The concept welcomed the proposal of Russia to implement on its territory a project of establishing an international center, under the IAEA safeguards, for providing enrichment services on the basis of an existing enrichment facility (in the city of Angarsk in Siberia). The concept also noted the U.S. announcement of converting 17 tons of excess HEU to LEU as a stockpile for providing assurance of supply.

The proposed six-power concept of an international mechanism required the *renouncement* by recipient states of indigenous "sensitive" activities in the field of nuclear fuel cycle. This condition, however, has received critical response on the part of many NNWS, in particular the non-aligned (NAM) states.

## SOLVING IRAN'S NUCLEAR PROBLEM

Iran's long-standing efforts to acquire a capability to enrich uranium without reporting these activities to the IAEA, as required by its safeguards agreement with the Agency, and its refusal early this year to abide by the Additional Protocol have caused much concern and an animated international debate. Iran has been asserting that its efforts are intended only to give it an indigenous source of low-enriched uranium fuel for its planned nuclear power industry. Many states, however, suspect that the country might use its enrichment capability for producing highly enriched uranium for nuclear weapons. They consider that this possibility must be halted sooner rather than later.

Russia is expected to finish the construction of the Bushehr power plant in 2007, and has pledged to supply it with nuclear fuel and to take back the spent fuel. It also proposed to Iran to set up a joint enrichment venture on the Russian territory for supplying Iran with LEU.

At present, Iran has a modest enrichment capability, but is planning to build a large enrichment plant in future. It has been reported, though, that so far Iran has been encountering serious technological problems with its enrichment capability .

Findings by the IAEA inspectors have confirmed that Iran has repeatedly breached its nuclear safeguards agreement by not reporting the clandestine acquisition of uranium enrichment technology from Pakistan through the A.Q. Khan supplier network.

For the last several years, the IAEA Board of Governors many times considered the Iranian nuclear issue, but was unable to get any definite response by Iran to its calls to fully abide by the requirements set out by the Board. Early this year the IAEA Board, under its Statute, had to refer the issue to the UN Security Council. The Council, having noted with serious concern Iran's decision to resume enrichment-related activities and to suspend the cooperation with the IAEA under the Additional Protocol, requested a report from the IAEA Director General on the process of Iranian compliance with the steps required by the Board, to the IAEA Board of Governors and in parallel to the Security Council.

In the meantime, the five permanent members of the UN Security Council and Germany, after meeting in Vienna on June l, 2006, have declared their willingness to undertake negotiations with Iran with the aim of reaching a negotiated agreement based on cooperation, should Iran suspend all enrichment related and reprocessing activities as required by the IAEA. In such a case they would suspend action in the Security Council. They have also declared that if Iran decides not to engage in negotiation, further steps would have to be taken in the Security Council. The proposal was presented to the Iranian authorities on June 6th. However, negotiations have not yet started due to the unwillingness of the Iranian side.

Under the circumstances, the Security Council on July 31st adopted Resolution 1696(2006), in which, acting under Article 40 of Chapter VII of the UN Charter, it:

- Demanded that Iran suspend all enrichment-related activities, including research and development, to be verified by the IAEA;
- Endorsed the six-power proposals of June 6, 2006;
- Requested by August 31st a report by the IAEA Director General on whether Iran has established full and sustained suspension of all activities mentioned in the resolution; and
- Expressed its intention, in the event that Iran has not by that date complied with the resolution, to adopt then appropriate measures under Article 41 of Chapter VII of the UN Charter to persuade Iran to comply with the resolution and underlined that further decisions will be required should such additional measures be necessary.

Both Russia and China voted for the resolution. The Russian representative said in the Council that the resolution expressed the need for Iran to establish full cooperation with IAEA to clarify outstanding questions and for restoring confidence in its nuclear program. The main purpose of the text was to support IAEA's efforts to resolve Iran's nuclear problem, and the Agency should continue to play a central role in resolving non-proliferation issues in the context of Iran's nuclear program. He expressed hope that, with the Council's support, it would be easier for the IAEA to do that job. By acting under Article 40 of the Charter, the Council has rendered mandatory suspension of all uranium-enrichment activities. If Iran did not comply, members had expressed the intention to take appropriate action under Article 41 (sanctions). As followed from the resolution, any additional measures that could be required to implement the resolution ruled out the use of military force. Russia said the resolution should help to clarify outstanding issues. This measure should be viewed as interim, for the period necessary for resolving the issue. If Iran complied with the resolution, members would be prepared to refrain from any further action, and if negotiations yielded a positive solution to the problem, no additional steps against Iran would be taken in the Council. The resolution also established a provision for Tehran's broad cooperation to meet Iran's energy requirements. The Russian delegation hoped that Tehran would seriously view the contents of the resolution and would take the necessary steps to redress the situation.

The initial draft resolution authorized action under Article 39, but Russia insisted it be removed because it refers to the use of force. As a compromise, the above mandatory clause was inserted to imply the resolution is legally binding. However, to make a resolution binding, it is required to have an Article 39 finding of a threat to international peace and security. In the final version of the resolution, the Security Council removed language calling the Iranian nuclear program a threat to peace and security.

However, the resolution threatens sanctions if Iran does not comply by August 31st, and includes a paragraph that could be described as either limited sanctions or an embargo. This paragraph calls on all states to prevent the transfer of any items, materials, goods and technology that could contribute to Iran's enrichment-related and reprocessing activities and ballistic missile programs.

Iran had promised to respond to a package proposed on June 6th by the permanent five and Germany (called P5+1) by August 22nd. Qatar, the sole Security Council member to vote against the resolution and the sole Arab state on the Council, asked the Council to wait to take action until Iran responded to the package. Given the crisis in the Middle East, Qatar said acting now did not serve the region's security.

The representative of Iran declared that the resolution would not lead to any productive outcome and, in fact, it could only exacerbate the situation. The people and government of Iran have shown, time and again, their resilience in the face of pressure, threat, injustice and imposition. He said that because its nuclear program does not pose a threat to international peace and security, dealing with it in the Security Council was unwarranted and void of any legal basis or practical utility, and that the resolution violates the fundamental principles of international law, the NPT and IAEA resolutions. Iranian authorities rejected the resolution and threatened to cut off supplies of oil. Iran is the fourth largest supplier of oil and has the second largest reserves of oil and gas.

Whether the Security Council resolution of July 31st would help to advance the solution of the Iranian nuclear issue or not, one has to bear in mind that the Middle East for a long time has been one of the most sensitive regions of the world. The international community should very cautiously and prudently—and certainly in full compliance with international law—seek settlement of any issues that may pose a threat to its unsteady security which may inevitably produce dangerous ramifications not only for the region but for the world at large.

Iran has its own security problems that should also be taken into account. To name one, it may perceive itself threatened by a U.S. military presence in Iraq. And as a member of the NPT it has the right, as all other parties to the NPT—in keeping with Articles II and IV of the treaty—to engage in all stages of peaceful nuclear energy activities.

But a right to do something does not necessarily mean that this right must be exercised, and nothing prevents states, especially in sensitive regions, from suspending or deferring any fuel-cycle activities, if their pursuit may have negative consequences.

The Iranian nuclear issue can still be settled in a generally acceptable way by diplomatic means, granted there is sufficient good will and willingness to negotiate from all sides.

## SETTLING THE NORTH KOREAN NUCLEAR AND MISSILE PROBLEM

The Democratic Peoples Republic of Korea (DPRK) has been a cause of deep concern to the international community for almost 15 years, ever since the IAEA inspections showed that North Korea must have produced more plutonium than it had declared. This was reported to the IAEA Board of Governors, which referred the case to the Security Council as a breach of safeguards obligations.

After a long period of debates, discussions, some temporary agreements and new recriminations, at the end of 2002 IAEA inspectors were asked by the DPRK to leave the country. In January 2003 Pyongyang announced its withdrawal from the NPT, and the plutonium reactor, frozen under the 1994 DPRK-U.S. Agreed Framework, was again put into operation. It is widely believed that the DPRK has accumulated a certain quantity of plutonium for weapons purposes. North Korea stated that the country possessed nuclear weapons. And in February 2005, the U.S. claimed that the DPRK was developing a capability to enrich uranium based on technology obtained through the network of A.Q. Khan.

In the meantime, in August 2003, talks were instituted within a six-party group consisting of China, Japan, North Korea, Russia, South Korea and the United States attempting to reconstitute the previous détente. In September 2005, a Joint Statement by the six was agreed upon which called for the verifiable denuclearization of the Korean Peninsula. Under it the DPRK committed to abandoning its nuclear weapons and existing nuclear programs and returning to the NPT and to IAEA safeguards. The U.S. affirmed that it has no nuclear weapons on the Korean Peninsula and has no intention to attack and invade the DPRK with nuclear or conventional weapons. The other parties of the six-power forum agreed to discuss at an appropriate time the subject of providing a light-water reactor to the DPRK.

The six parties also agreed to take coordinated steps to implement the above consensus in a phased manner in line with the principle of "commitment for commitment, action for action." However, since September 2005, no meetings of the six-party forum took place.

A new flare-up of tension around the DPRK nuclear and missile programs occurred in July 2006, when North Korea broke its 2000 pledge to maintain a moratorium on missile launching, and on July 5$^{th}$ committed multiple launches of ballistic missiles. On July 15$^{th}$, the UN Security Council in resolution 1695 unanimously condemned this act and demanded that the DPRK suspend all activities related to its ballistic missile program and re-establish its pre-existing commitments to a moratorium on missile launching. The resolution also urged the DPRK to return to the six-party talks in order to work towards the expeditious implementation of the September 2005 Joint Statement. The North Korea's UN ambassador, however, rejected the resolution after it was passed.

Under the current circumstances, it does not appear that the six-party talks would be resumed any time soon.

## COMPREHENSIVE NUCLEAR-TEST-BAN TREATY

The adherence of all states to the Comprehensive Nuclear-Test-Ban Treaty (CTBT) would serve a vital objective of promoting nuclear non-proliferation. The preamble of the NPT recalls the determination expressed by the parties to the 1963 Partial test ban treaty "to seek to achieve the discontinuance of all test explosions of nuclear weapons for all time." One of the key components of the package deal that led to the indefinite extension of the NPT in 1995 was a call for the completion of negotiations on a CTBT by 1996.

However, while the CTBT was indeed concluded and opened for signature in September 1996, it has still not entered into force. As of August 2006, the number of signatories has grown to 176 states, with 134 ratifications. But the treaty will only enter into force after 44 states designated in its text as involved in significant nuclear activities have ratified it.

Of these 44 states, only 34 have ratified the treaty so far. Among the ten that have not, seven states—U.S., China, Israel, Iran, Egypt, Indonesia, and Colombia—have signed but not ratified it. Three states have neither signed nor ratified the treaty: India, Pakistan, and North Korea.

Though the CTBT has not entered into force, NWS continue to maintain a moratorium on testing and refrain from conducting them: Russia, since 1990; UK, since 1991; U.S., since 1992; China and France, since 1996; and India and Pakistan, since 1998. North Korea and Israel have not conducted tests either. However, the U.S., Russia and UK continue to carry out the so called subcritical experiments at test sites in Nevada and Novaya Zemlya.

A U.S. decision to ratify the CTBT would strongly influence other countries to follow suit. While no nuclear explosive tests have been carried out for many years, leaving the treaty in limbo is a risk to further efforts in nuclear arms control and nuclear non-proliferation.

The global verification regime of the CTBT is already 70% operational, comprising facilities for seismological, hydro-acoustic, infrasound and radio-nuclide monitoring. However, the CTBT Organization (CTBTO) has had difficulties in collecting the annual dues owed to the organization. The U.S. refuses to pay its contribution for the CTBTO work on designing a system for on-site inspections. There is no doubt that the progressively installed monitoring system is essential to the continued credibility of the CTBT.

## FISSILE MATERIAL CUT-OFF TREATY (FMCT)

Prohibition of the production of fissile material for use in nuclear weapons has long had broad support in the world community. It has an explicit non-proliferation value and was included as a goal in the package deal that led to the indefinite extension of the NPT in 1995. While not alone sufficient to bring about nuclear arms control and disarmament, ending fissile material production for nuclear explosives would serve important non-proliferation objectives and halt fresh supply of Pu and HEU for weapons.

Of the NPT NWS only China has not yet officially declared that it is no longer producing such materials for weapons. The Conference on Disarmament (CD) agreed on a negotiation mandate for the FMCT. However, there have been a number of difficulties that have so far prevented the CD from producing such a treaty.

Even if fresh production of fissile materials for weapons were to be stopped, states could still make new weapons from stockpiled material. Because such stocks are quite large in several states (e.g., some experts estimate that U.S. possesses 100 tons of Pu and Russia up to 150), many NNWS have maintained that that FMCT should cover such stocks. NWS, which is not unexpected, oppose this idea. Pakistan and Arab states in the Middle East want stocks to be included, while India and Israel do not.

Another problem is verification of the FMCT. Nevertheless, in July 2004, after having supported it as a key element in a treaty, the U.S. reversed its policy and declared that "realistic, effective verification of an FMCT is not achievable." However, the IAEA has a long experience of verifying the peaceful use of relevant facilities in Brazil, South Africa and Japan, and a vast majority of states continue to support verification as an important component of FMCT.

On May 18, 2006, the U.S. submitted to the CD a draft Treaty on the Cessation of Production of Fissile Material for Use in Nuclear Weapons or Other Nuclear Explosive Devices. The draft commits its parties not to produce fissile material for use in weapons, or "use any fissile material *produced thereafter* in nuclear weapons or nuclear explosive devices." It provides for no verification. After the submission of the American draft it was immediately rejected by the delegation of Egypt.

The G-8 Summit in Saint-Petersburg, which met in July 2006, expressed its "support of the early commencement of negotiations on the Fissile Material Cut-Off Treaty in the Conference on Disarmament," without, however, mentioning the U.S. draft treaty.

## IMPLEMENTING NPT ART. VI (NUCLEAR DISARMAMENT)

Since the conclusion of the NPT in 1968, the U.S. and USSR/Russia have reached a number of agreements and decisions to cut down their strategic (as well as intermediate and tactical) nuclear weapons.

According to the July/August 2006 issue of the *Bulletin of the Atomic Scientists,* NWS have reduced the global stockpile to its lowest level in 45 years. The total global nuclear weapons stockpile is estimated to be substantially smaller than the 1986 Cold War high of 70,000-plus warheads. In the same period, the number of nuclear-weapon states has grown from six to nine. It is estimated that these nine states possess now about 27,000 nuclear warheads, of which 97 percent are in U.S. and Russian stockpiles. About 12,500 of these warheads are considered operational, with the balance in reserve or retired and awaiting dismantlement. Estimating the arsenal sizes of the smaller nuclear powers—Israel, India, Pakistan, and North Korea—it is considered that India and Pakistan have about 110 nuclear warheads between them and the North Koreans could have around 10. Though Israel has not acknowledged that it possesses nuclear weapons, the U.S. Defense Intelligence Agency estimates that it has between 60 and 85 warheads.

However, since the signing of the 2002 Moscow SORT Treaty requiring the U.S. and Russia to each reduce their deployed strategic nuclear warheads to 1,700–2,200 units, there have been no formal talks between Washington and Moscow on further nuclear cuts. This four-year hiatus has generated the impression in many countries that the NWS are derelict in pursuing the "cessation of the nuclear arms race at an early date and nuclear disarmament" as required by the NPT Article VI.

If the U.S. and Russia would allow the 1991 START-1 Treaty to expire in 2009, and the Moscow Treaty to lapse when its proposed ceilings enter into force in 2012, after that Washington and Moscow will be under no obligation to limit their nuclear-weapon arsenals, and this would be considered by other NWS, including China, as well as by

NNWS as an end of nuclear arms control and as a signal to feel free to develop their existing or start new nuclear-weapon programs.

President Putin stated that key disarmament issues are all but off the international agenda, and that the arms race has entered a new spiral with the achievement of new levels of technology that raise the danger of the emergence of a whole arsenal of so-called destabilizing weapons. The Russian president announced that Moscow was calling for the renewal of dialogue on key weapons-reduction issues, first of all, negotiations on replacing the START Treaty by some new arrangement, adding that it was necessary to help reverse a period of "stagnation" in disarmament.

Moreover, the possibility of nuclear arms in the hands of more nations in volatile regions of the world raises the possibility of another arms race: an unenviable choice that might be precluded by more energetic nuclear arms control diplomacy.

No one should be under any illusion that progress on nuclear disarmament can be achieved easily. This subject was discussed by the U.S. and Russian presidents last July on a bilateral basis on the eve of the G-8 Summit in Saint-Petersburg. According to reports, both leaders have agreed on a program of action in this area, and as far as the START-1 is concerned, they instructed respective experts to review the treaty and report to them, whether and to what degree the provisions of the treaty have become obsolete, which of them should still be kept in force and which other provisions need clarification.

As can be seen from the above brief review of the functioning of the NPT, the nuclear non-proliferation system, throughout its lifetime, has been the object of continuous debate, sometimes quite heated, and is constantly challenged by various proliferation threats. Nonetheless, the treaty continues to provide a solid international legal foundation for the non-proliferation regime, and there is no substitute for it. One has to accept and appreciate this fact, and the only approach to keep the regime working is to persistently seek ways of improving it and to search and find methods for counteracting any challenges and threats to it. There does not seem to be any other option.

# HOW SERIOUS IS THE CRISIS OF THE INTERNATIONAL NUCLEAR NON-PROLIFERATION REGIME?

JOACHIM KRAUSE
Institute for Social Sciences
University of Kiel, Kiel, Germany

*Summary: The nuclear non-proliferation regime is in a crisis, but it is definitely not as severely damaged as proponents of the liberal arms control school are suggesting. Their main argument is that contractual breaches (first and foremost) by the nuclear weapons states as well as by non-nuclear weapons states (Iran, North-Korea; Iraq and Libya in the past) and the ongoing abstentions of India, Israel and Pakistan from the regime are the main causes for the pending collapse. It is argued here that the main factor in preserving the nuclear nonproliferation regime has been the relative success of the rule of non-use of force in interstate relations. It is more important to see to that this rule will be maintained—for which the role of the U.S. as a steward of international peace is crucial—than in making assumptions whether there was a basic deal between nuclear weapons states and non-nuclear weapons states and about who was responsible for the alleged unraveling of that deal*

Concerns about nuclear programs in North Korea and Iran, along with the controversies surrounding the Indian-American Nuclear Agreement of the summer 2005 have generated a deep pessimism about the prospects of the nuclear non-proliferation regime. Assertions that the regime is broken and world order itself is in danger have become increasingly frequent. But it is nonetheless not yet clear just how serious the crisis is and how the two are interconnected. At present at least three different interpretations can be identified as to why and how gravely nuclear non-proliferation policy is endangered and what the consequences for world order will be:

- First, the widespread theory of the liberal school of arms control cites three threats to the nuclear non-proliferation regime: (1) the failure of nuclear states to disarm, (2) the continued existence of loop-holes in the regulations of the Nuclear Non-Proliferation Treaty (NPT) of 1968 as well as (3) terrorism. The liberal school of arms control assumes that all arms represent a risk and that nuclear arms are particularly menacing;[1] it emphasizes the dangers of arms races and considers the greatest risk potential to originate with nuclear weapons states that have set a bad example for the others by refusing to reduce their own arsenals. Proponents of this school argue that the nuclear and non-nuclear weapons states entered into a firm agreement on nuclear disarmament in the sixties and that since the non-nuclear states have renounced nuclear weapons of their own, it is now high time that the nuclear powers completely destroy their stockpiles. They consider the difficulties in dealing with actual or presumptive treaty breakers to be primarily a consequence of the misguided policy of those states with nuclear weapons, in particular the USA.[2]

- The opposite view is being held by the "realistic" school. Its adherents proceed from the assumption that the non proliferation regime was an anomaly: they argue that states cannot be permanently denied the right to maintain their security by whatever means they deem to be necessary. According to their assessment, a world with many nuclear weapons powers was, in principle, more stable than one in which only a few have such weapons.[3] The present nuclear non-proliferation regime reflected the hegemonial role of the USA in the international system. And should this hegemony be called into question, the non-proliferation regime would automatically collapse.
- A third school of thought, at present most prominent in the U.S. Administration and Congress, asserts that the nuclear non-proliferation regime is in principle viable but that it is confronted with numerous challenges that can no longer be adequately mastered with the classical means of multilateral diplomacy. On the contrary: the established mechanisms of multilateral, global diplomacy can often actually pose obstacles, since debates in this context tend to circle endlessly around relatively insignificant problems while the true issues are left practically unaddressed. Unilateral or multiple measures should, therefore, also be undertaken, up to and including military intervention and where necessary preventive measures.

All of these schools contain a kernel of truth, but they all remain ultimately unsatisfactory. The arguments of the first school of thought are weak, because they are based on the assumption that a natural division exists between those states with and those without nuclear weapons that determines their respective security interests. But in reality no state can base its security strategy principally on its membership in the one of these groups. It is, rather, more likely that their strategies will depend on how they perceive their situation, its risks and threats at any given time. Hardly any cases (with the possible exception of India) exist in which states were motivated to acquire nuclear weapons because of the supposed bad example of the five original nuclear weapons states. It would be equally difficult to identify states that assume they have a fundamental right to nuclear arms and are only waiting for the non-proliferation regime to collapse or for the nuclear weapons powers to offer them something as compensation for continuing to renounce nuclear weapons. The overwhelming majority of states do not wish to acquire nuclear weapons: a fact that would appear to contradict the theoretical assumption of the realistic school. Moreover, most states accept the more or less permanent inequality between states that possess nuclear weapons and those that do not, as long as no tangible disadvantages arise for their security interests. Furthermore, many states have in the past perceived and continue to view the nuclear weapons potential of the USA as the guarantor of their security, as was certainly the case in the Federal Republic of Germany during the East-West conflict. Granted, the voting behavior of many of the non-nuclear weapons states during the Review Conferences on the NPT would seem to corroborate the thesis that there are various camps. But it does not reveal the existence of any united front of non-nuclear weapons states. Even those governments that were the most radical critics of the nuclear weapons states during these conferences (Mexico, Malaysia, and Nigeria) did not imply that their discontent over the behavior of the nuclear weapons powers would lead them to seek their own nuclear weapons. The few states that are actually suspected of developing secret nuclear weapons programs usually kept a low profile during such debates.

## WHY HAS NUCLEAR NON-PROLIFERATION SUCCEEDED?

In understanding the nature of the crisis, one first has to ask for the reasons for the successes of the nuclear non-proliferation regime during the past 35 years. The fact that so many states that were supposed to have become nuclear weapons states have rather have chosen the non-nuclear weapons status still has to be registered as an outstanding success. Why have the 182 non-nuclear weapons states that signed the NPT—with few exceptions—been satisfied with the nuclear status quo in the past? Two phenomena that shaped the past decades explain this acceptance of the inequality between states with nuclear weapons and those without: respect for the principle of the prohibition of the use of force between states and the occurrence of structural changes within the states of the developed, western countries and the threshold countries of Asia and Latin America.

The prohibition of the use of force between states was established in the UN Charter and, judging by the last 60 years, can be considered to have been relatively successful. But the continuous decrease in the use of force between states cannot be explained by the UN Charter alone. It was and is much more crucial that institutions and states exist that take responsibility for ensuring that the principle is upheld. In the more than 60 years since the UN was founded, it has typically been the U.S. administration rather than the UN Security Council that successfully committed itself to uphold this principle, either through the vehicle of the UN, through NATO, either in cooperation with allies or as sole intermediary, as guarantor of peace agreements or of the security of its allies. U.S. advocacy of the prohibition of the use of force marks a fundamental difference to the period between the world wars when there was no power willing and capable of guaranteeing the international order of collective security. Without American security guarantees and the repeated endeavors of Washington to solve conflicts in a preventive, diplomatic manner, to intervene in crisis situations and if necessary apply massive pressure in order to bring regional wars (such as in the Middle East or Southern Asia) to a quick conclusion, the renunciation of force proclaimed in the UN Charter would have had no more effect than the Briand-Kellog Treaty of 1928. That is, without the effectiveness of the prohibition of the use of force, the nuclear non-proliferation regime could never have been successful.

The other development that decisively contributed to the success of this regime was the structural change in the nature of the state in western industrial countries as well as in the industrial and threshold countries of Asia and Latin America. As the new world order emerged after World War II, a shift occurred in the functions of the state toward more intervention in the economy and modernization of infrastructure, as well as expansion of the welfare state and redistribution of wealth. Political success was no longer defined in categories of territorial expansion and security, but rather by measures such as creation and securing of employment, through the ability to compete in international markets and through greater social security. In the wake of globalization, this model calling for a primarily economic role for the state aimed at satisfying domestic needs has become attractive to other states outside the western world. The British political scientist, the late Susan Strange, attributed this trend to the influence of the USA which used its preeminence in the international system after World War II to define the rules of the international economic system and brought the states of Western Europe and Northern Asia into the fold of a free trade economy. This movement has since developed such momentum that the power of the states has begun to recede as impersonal market dynamics gain sway.[4]

These functional changes and the resultant loss of power of the state have repercussions on nuclear proliferation: states that assign great value to a functioning economy, where economic well-being depends on access for their firms to international markets and their capacity to attract foreign investors can today no longer afford to acquire nuclear weapons. In the nineties, Erwin Häckel and Karl Kaiser presented an analysis of opportunity costs of a hypothetical nuclear option for the Federal Republic of Germany. The conclusions were clear: the political and economic opportunity costs were so high that they clearly precluded such a decision.[5]

Similar calculations can surely be made for almost every state—around 50 today—with appreciable nuclear capabilities. There a few exceptions, but they tend to confirm the rule. This applies not only to those countries that have not joined the NPT (Israel, India, Pakistan) but to those that have broken the treaty as well. Israel is one of the few countries that actually have a massive security problem: nuclear weapons represent an existential guarantee for its survival. India is the only country to follow the example of the USA, China, France and Great Britain in an effort to underline its pretensions as a world power in the manner the liberal arms control theory has described. But as India has become more aware of its increasing interdependence within the world economy, it has adopted a more reserved approach. The conclusion of the treaty on cooperation in the field of civilian nuclear energy with the USA suggests that New Delhi has come to recognize the signs of the times. Pakistan, on the other hand, became a nuclear weapons power because it saw no other way of dealing with the India's superior power. Iraq (under Saddam Hussein), Libya and Iran are rentier-states that share the advantage of oil producers that do not necessarily have to worry about cooperative standards. The regular flow of gigantic revenues has made it possible for adventurers, criminal family clans, religious fanatics and eccentrics to maintain power there. These states with huge assured incomes can become potential buyers of nuclear weapons should they channel internal problems into international aggressiveness or seek to avoid international sanctions or interventions.

Being not a rentier-state, North Korea represents the special case of a state that has gone bankrupt due to its international isolation and believes that it can overcome the crisis through nuclear blackmail.

## NUCLEAR ORDER AND THE PROHIBITION OF THE USE OF FORCE

Thus one might be tempted to agree with the third school of thought that the world nuclear order is not facing such a fundamental threat after all. It will, indeed, remain secure as long as the principles of the international political order sketched above (continued prohibition of the use of force either through the UN Security Council or the USA as well as the primacy of economic and welfare considerations) are upheld. There is some question, however, whether or not nuclear non-proliferation could be eroded anyway as a consequence of the erosion of the prohibition of the use of force. And, in the past 15 years, a number of developments have arisen that suggest that this principle of prohibition is in crisis. There appear to be two main reasons:

- The increasing level of violence in domestic social conflict observable primarily in failed states has become a real factor in politics today. In most cases the universally valid principles of international law that constrain the use of force are being violated on a massive scale without triggering any appreciable intervention by the community of states.
- The failure of the central organ of collective security, the Security Council of the UN, in the face of the international crises of the past 15 years (Bosnia-Herzegovina, Kosovo, Ruanda, Congo, Sudan, Iraq, North Korea, Middle East) has contributed in a major way to the erosion of the prohibition of the use of force in various regions of the world. Africa is the most prominent example. Where ever the USA, NATO or other alliances of western states did intervene, with or without a mandate from the UN, this erosion was stopped.

Furthermore, the increasing acceptance of the incendiary slogans of political Islam in the Islamic world should be cause for considerable concern. If they were ever to become an integral part of the political programs of existing governments, they could potentially become a fundamental threat to the international prohibition of the use of force. Just how closely the nuclear order and the international political order are interconnected becomes apparent when one considers that if representatives of radical political Islam were to gain control of nuclear weapons the entire prohibition of the use of force regime could be overturned. If Iran were to acquire nuclear arms and the otherwise rhetorical threat of eradicating Israel became a real option, nuclear conflict in the Middle East would become a distinct possibility: Given its small size, Israel could be "eradicated" with a relatively small number of nuclear explosions.

## THE PRECARIOUS ROLE OF THE USA

Without the repeated U.S. advocacy (alone or together with the Europeans and other states of the western world) for the prohibition of the use of force and adherence to the NPT, both the international political order as we know it and the non-proliferation regime would be barely existent today or limited to the western world. In this sense the argument of the proponents of the third school of thought that the USA is the guarantor of the nuclear non-proliferation regime and the international prohibition of force is logical. There is, however, one problem: the more the USA is willing to compensate for the deficits of multilateral institutions, the more resistance it generates to its efforts.

There are two reasons for this resistance: first, unilateral action on the part of a super power like the USA—no matter how justified—often triggers counter movements that develop out of a general defensive stance and an instinct to resist that reflect prejudices and animosity vis à vis that larger power. Second, American policy has never been without flaws and imponderability, and strong doubts as to the quality and professionalism of those acting in the name of the USA have often been justified. This was and is the case in other fields as well,[6] but the problem has never been as clear as under the present administration. The dilettantism with which it prepared and executed the invasion of the Iraq War (that was supposed to restore the authority of the UN Security Council but was then substantiated in detail with hair-raisingly false assertions) and the catastrophic PR policy of the Bush administration have caused many countries to view the USA as a greater threat to international security than Iran with its nuclear ambitions. This clearly demonstrates the fundamental dilemma involved in upholding the international political order (defined as the prohibition of the use of force) and the nuclear non-proliferation regime. The more the weakness of multilateral institutions causes the USA to take over these tasks, the harder it gets to win international acceptance. On the contrary: the more the USA acts unilaterally, the stronger the resistance becomes, thus creating a situation that opens up undreamt of opportunities for those states that are mounting a massive challenge to this very order.

The Iranian leadership has recognized this opportunity and is exploiting the situation to create the capabilities necessary to get as close as possible to building a nuclear weapon. Most remarkably, after the exposure of its secret enrichment programs in 2002, Iran chose the political offensive and became a vocal advocate of the right of all Third World states to nuclear enrichment. The Islamic Mullah regime in Tehran has used the divisions that surfaced between the USA and its allies since 2003 to stage a confrontation with the USA and the UN. This, in turn, has helped it shore up its domestic power base. The leadership of North Korea has taken a similar tactic, which suggests that it shares this assessment of the international situation.

The battle over nuclear non-proliferation and the international order could be lost, if this trend is allowed to continue. No less an authority than the former U.S. Secretary of State, Henry Kissinger, has warned that both crises could mark an historical turning point. As in the 1930's, the entire international order could collapse, if those powers responsible for its preservation no longer support it. "A failed diplomacy," Kissinger asserts, "would leave us with a choice between the use of force or a world were restraint has been eroded by the inability or unwillingness of countries that have the most to lose to restrain defiant fanatics."[7]

## OUTLOOK

In dealing with the crisis of nuclear nonproliferation, a paradigm shift is needed. The dominant scholarly paradigm – the liberal arms control school – is not (or no longer) helpful in addressing nonproliferation issues. On the contrary, it has become part of the problem we face in dealing with problematic states. Their proponents' main concern is disarmament but not security under given circumstances with lesser nuclear weapons. Hence, their arguments are being used by challengers of the regime—such as the Iranian president—in order to further their case. These challengers basically want to defy an international order that is based on U.S. stewardship. What is often overlooked is that without that stewardship, the order of the non-use of force between states would collapse, as well as the nuclear nonproliferation order. Hence, the stakes are higher than just nuclear nonproliferation. However, the problem is not just being posed by the challengers; it is also how the U.S. is living up to its stewardship. The past years have been marked by growing doubts as to the ability of the current U.S. administration to meet this goal. In this regard it is of growing importance whether and how the U.S. is supported or even substituted in its stewardship role by the member states of the European Union.

The inequality between nuclear-weapons states and non-nuclear weapons states will continue—and it will most likely pose no major problem as long as it does not go along with tangible security disadvantages for non-nuclear weapons states. Indeed, many non-nuclear weapons states do not consider the nuclear weapons option because they are under some nuclear umbrella or under a broader security guarantee given by a nuclear weapons state. The danger of a collapse of the nuclear non-proliferation regime is there; but it is closely related with the way non-nuclear weapons states perceive their respective security environment and how strongly they are trusting existing mechanism of guaranteeing the rule of the non-use of force in international relations.

In the long perspective, the most likely danger for the nuclear non-proliferation regime is the combination of a political ideology that is defying the norm of non-use of force with the quest for nuclear weapons. In this regard, the most likely danger comes from extremist versions of the ideology of political Islam (Islamism). Radical Islamism is adamantly opposed to the norm of non-use of force. In case the current radical Islamist leadership of Iran would be in possession of nuclear weapons, the main problem would not be the emergence of a nuclear arms race, but the outbreak of a nuclear war in the Middle East. A similar danger is associated with Pakistan, where a takeover by Islamist forces might result in a severe international crisis with the danger of a nuclear war.

REFERENCES

1. For a typical example of this school, see the report of the Weapons of Mass Destruction Commission (chairman: Hans Blix): *Weapons of Terror: Freeing the World of Nuclear, Biological, and Chemical Arms*, Stockholm June 2006, f.i. pp. 62-66.
2. Compare William Walker: *Weapons of Mass Destruction and International Order.* London (IISS – Adelphi Paper 370) 2004.
3. Kenneth N. Waltz: *The Spread of Nuclear Weapons: More May be Better*, London (IISS Adelphi Paper) 1981.
4. Worth reading in this context is Susan Strange: *The Retreat of the State. The Diffusion of Power in the World Economy* (Cambridge 1996) as well as Philip Bobbit: *The Shield of Achilles. War, Peace, and the Course of History* (New York 2003).
5. Erwin Häckel and Karl Kaiser: "Kernwaffenbesitz und Kernwaffenabrüstung. Bestehen Gefahren der nuklearen Proliferation in Europa?" in: Joachim Krause (ed.): *Kernwaffenverbreitung und internationaler Systemwandel* (Baden Baden 1994), pp. 239-262.
6. See Susan Strange: "Reaganomics, the Third World and the future", in: Altaf Gauhar (ed.): *Third World Affairs* (London 1986) pp. 65-72
7. Henry A. Kissinger: "A Nuclear Test for Diplomacy", in: Washington Post 16. May 2006, page A 17.

# ROGUE STATE HELVETIA ? SWITZERLAND AND THE ATOMIC BOMB 1945-1988

CHRISTIAN BÜHLMANN
Military Federal Department, Bern, Switzerland

## INTRODUCTION

The idea that neutral Switzerland—a peaceful country, home of the United Nations Office in Geneva, home of the international Red Cross and of the Geneva Convention—had tried to develop an atomic bomb may seem absurd.[1]

Therefore, it may come as a surprise to discover that Switzerland indeed had secret plans for the development of nuclear weapons.

This article presents the main reasons why Switzerland considered developing a nuclear capacity in the 1950s and 60s and why it finally decided to sign the Treaty on the Non-Proliferation of Nuclear Weapons (NPT).

I will conclude by showing that Switzerland, in its quest for nuclear weapons, was far from being a rogue state and how those past developments paved the way to Switzerland's current policy of non-proliferation.

### State of the sources

The Swiss quest for nuclear weapons is complex. Its history is still fragmentary, even though a couple of studies and articles have been published, most of them in German (Stüssi-Lauterburg 1997; Jorio 2001; Breitenmoser 2002; Neval 2003; Wollenmann 2004; Braun 2006).

This article is based on those chronological studies. However, it presents and summarizes them within a topical context. In order to simplify the text, the Swiss political structures have been anglicised.

## IN SEARCH OF NUCLEAR WEAPONS

### First reactions after WW II

The bombing of Hiroshima and Nagasaki made a great impression on the Swiss military leaders. A couple of days after the second explosion, the armed forces chief of training, Lieutenant General Hans Frick, wrote a letter to the minister of defence, Federal Councillor Karl Kobelt. Protection against atomic weapons was his main concern, but he also wondered whether Switzerland might be able to develop nuclear weapons on its own (Braun 2006, 748-749).

In the middle of the 1950s, tactical atomic weapons appeared in the European theatre. This led to considerations on a redesign of Swiss defences: until that time, strategic nuclear weapons had not been regarded as a direct threat for its armed forces, because of their huge destructive power. Tactical nuclear weapons, however, could be used as a theatre weapon to selectively destroy military reserves without 'significantly'

---

[1] *The author would like to thank Dr. Peter Braun and Anthony Gygax for their helpful comments on this text.*

destroying civilian and industrial infrastructure. Should, therefore, Switzerland's rather linear defence be replaced by a less vulnerable, but more expensive mobile one or should a more affordable area defence be chosen? This question lasted for years, until it was settled in 1966 (Ernst 1971; Braun 2006).

During this period, the Swiss defence community assessment had mainly four reasons for developing atomic weapons:

1. Tactical level nuclear weapons were considered 'better bullets'
2. Operational level nuclear weapons could act as theatre deterrent
3. Strategic level nuclear weapons would allow to counter any Soviet nuclear blackmail
4. Atomic weapons would help to balance a threatening German proliferation.

Tactical level

Feeling that nuclear weapons were increasingly gaining the status of 'normal' weapons, several Swiss military authors openly advocated the acquisition of atomic weapons. In 1957, the Swiss Officers Society published a report that called for the procurement of nuclear weapons. The Federal Council discussed this topic in 1955 and came to the conclusion that, although nuclear weapons were morally repulsive, it might be appropriate for Switzerland to purchase them. Finance minister Streuli, hoping that atomic armaments might be less expensive than conventional ones, was in favour of nuclear ones. A cooperation with Sweden, which at that time was proceeding with similar research, was envisioned.

Warfare with tactical nuclear weapons was considered manageable. As such, it was assumed that atomic bombs could help balance the conventional gap between small states and superpowers. However, the strategic consequences caused by their possession were not anticipated at that time; the political, social, psychological and ecological consequences were not taken into account. Nuclear bombs were then understood only as a more potent ammunition and not as a strategic way of sanctuarising territory.

Operational level

For most Swiss military analysts, there was a major risk that the armed forces of the Warsaw Pact would use the so-called "atomic void" of Switzerland in a move to outflank the NATO forces. For that purpose, the eastern forces would use tactical nuclear weapons to destroy the Swiss defence forces and then proceed through the country. However, NATO or France would not stay idle and engage the Soviet divisions with nuclear weapons.

To avoid this double threat and enhance its neutral stance, the Swiss strategy of dissuasion would require operational-level nuclear weapons, in order to threaten WAPA's rear bases and logistic lines and thus deter the eastern countries.

Strategic level

A further aspect that was discussed in the middle of the 1960s was the risk that the Soviet Union could coerce Switzerland with nuclear blackmail. Switzerland would be threatened by atomic destruction, should the country not act according to Soviet will.

According to military strategist Major General Gustav Däniker (Däniker 1966), the only solution to counter that threat would have been the possession of strategic nuclear weapons as well as delivery platforms to threaten the Soviet heartland in return.

Proliferation balance

The last and more potent reason for possession of atomic weapons was the fear of nuclear proliferation.

At the political level, the acquisition of atomic weapons was mainly an answer to proliferation in the event of the monopoly of the USA, UK and the USSR being broken. The strategists were concerned that the European military equilibrium might be destroyed if France, and above all Germany, were to procure atomic weapons.

It would place Switzerland back in a geopolitical position similar to the one it had during the 19th and the first half of the 20th century: a very delicate position between possible foes that might use Swiss territory as a theater of operations.

Swiss nuclear weapons would help conserve the equilibrium within Europe as well as balance the conventional gap between small states and superpowers.

For most of the time, technological reflections on nuclear weapons dominated. The strategic approach came rather late and was influenced by the French general and author André Beaufre.

Approaches to nuclear development

There were three major approaches to nuclear weapons: two rather theoretical, one more practical.

*SKA—The first study group*

It was already clear from the outset that the armed forces would not be able to develop atomic weapons on their own. Civilian resources and expertise would be required. A nuclear energy study group (*Studienkommission für Atomenergie*—SKA), integrating scientists and military personal, was therefore created in 1946. It was chaired by Professor Paul Scherrer, at this time the leading Swiss nuclear scientist. The secret study group's tasks were primarily to investigate protective measures. However, the development of weapons was also an option: the draft research order envisions atomic land mines for destruction and sabotage, nuclear artillery shells as well as atomic air-to-surface bombs (Breitenmoser 2002, 91).

In order to finance the studies, money had to be supplied by Parliament. The message from the government stressed the need for civilian research, but made no mention of the secret military mission. However, some members of Parliament condemned financing the project. They feared that, by not explicitly banning the development of atomic weapons, Switzerland might be perceived as a threat by its neighbours (Braun 2006, 752-756).

During the same period, interest arose in the industry for nuclear energy. It brought welcome capabilities to the study group: the armed forces would not have been able to provide the human resources, let alone finance the research on their own. In 1953, after Eisenhower's "Atoms for Peace" initiative, Swiss industry began to work in this direction. A joint civilian-military research reactor was built. Very soon, however, private industry lost interest in it, when the USA sold two civilian reactors at very low prices, with constraints that the radioactive material should not be used for military purposes.

At the end of the nineteen-fifties, the SKA was merged with the civilian nuclear commission within the energy department and the military department lost its lead in nuclear energy.

*The MAP Study*

Towards the end of the fifties, the department of defence created a new expert group, on part time basis, to assess the possibilities of developing atomic weapons in Switzerland. In 1963, the group gave a report, *Möglichkeiten einer eigenen Atomwaffenproduktion* [Nuclear weapon production options] (MAP), stating that Switzerland could produce atomic weapons autonomously. In order to develop a plutonium-based bomb—the more expensive type—it was estimated that 750 experts and 2.1 billion Swiss Francs would be required over 30 years. For more extensive information, the group requested a further study over 3 years, requesting 20 full-time experts and 20 million Swiss Francs. Still, the government did not permit the formation of the expert group: it seemed that it only wanted to keep the window open for the procurement of atomic weapons, but not to procure them.

*Obtaining fissile material*

The main practical problem was obtaining radioactive material in order to proceed with developments. Three approaches were to be pursued:

1. Search for uranium in the Swiss Alps
2. Study the use of other radioactive elements that might be found in Switzerland
3. Purchase of uranium from abroad.

The first two approaches were unsuccessful. The third effort, cautious enquiries, first with the USA, then with east-European countries, the Popular Republic of China and India, failed too, partially because the USA had purchased most of the available material to avoid proliferation. In 1954 and 1955, it was finally possible to obtain 10 tons of uranium from the Belgian-Congo through a contract with the United Kingdom and Belgium, under the condition that it shouldn't be used for any military purpose (Braun 2006, 759-763). This amount would have only been sufficient to build a single atomic bomb. Some small samples were used for experimental purposes and the rest was stored in an underground facility and almost lost.

## TOWARDS THE NPT

Wthin the context of the Cold War, the Swiss nuclear strategy had some rationality. Why did Switzerland change it in order to embrace the NPT? It is not, as is sometimes assumed, that the Swiss population had voted against atomic weapons.

Initiatives against nuclear weapons

Indeed, at the end of the 1950s, a civilian group against atomic weapons (*Bewegung gegen den Atomtod*), assembling representatives from churches, universities and left-wing groups, began to voice its concern. It called for a nuclear ban in Switzerland.

The Swiss government was upset by that attitude and intended to counter this 'defeatist propaganda.' Therefore, in July 1958, it officially declared that atomic weapons were not only offensive, but also defensive weapons. For a neutral state which has to use the best weapons to defend itself, nuclear weapons were definitely a possible option. This report described a long-term vision but was misunderstood and led to the impression that Swiss Armed Forces would immediately begin with procuring atomic bombs.

There was some concern in western foreign countries that this decision might encourage proliferation. Since this inception, the Warsaw Pact questioned Swiss neutrality and is assumed to have included Switzerland in its offensive plans for that reason. At the national level, two popular initiatives were submitted to the Swiss people. The first one, sponsored by pacifists, intellectuals and religious movements, sought to prohibit any development, procurement, construction, storage or use of atomic weapons on Swiss territory. The second one, endorsed by the Swiss socialist party to avoid a split between its left wing in favour of the first initiative, and its right wing against it, sought to achieve a lesser goal: any action in atomic weapon development was to be submitted to the people. Both initiatives were rejected at a rate of about two to one: the population did not veto possible acquisition of atomic weapons by Switzerland.

The reasons for the change in strategy were indeed both domestic and external:

Domestic reasons

*Human resources problem*

Firstly, the MAP report stated that, in order to develop a plutonium-based bomb—the more expensive type—750 experts and 2.1 billion Swiss Francs would be required over 30 years. It would have been very difficult to train and recruit those scientists.

*Lacking political will*

Secondly, a gap had developed between the defence and the foreign departments. Since the beginning of the 1960s, the foreign department had been considering integration into the international community through non-proliferation to be a safer strategy than autonomy in nuclear weapons. The foreign department did not support nuclear weapons development any more. The military did not share that opinion and pushed forward—alone.

*Mirage affair*

Finally, in 1964, a setback in the procurement of a French fighter aircraft—which could have been used as atomic bomber—further closed the nuclear window of opportunity. What is known as the 'Mirage affair' erupted when the procurement costs for 100 aircraft exceeded the original budget by more than 50%. Parliament refused to endorse the budget increase and cut the number of airplanes down to 57. The armed forces were thus denied a delivery platform for an atomic bomb. Moreover, the department of defence lost a lot of political support. This affair also clearly revealed that the armed forces had been unable to manage a complex project. The nuclear development programme would have been much more demanding and political faith in Swiss military capabilities had diminished.

External reasons

There are also three central external reasons that induced the stop of the nuclear development of atomic weapons:

*Non proliferation deals*

Firstly, because of American dumping measures to support the development of a civilian nuclear industry, such as cheap heavy water or the discounted sale of civilian nuclear power reactors, it was possible to avoid a civilian-military joint venture and thus render autonomous development much more costly and difficult. This excluded a dual-use civilian military approach.

*Diplomatic and economical pressure*

Secondly, pressure from the USA on non-NPT signatory states was rising: should Switzerland not sign the NPT, its civilian industry might not receive fissile material anymore.

*Security through non proliferation*

Finally, the prospects of signing the NPT and thus of achieving greater integration within the international community, avoiding pressure and keeping civilian fissile material would bring Switzerland more security in general than atomic weapons.

As Germany decided to sign the NPT, nuclear stability in Europe would remain unchanged. It was an easier and cheaper way of achieving equilibrium than by developing atomic weapons.

The NPT Track

The path to the NPT had two steps: Firstly, the signature of the Treaty Banning Nuclear Weapon Tests in the Atmosphere (PTBT), then the signature and the slow ratification of the NPT.

*Signature of the PTBT*

It was, however, Switzerland who signed the PTBT in Moscow in 1963. The gap had widened between the Department of Foreign Affairs and the military authorities. The latter wanted to keep the atomic weapon procurement window open. The former wanted to join the Treaty on the Non-Proliferation of Nuclear Weapons. Sweden, fully embracing the NPT path, was no longer a research partner. On the one hand, it had rendered autonomous development even more difficult. On the other, it gave a role model to follow.

In 1969, the Swiss government signed the NPT. The military was strongly opposed to the idea and tried to influence the legislative power. Ratification by parliament took a long time. Firstly, because the government wanted to be sure that the NPT was effective before issuing its message to the legislative. Then, Parliament wanted to wait for results. Therefore, the treaty was only ratified in 1977. There was subsequently no further opportunity for the military to carry out atomic studies beyond the laboratory level.

*The AAA study group*

When the Swiss government signed the treaty in 1969, the military authorities wanted to maintain as much freedom of action as possible by retaining a nuclear study capability and by preserving the status of a threshold country. A further study group, *Arbeitsauschuss für Atomfragen* (AAA) was therefore constituted. Its task was to counsel the military leaders on any nuclear-related issue. However, as it did not convene on a regular basis, the study group's goal was rather to prepare options, should Germany leave the NPT.

CONCLUSION

In 1981, after the Swiss government came to understand that the NPT was functioning, it placed the Congolese uranium under the supervision of the IAEA. In 1988, the AAA was finally dissolved.

Current Swiss policy

In 1995, Switzerland signed indefinite and unconditional extention of the NPT . In 1996, it also signed the Comprehensive Test Ban Treaty (CTBT). Switzerland's official long-term goal is the universal and verifiable elimination of nuclear weapons. Some of it's mid-term objectives include the preservation and reinforcement of the NPT, the definition of a Fissile Material Cutoff Treaty (FMCT) as well as support for the CTBC. Officially, it is believed that international agreements on the limitation or reduction of nuclear weapons stocks contribute to transparency and confidence-building. They thus increase the security of the international community and of Switzerland.

An example for today?

Can the example of Switzerland's past path to the NPT constitute a solution for today's concerns with proliferation ? I think not.

1. The Swiss approach to nuclear armament was rather theoretical. The approach to the development of nuclear energy was based on a technological mindset, not on a strategic one. Therefore, Switzerland never intended to become a rogue state.
2. Even though its population rejected a ban on the development of nuclear weapons twice, Switzerland chose the path of non-proliferation. It changed its paradigm by realising that the nuclear approach would bring less security and be much more expensive than the one of non-proliferation. This was due to incentives brought by the international community in terms of security and the threat of arresting civilian uranium deliveries. In today's world, proliferating countries may believe that nuclear weapons bring greater security through sanctuarisation of their territory.
3. What is more, during the Cold War, Switzerland's conventional strategy benefited indirectly from NATO's nuclear weapons protection. There was no need to purchase them.

The real lesson might be that in its search for security, Switzerland chose diplomacy rather than defence.

REMARKS

The views expressed here are exclusively the author's own. They are not necessarily those of the Swiss Federal Department of Defence, Civil Protection and Sport.

REFERENCES

Braun, Peter. 2006. *Von der Reduitstrategie zur Abwehr: Die militärische Landesverteidigung des Schweiz im Kalten Krieg 1945-1966*. Baden: Hier und Jetzt.

Breitenmoser, Christoph. 2002. *Strategie ohne Aussenpolitik: Zur Entwicklung der schweizerischen Sicherheitspolitik im Kalten Krieg*. Vol. 10, *Studies in Contemporary History and Security Policy*. Bern et al. Loc: Peter Lang.

Däniker, Gustav. 1966. *Strategie des Kleinstaats: Politisch militärischer Selbstbehauptung im Atomzeitalter*. Frauenfeld und Stuttgart: Huber.

Ernst, Alfred. 1971. *Die Konzeption der Schweizerischen Landesverteidigung 1815 bis 1966*. Frauenfeld und Stuttgart: Verlag Huber.

Jorio, Marco. 2006. *Armes atomiques*. Dictionnaire historique de la Suisse (DHS) 2001 [cité 29.07. 2006]. [Internet] http://hls-dhs-dss.ch/textes/f/F24625.php.

Neval, Daniel A. 2003. *"Mit Atombomben bis nach Moskau": Gegenseitige Wahrnehmung der Schweiz und des Ostblocks im Kalten Krieg 1945-1968*. Zürich: Chronos Verlag.

Stüssi-Lauterburg, Jürg. 1997. Historischer Abriss zur Frage einer Schweizer Nuklearbewaffnung. In *Travaux & recherches / Beiträge zur Forschung 1997*, édité. Bern: Schweizerische Vereinigung für Militärgeschichte und Militärwissenschaft / Association suisse d'histoire et de sciences militaire.

Wollenmann, Reto. 2004. *Zwischen Atomwaffe und Atomsperrvertrag - Die Schweiz auf dem Weg von der nuklearen Option zum Nonproliferationsvertrag (1958-1969)*. édité par A. Wenger. Vol. 75, *Zürcher Beiträge zur Sicherheitspolitik und Konfliktforschung* Zürich: Forschungsstelle für Sicherheitspolitik der ETH Zürich.

# 4. AIDS & INFECTIOUS DISEASES
# FOCUS: AVIAN FLU — GLOBAL HEALTH

# THE NEED OF A GLOBAL TASK FORCE FOR INFLUENZA

ALBERT D.M.E. OSTERHAUS
Department of Virology, Erasmus Medical Center
Rotterdam, The Netherlands

Influenza A-viruses are divided into two distinct groups based on their ability to cause disease in birds: low pathogenic and highly pathogenic avian influenza (LPAI and HPAI). Only viruses of the H5 and H7 subtype have been shown to cause HPAI. Migratory birds are the reservoir of influenza A-viruses and are the source of LPAI viruses that may mutate into HPAI viruses in domestic poultry, causing severe disease outbreaks with high mortality in these animals. Since 1997, it has become clear that avian influenza viruses may infect humans: in Hong Kong 18 people became infected with the HPAI H5N1 virus and 6 of them died. In 2003, a massive outbreak of HPAI amongst poultry in The Netherlands, caused by a H7N7 virus, resulted in 89 clinical cases among farmers and poultry workers, one of which was fatal. The ancestors of this H7N7 virus had been found in migratory ducks prior to the outbreak. The virus had probably spilled over as a LPAI H7N7 virus to free-range chickens, in which it mutated to an HPAI virus.

In the past three years, outbreaks of HPAI in domestic poultry in Asia, caused by HPAI H5N1 viruses were associated with about 200 severe human infections, more than half of which were fatal. From Asia, HPAI H5N1 virus infections spread to the Middle East, Europe and Africa; affecting domestic poultry, wild birds and a number of mammalian species. Also from Turkey and the Middle East, human cases—some with fatal outcome—were reported. The pattern of spread largely coincided with the flyways of migratory birds.

The respective zoonotic events constitute a severe warning for the looming threat of an influenza pandemic. Avian influenza A-viruses may adapt to efficient human-to-human transmission, either directly by mutation or by reassortment with mammalian influenza A-viruses. Since little or no immunity is likely to exist against such a virus in the human population, this may result in a pandemic outbreak of influenza. Although several countries are currently preparing for such an event by developing "pandemic preparedness plans" according to WHO recommendations, the world at large is not sufficiently prepared for such a catastrophe. Better collaboration and coordination between all the stakeholders is urgently needed to establish early warning systems and effective global pandemic preparedness plans. One way forward would be the establishment of a global task force for influenza.

# CREATING CHANGE IN GLOBAL HEALTH

AHMAD KAMAL
Senior Fellow, United Nations Institute for Training and Research
New York, USA

## INTRODUCTION

We pride ourselves in our belief that the public health situation of much of the world's population shows so many signs of considerable improvement over the last decades. Improved economic conditions, better physical circumstances of life and of improved nutrition, improvements in access to the fundamental requirements of public health such as clean water and adequate sanitation, better access to essential health services particularly at the primary care level, and recent technological advances in medical and surgical management, have all combined together to result in greater life expectancy in many countries. There have been great headline advances, such as the complete eradication of a killer disease like smallpox almost thirty years ago, the forthcoming elimination of poliomyelitis, and the manageable control of several other communicable diseases using immunization.

However, whilst the world's populations perhaps experience better public health now in aggregate than in previous eras, there are devastating differences and deficiencies. Each country has a unique health profile and, although the diversity is extreme, there are many common issues. Factors such as stable economies, strong health systems and supportive environments are associated everywhere with well being and security. But there are still far too many areas where, in spite of the great potential, such improvements are just not happening. Instead we see wasted opportunities, instability and exclusion from the benefit of progress.

This paper looks at the opportunities and constraints which impact all efforts at international intervention to improve global public health, and presents some suggestions as to how better results might be achieved.

## THE DETERMINANTS OF GLOBAL PUBLIC HEALTH

The main determinant which explains much of these differences in results is poverty. Poverty is not just a statistical number of the poor and the very poor. It is a very real and concrete situation, which has a major impact on all aspects of public health. It limits the access to primary health care, without which life starts out limited at birth, with dramatic increases in the figures of maternal and child mortality. It results in inadequate access to clean water, which then in turn increases the incidence of water-borne diseases. It has a serious effect on the degree of sanitation that can be provided, particularly to the most vulnerable parts of the population, namely, the very young and the aged. It reduces the chances of corrective education even in the basics of sanitation and preventive health; even in some middle-income developing countries, poorer women still do not understand the link between diarrhea and dehydration. Poverty is thus a multitude of ills: poverty of income, poverty of knowledge, poverty of access to health care, poverty of safe motherhood and reproductive health, poverty of nutrition, and poverty of opportunity.

Overall, it condemns the poor to early death, or to shorter lives passed in the travails of agonized survival. Poverty thus remains key to understanding public health.

Whilst the numbers of people living in poverty (commonly defined as living on less than $2 per day) and absolute or extreme poverty (living on less than $1 per day) has been reducing worldwide, there still remain over two billion in the first category and over one billion in the second. That is more than one-third of the population of the world. Even the much touted reductions in poverty are geographically skewed, because of the results obtained in only a couple of countries, namely, in China and in some parts of India.

These are profoundly disturbing totals, with very serious and negative consequences for global health. Further, poverty itself is badly mal-distributed amongst populations. The huge inequity between regions in the share of the population living in extreme poverty, defined as an income of less than $1 per day, may be seen in Figure 1 below.

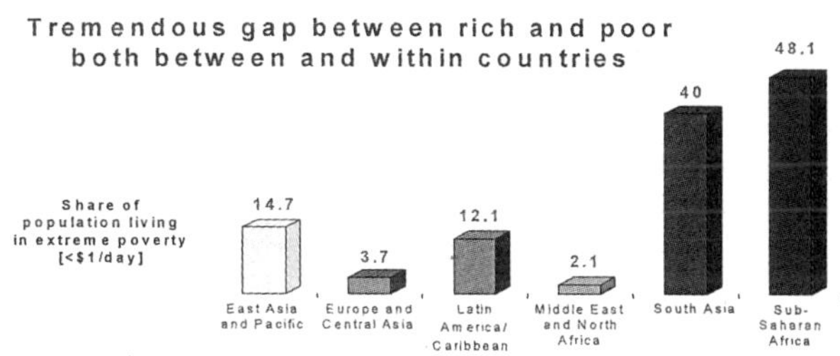


Source: World Bank

*Figure 1.*

While there is an obvious differential in the health experiences of different national populations, there are also powerful differential health effects between groups within a single national population. These factors include early life experiences; stress; social exclusion, unemployment; the lack of availability of social support structures; poor work experiences; food insecurity, food availability and quality; uncertain transport systems, and addictive lifestyles.

Lifestyle determinants are of the greatest importance. As lifestyles become increasingly globalised, there is an increase in the relative significance of chronic diseases, including in the developing countries. Factors here include tobacco, alcohol, violence, road traffic accidents, food and nutritional factors (particularly obesity) and mental health problems. Food and nutritional factors are particularly affected by agricultural and trade negotiations and agreements, and disputes, although a detailed discussion here is outside the scope of this chapter.

Super-imposed on these vitally important factors are others, of lesser but still significant importance. Three will be mentioned.

The first is migration. There is a growing body of evidence that in the short- and medium–term, migrations can affect the health of the peoples who move and that of the communities they move into. Migration is a formative and historic determinant in all populations around the world, and even in its currently restricted form, it is an obvious vector in the transmission, not just of communicable diseases such as HIV/AIDS and SARS, but also in the genetic, cultural, and lifestyle effects which influence the manner in which diseases spread.

The second factor lies in the health effects of natural and man-made disasters, such as earthquakes, floods, and tsunamis, on the one hand, and radiological and chemical technological disasters such as Chernobyl and Bhopal on the other. Reported disasters of the former type appear to have increased significantly in recent years, and are possibly associated with global warming. According to the United Nations Environmental Programme, such natural environmental disasters have increased three-fold over the past decades, with an eight-fold increase in economic costs, and a sixteen-fold increase in insurance claims.

The third results from the visible growth of wars and ethnic conflicts. These have swollen the numbers of refugees and displaced, as well as of those who are affected by the serious deterioration in health services as a consequence of war and complex emergencies, and made them exceptionally vulnerable to communicable and other diseases.

Health expenditure distribution both at national and individual level is also a critical factor. Almost a thousand years ago, Al-Asuli, a great physician writing in distant Bokhara, divided his classic pharmacopoeia into two volumes entitled, "Diseases of the Rich" and "Diseases of the Poor." It is easy to apply that very same classification to the division of actual expenditures between secondary and tertiary health on the one hand, and primary health on the other.

Overall expenditures on secondary and tertiary health, or on the "diseases of the rich," are almost 80% of the total health expenditures in the world. This leaves just one-fifth of the available funding for primary health, or the "diseases of the poor," even though far more die from these primary diseases than from the secondary or tertiary ones. The poor results that are being registered in global health statistics are a direct result of these statistical priorities. Health improves largely in the developed countries, where secondary and tertiary health is important, and continues to deteriorate in the poorer countries where primary health is the fundamental objective.

Another important factor lies in the general inattention that is given to preventive health education. Preventive health costs far less than curative health, but is heavily dependent on education. The allocation of adequate resources to preventive health education would thus go a long way to resolving a large part of the health-related problems in developing countries.

The result of all these factors is that many societies remain afflicted with traditional primary health diseases such as diarrhea, malaria and tuberculosis. These are the bane of the least developed states, and represent an unacceptable situation in a globalised world. Maternal and infant mortality remains unacceptably high, reproductive health services are woefully deficient, with new epidemics such as HIV/AIDS adding even more to the burden of disease. HIV/AIDS alone has dramatically reversed public health improvement in many societies, particularly in Sub-Saharan Africa, as is clearly illustrated in Figure 2 below.

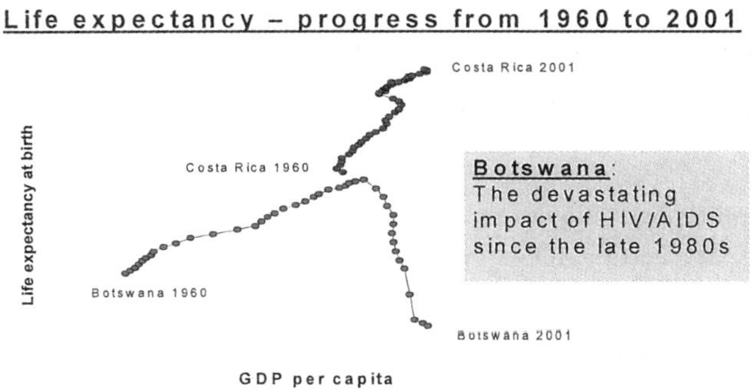


Source: Figure generated by GapMinder, based on World Bank data. *Figure 2.*

What are the relative contributions of all these factors? These will differ from population to population. Across most of the world, the fundamental determinants of health—peace and security, nutrition, economic and social circumstances, public institutional capacity concerning water, sanitation, lifestyle factors and the environment—far outweigh the contribution from individual clinical care.

## A GLOBAL FRAMEWORK FOR ACTION: GOALS PRIORITIZATION AND PROGRESS MEASUREMENT

Health development plays a vital contribution in the overall human economic and social development. Whilst always dependent upon increased financial resources and developments in science, public health action is also fundamentally about creating change in the many circumstances of peoples' lives. It estimated that timely and bold health action could save 8 million lives each year, extending the life spans, productivity and economic well being of the poor. Such a strategy would require two global initiatives: a significant scaling up of resources for health in poor countries, and tackling the non-financial obstacles in poor countries to the effective delivery of health services.

There is general agreement about the objectives towards which we all aspire. However it is absolutely clear from the earlier analysis that there are serious gaps between these goals and present results. As the technological and communications revolution continues to shrink and globalize our world, it is obvious that poverty remains the root cause of most of the problems of health, as well as in other areas of development, in the developing countries. Only after the levels of poverty have been addressed do the other determinants of health make their impact. These include the quality of governance; the capacity of public administration to provide basic public health provisions such as clean water, sanitation, immunization, maternal and child health care; lifestyle factors such as smoking, alcohol consumption and sexual behavior; environmental circumstances and the extent of environmental pollution, and the overall capacity of health services and whether these are available to all.

Goals, targets and standards for international public health improvement have long existed but are increasingly recognized as required for real progress. For the past six years, the Millennium Development Goals (MDGs) have provided a framework for international cooperation in development, against which improvements in global public health may be assessed. These MDGs are shown in Figure 3 below.

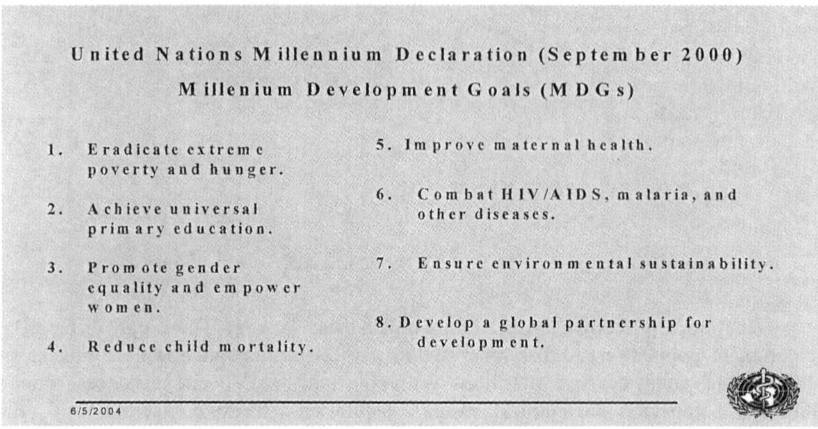


Extracted from the Millennium Declaration.

*Figure 3.*

The eight objectives of the MDGs are framed as a compact between rich and poor countries, and are supported by numerical targets and indicators leading to the target year 2015. Goals 1-7 focus on key elements of human development relevant to health, including poverty, education, and the environment. Goals 4, 5 and 6 are specific to health (Goal 4: reduce child mortality by two thirds; Goal 5: reduce maternal mortality by three quarters; and Goal 6: halt and reverse HIV/AIDS, malaria and other diseases). Goal 7 refers to ensuring environmental sustainability, including through the provision of potable drinking water. Finally, Goal 8 refers to the obligations of developed countries to increase external aid and debt relief, to establish a fair trade and financial system, and to improve the transfer of technology, including the provision of access to affordable essential drugs in developing countries.

The MDGs provide a simple understandable framework for improving human development, including health, and for monitoring and assessing international action and performance. The global MDGs are supported in many countries by national MDG monitoring and reporting. In 2005, a global review of progress towards the MDGs was conducted and reviewed during a summit meeting of heads of state and government. At the current rate, it seems highly unlikely that these goals will be achieved. The prospects for health appear exceptionally bleak. At current trends, the child mortality goal will not be met in Sub-Saharan Africa until 2065 (source: UNDP Human Development Report for 2003).

## CONSTRAINTS ON IMPLEMENTATION OF A GLOBAL STRATEGY

The state and determinants of health amongst the global population are well understood. Whilst the technology does not perhaps exist to deal effectively with all health problems, the knowledge is there to improve global health very significantly. The question remains why our world, with its powerful understanding, resources and technology does not do a better job. Why has the world community been unable or unwilling to produce better results?

Part of the reason may be the obvious if regrettable fact that health attracts, by and large, despite improvements in rhetoric, a relatively low priority on the international agenda, and in most international institutions. The United Nations, (including the Security Council, the General Assembly and the Economic and Social Council) are preoccupied with other political and economic issues, as are the Bretton Woods Institutions. As for the World Health Organization, despite its specificity in dealing with this subject, it has highly limited resources, particularly in health policy development assistance to developing countries.

At national level the same depressing conclusion about the priority given to public health is relevant. In developed societies political, public and media attention all too often is preoccupied with technology-based care services, with a much lower visibility allocated to public health action. Even developing countries themselves often give visibly low priority to health, both in the percentage of financial resources devoted to this sector, and also in the relegation of their Ministries of Health to positions of lesser influence within government hierarchies. This is surprising, given the uniform results of public opinion surveys, which show that populations generally give very high priority to health issues.

One reason for this poor treatment of the health sector by governments may be that health-related expenditures are generally perceived politically as no more than a "consumption" item. Nothing could be further from the truth. The development potential of expenditure on health makes it as an "investment" item. Any parents anywhere in the world would say the same when talking about the health of their children.

The challenge is to capture this concern and commitment to health amongst populations and turn this into effective at global and national levels. There is now a much wider realization than before that investment in health is an essential part of social and economic development, and that a healthy work force is a more productive and committed one. It is interesting that realizing this fact, some of the most important recent initiatives to provide health care and HIV/AIDs treatment in sub-Saharan Africa have come from the private foundations established by philanthropists.

Even if countries, or international organizations for that matter, make political declarations about the importance of public health, these would have to be matched by financial outlays. For the present, these are woefully missing.

For the foreseeable future it will be very difficult for developing countries to produce the necessary absolute amounts of financial resources to resolve the problem.

Most of them have severely limited resources, and just cannot to produce adequate finances for effective interventions. So the problem remains largely intact. The question then is, from where will these development resources be obtained?

International development assistance will remain vital. It is estimated that around eight million lives per year could be saved by the year 2015, mainly in the low-income countries, by a range of simple but essential preventive and therapeutic health interventions. Taking account of domestic resource utilization in the countries concerned, this would cost donors around $27 billion per year. Unfortunately, such a sum is more than triple the total current official development assistance for health (DAH) of around $8 billion annually. Nevertheless the requisite sums are eminently affordable; they represent only around 0.1% of total donor GNP. Such outlays on health would also yield economic benefits vastly greater than the costs.

Another set of constraints which prevent due importance being given to public health issues arises from the fact that most country-based legislation on the subject is limited to communicable diseases, food safety, and some child health functions. There is usually very little legal structure supporting the implementation of a public health strategy, or for any multi-sectoral public health action. There have been attempts in some countries to develop and implement comprehensive strategies for health improvement, but success has been partial and limited.

A factor to which very little public debate has been devoted so far arises from the nature and role of the global pharmaceutical industry. Much of this industry is based on its outright search for profits, as is perhaps only normal. What is less so is the size of the profits generated by this industry, even at a time of recession affecting almost all other economic and commercial sectors. Drug prices are at very high levels, generic drugs are largely kept out of the market, and the common man is expected to accept this as a normal part of the market economy. Hippocrates would be turning in his grave.

There is also a well-documented imbalance between the R&D investments of the pharmaceutical industry geared to the rich and the poor. It has been estimated, for example, that out of the 1223 new chemical entities marketed worldwide between 1975 and 1996, only 13 were developed specifically for tropical diseases. This disparity is primarily due to the low level of profits which the pharmaceutical industry would expect in developing countries.

Some of this came to the forefront of international public debate when an effort was made to keep generic equivalents out of the hands of sub-Saharan Africans suffering from HIV/AIDS. A similar effort came to light when the United States ran short of an antidote drug during the anthrax scare, and had to face an attempt by drug companies to prevent identical drugs being imported from Canada where they were being sold at much cheaper prices.

To sum up, we live in a sad world where in spite of our very often knowing what to do, there is no effective global or national public health vision, and even more limited capacity for strategic public health management. Crisis management (or crisis creation) dominates our thinking and actions, and in this order of things, health ranks low on the scale of preoccupations.

## THE WAY FORWARD

How can more effective public health action be created? We must recognize that health development in modern societies is a complex endeavor. No conceivable scientific or technological breakthrough could bring such a large improvement in global health status as a reduction in the current global inequities, above all of poverty. Between and within societies, policies are needed that increase the income, the educational background, the living conditions and the social environment of the poor.

Improving the effectiveness and quality of health care is a vital component of a policy for public health improvement. Primary health care provides the institutional base for the contribution of health care to public health improvement. The primary care team must work closely with the community mobilization mechanisms that are an essential part of the primary health care approach. In the secondary and tertiary hospital sectors, the rapid development of medical and communication technologies will have major management implications and challenges for hospitals and other health facilities. A vital development will also be a much more comprehensive system to assess the outcomes and quality of individual clinical care.

One particularly important contradiction that must be recognized often arises between the issues of employment and those of health. That is why the tobacco industry, despite its known adverse impact on public health, survives with constant employment levels, and in fact continues to make handsome profits. Some of this may also be due to the fact that the decreased sales in Europe and America are being compensated by aggressive sales in developing countries. To counter such influences requires courage on the part of both politicians and public health managers.

One very important component of public health management and action is changing the nature of the debate. Change cannot just be left to professionals, who may not have the change management skills required. Nor can it just be left to politicians and administrators, who may lack the necessary courage to promote health as a priority alongside all the other interests clamoring for their favors. That is why the influence of civil society, the media, and populations themselves is so crucially important in creating advocacy and changed public perceptions.

Another important approach is to better educate the public also about the importance of public health as an issue of community development, improving the overall economic performance of populations and creating social and individual well being. Better and regular public health reporting to populations, as well as politicians and professionals, is vital. Taken together such developments can help decision-makers feel braver and more comfortable in promoting health and tackling influences inimical to health. Health needs a better voice at the decision-making table. The aim is to hold governments responsible and accountable for reflecting the importance of health across all levels of society. But not just governments: this same sense of responsibility and accountability for health needs to be felt across all sectors and actors within society whose decisions and actions have health effects.

Civil society organizations, many of which have considerable experience in public health projects, have a particularly important role to play. Such organizations can not only ensure deeper grass-roots presence at the local and rural levels, but can also be able to help in improving public health reporting, while at the same time, informing and educating the public about health problems. So far, regrettably, such civil society organizations have been perceived by governments as being largely adversarial in nature. It is essential to correct this impression, which has slowed down results in many countries. Health is a public service, not a power-sharing issue.

Nothing is ever possible without leadership and strategic vision. For all the reasons listed earlier, this has been sorely lacking in this basic sector of health. It has been difficult for leaders to shed their preoccupation with politics in the interest of the health sector, or to understand the long-term contribution of actions in this sector as contributions to peace and security. Much use is made of the slogan about democracies not going to war. Perhaps the time has come to point out that countries that put the proper emphasis on public health do not go to war either.

Nor should it be too difficult to point out to international organizations and their governing bodies that, in their case too, their current preoccupation with the political crises of the day are pushing back the aspirations of peoples into an ever distant future. Statistics abound about the costs of wars, about the damage to societies and their infrastructures, but also about the results which would have been achieved by the same expenditures were they to be diverted to social welfare objectives in the health and education sectors.

## CONCLUSION

Much of what needs to be done is clearly visible in the Millennium Development Goals. What they do is to set out the objectives as loftily as is possible. What they do not do is to chart out the methodology by which these objectives can actually be implemented within the prescribed time frames. The ideals of the first seven goals, which talk of a brave new world, will depend entirely on the eighth goal which sets out the funding needs. More than anything else, it is this last goal that will determine the success or failure of the MDGs as a whole. So far, it remains largely a paper promise.

Can the leaders of our times agree that health and education and human rights and gender equity are as important, if not more important, as their preoccupation with issues of peace and security, or of achieving it through wars? Could they agree that these social issues may well determine whether we will see peace and security in our times? And could they finally agree, as asked for in the last of the Millennium Development Goals, that nothing would be achievable unless we back the high objectives of the Goals with the necessary financial resources?

# 5. CLIMATOLOGY

## FOCUS: GLOBAL WARMING/AEROSOLS & SATELLITES

# AEROSOLS, AIR QUALITY, AND INTERNATIONAL POLICY

LAWRENCE FRIEDL
NASA Applied Sciences Program
NASA Headquarters, Washington, DC USA

## ABSTRACT

Aerosols and particulate matter pose significant health and environmental impacts to humans and ecosystems. Environmental and public health organizations have developed methods to identify aerosol impacts, develop policy, and create aerosol air quality forecasts. Research satellite observations and atmospheric models have provided new insights on aerosol distribution, interactions with climate change, and the regional-to-global transport of aerosols and other pollutants. In addition, the research, observations, models, and analytic techniques to characterize aerosols have also provided national-to-local air quality managers with advanced methods to identify aerosol transport and forecast pollution episodes. While aerosols were once thought to be a local air quality issue, there is greater awareness of the regional, continental, and intercontinental transport of aerosols. Internationally, the United Nations Convention on Long-Range Transboundary Air Pollution has established a task force to characterize air pollution transport and establish scientific guidance for international policy and protocols. Given the international sources of aerosols, long-range transport, and climate change impacts, international solutions are needed to help achieve national air quality goals and protect public health.

This paper will present the use of satellite observations to characterize aerosols, complement ground measurements, and inform air quality forecasts. It will discuss aerosol transport in air quality forecasting and issues related to air quality, aerosols, and climate change. Finally, the paper will describe the use of satellite observations and models to support the United Nations task force and international policy efforts to address long-range transport of air pollution.

## INTRODUCTION

Earth system science research has focused significant attention on the measurement of aerosols and their role in atmospheric radiative forcings and role in climate variability. Earth science satellites have contributed large-scale measurements and atmospheric models that provide new insights to understand the global to regional aerosol distributions and transport. The satellite observations have provided air quality managers with new measurements to forecast pollution episodes and international efforts to support scientific assessments for international policy.

Aerosols are both of natural sources, such as entrained dust, wildfires, or sea salt, and human sources, such as biomass burning and industrial pollution. Some aerosols, such as sulfates, have cooling effects on the atmosphere, and other types have warning effects, such as black carbon. Aerosols are also categorized by their size: aerosols and particulate matter between 2.5μm and 10μm in diameter are referred to as coarse, such as dust and sea salts. Particulate matter less than 2.5μm in diameter are referred to as fine (aka, PM2.5), such as combustion products. Environmental and public health organizations have identified that PM2.5 pose significant threats to human health and ecological systems.

EARTH SCIENCE SATELLITE OBSERVATIONS

Earth system science organizations have pursued significant ground, airborne, and space-based measurement efforts to answer fundamental sciences questions about the Earth system, including how aerosols effect the climate and generate other pollutants. Satellite measurements (polar-orbiting and geostationary) provide unique spatial, spectral, and temporal measurements of aerosols, especially global distributions and measurements in areas without ground monitors.

Several Earth science satellites fly in a constellation formation, which the Earth science community refers to as the "A-Train" (Figure 1). The A-Train consists of seven international and U.S. satellites that fly within 20 minutes of one another, providing coordinated science and multiple measurements of the same underlying environment and atmosphere.

*Figure 1. The A-Train Constellation of Satellites.*

The following is a brief summary of some sensors on NASA Earth science research satellites that collect aerosol measurements (several of the sensors are joint international activities); each of these sensors is represented in the A-Train.

<u>Moderate Resolution Infrared Spectroradiometer (MODIS)</u>

MODIS sensors fly on NASA's *Terra* and *Aqua* satellites. MODIS has 36 spectral channels from 0.41-15um representing three spatial resolutions—250m, 500m, and 1km—and the sensors gather information about the location and thickness of dust and haze (among numerous other geophysical parameters). *Terra* and *Aqua* are both polar-orbiting, sun-synchronous satellites. *Terra* (launched in 1999) has a 10:30am equatorial crossing time and *Aqua* (launched in 2002) has a 1:30pm crossing time.

Two standard MODIS products include the Aerosol Optical Depth (AOD), which is a measure of the total extinction by aerosols from the satellite to the ground, and the Cloud Optical Thickness (COT). MODIS AOD values retrieved over land use 500m bands at 0.47 and 0.66 μm and interpolated to 0.55 μm in order to be mapped with those retrieved over ocean (Chu et al., 2005). Daily AOD/COT products are produced at a spatial resolution of a 10×10 1-km (nadir)-pixel array. Figure 2 provides a visible image and a processed image for AOD and COT for a pollution event in North America in 2002; the comparisons highlight Hurricane Gustav off the East coast of North America and high aerosol/haze conditions in the Midwest United States. For AOD, which is a unitless parameter, lower values (color coded blue) represent lower levels of aerosols.

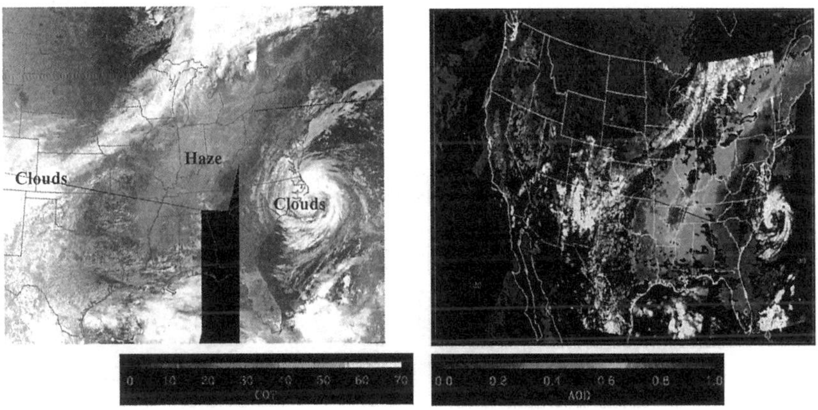


Cloud Optical Thickness

Aerosol Optical Depth

*Figure 2. MODIS visible image of haze and clouds (left) and MODIS processed image of aerosol optical depth and cloud optical thickness (Sept. 10, 2002). Images from NASA-GSFC & LaRC.*

Aerosol optical depth is a satellite-derived measure of light extinction through the atmosphere, which is proportional to the number of particles in the atmospheric column. MODIS AOD is a columnar measurement, and it does not provide vertical resolution to indicate the height of the aerosols in the atmosphere. The algorithm accounts for reflectances over land or oceans, aerosol types, size and scattering parameters, and other factors. For example, the algorithm in Europe and the eastern North America uses an urban/industrial aerosol model, and in western North America it uses a smoke-based aerosol model. Remer et al., (2005) provide a detailed description of the MODIS AOD algorithm. Summarizing some of that description, AOD is most reliable over ocean due to its flat, dark, uniform surface, and, over land, AOD is most reliable over flat terrain and dense vegetation. AOD is not available with cloud-cover, and it is least reliable over deserts and snow-covered areas due to the bright background reflectance. In addition, the AOD does well at detecting sulfates and other spherical particles that scatter light well, and it does less well at detecting irregularly shaped particles that do not scatter light well. Work is in progress to improve the AOD retrievals over bright surfaces.

Multi-Angle (MISR)

MISR is also on the *Terra* satellite. MISR is a 9-camera instrument with fore, nadir, and aft viewing cameras at different view angles (0, 26.1, 45.6, 60.0, and 70.5 degrees). Each camera has four spectral bands (blue, green, red, and near-infrared), and the detectors are charge couple devices. Each camera has a 360km crosstrack swath width, and the global cover is every 9 days (some latitudes receive more frequent coverage). The multi-angle cameras provide unique insights on aerosol distributions, including some estimates of smoke plume and aerosol heights.

CALIPSO

Launched in April 2006, CALIPSO is a joint U.S.-French (NASA and CNES) satellite. The satellite has a 3-channel LIDAR (CALIOP-Cloud-Aerosol LIDAR with Orthogonal Polarization) and two passive instruments to obtain coincident observations of radiative fluxes and atmospheric conditions. The LIDAR provides insights into the vertical dimension of clouds, dust and aerosols. Figure 3 provides an pre-validation view of the 532nm total attenuated backscatter from 7-June-2006.

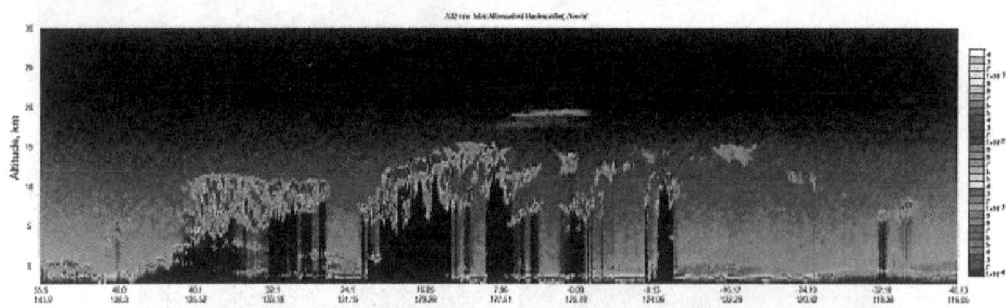

*Figure 3. CALIPSO Total Attenuated Backscatter 532 nm (7 June 2006). Vertical structure of aerosols and clouds visible. Light, horizontal streaks in center of image are likely sulfate aerosol layer resulting volcanic plume from the 20-May-2006 eruption of Soufriere Hills, Montserrat. Image courtesy of NASA-Langely.*

Ozone Monitoring Instrument (OMI)

The OMI sensor is one of four sensors aboard the *Aura* satellite, which launched in 2004. OMI is a Dutch-Finish instrument, and the U.S. is part of the joint Dutch-Finish-U.S. OMI science team. OMI is a nadir-looking solar backscatter spectrometer (280-500nm) with a 13x24km footprint and a 2.6km swath width. In addition to measuring ozone and other trace gases, OMI uses the UV to detect light-absorbing aerosols, such as Black Carbon and dust. OMI produces an Absorbing Aerosol Index (AAI), which is beneficial for geographic and source identification but less so for quantification. *Aura* has a 1:45pm equatorial crossing time (ascending node), and it follows *Aqua* in the same orbit by 15 minutes.

CloudSAT

CloudSAT, which is a cooperative mission with Canada, uses an advanced radar to "slice" through clouds to measure their vertical structure. The Cloud Profiling Radar on CloudSAT measures the structure and composition clouds, and scientists use the CloudSAT data to study clouds globally, characterize the effect of aerosols on the hydrologic cycle, and support climate models. CALIPSO launched in April 2006.

## AIR QUALITY FORECASTING

The AIRNow program is a joint partnership between the U.S. Environmental Protection Agency (EPA) and State and local air quality agencies to provide real-time air quality information in a visual format, such as the Air Quality Index (AQI). The AQI is derived using an extensive network of ground monitors, and the system alerts the public once pollution levels exceed a certain level so that sensitive groups may restrict their activities. Since 2002, a team of researchers from NASA, EPA, NOAA, and academia has developed the use of near real-time MODIS satellite data in AQI to help forecasters improve the next-day, regional forecasts of particle pollution. Forecasters now include the satellite-based data products and techniques in operational use.

The project team initially developed a "fused" product that combined EPA ground measurements with the MODIS satellite data and expanded this product to multiple-day, time-looped animations of aerosol levels and trends. These products helped forecasters visualize the relationship between MODIS AOD and PM2.5 ground monitor data. The tools provided the ability to identify and track the frequency and extent of particle pollution transport episodes. During a prototype test, the project team created three-day loops, each showing MODIS aerosol and cloud data over the entire continental United States, overlaid by either wind vectors or air parcel trajectories and hourly measured particle concentration data from ground monitors. The visualization loops depicted temporal and spatial relationships to provide a synoptic view of aerosol events across North America.

Figure 4 below provides samples of MODIS images and AQI measurements from a North American haze event from September 8-14, 2002. This multiple-day sequence shows the transport of a regional air pollution event. The color contours show MODIS aerosol optical depth, and the color posts show EPA ground-based PM2.5 measurements (AQI levels). The similarities in levels suggest that satellite data can assist air quality managers observe aerosol levels, location and transport as well as provide aerosol measurements where there are few or no ground monitors. In this episode, a large-scale aerosol event formed south of Lake Michigan (several hourly AQI readings above 100). Primarily driven by large-scale meteorological conditions (including Hurricane Gustav off the East Coast), aerosols moved north to Canada, south toward Texas, and then East along the Gulf of Mexico coast over the seven-day event.

The data fusion products for the 2003 prototype included MODIS aerosol optical depth, MODIS cloud optical thickness, hourly PM2.5 concentrations from ground, modelled winds fields, satellite-based fire counts, and modelled air parcel trajectories.

Sept 8, 2002
Large aerosol event forming in the Midwest U.S. Hurricane Gustav approaching East Coast.

Sept 10, 2002
AQI>100 in Midwest. Bands of high aerosols to Texas and Canada.

Sept 12, 2002
High aerosol levels over Texas and transport across much of Southeast US.

*Figure 4. MODIS AOD Indications of Regional Transport Event in North America.*

Figure 5 presents a plot of the hourly PM2.5 concentrations (black line) for a ground monitor at Baltimore, Maryland and the MODIS AOD levels (red dots). Although there were 82 days plotted on the chart for the ground measurements, there were only 15 occasions with concurrent MODIS AOD measurements over this period, which is likely due to cloud cover which prohibits AOD measurements. At some dates, such as August 10-11, the MODIS AOD levels tracks well with the hourly concentrations, suggesting the aerosols are within the boundary layer. On other days, such as August $12^{th}$ and $26^{th}$, the MODIS AOD levels are higher than the ground measurements, suggesting that the aerosols are aloft rather than in the boundary layer. Extensive studies at numerous locations have indicated strong correlations between the hourly PM2.5 surface monitors and AOD levels in coincident pixels (10km x 10km) (Engle-Cox et al., 2004).

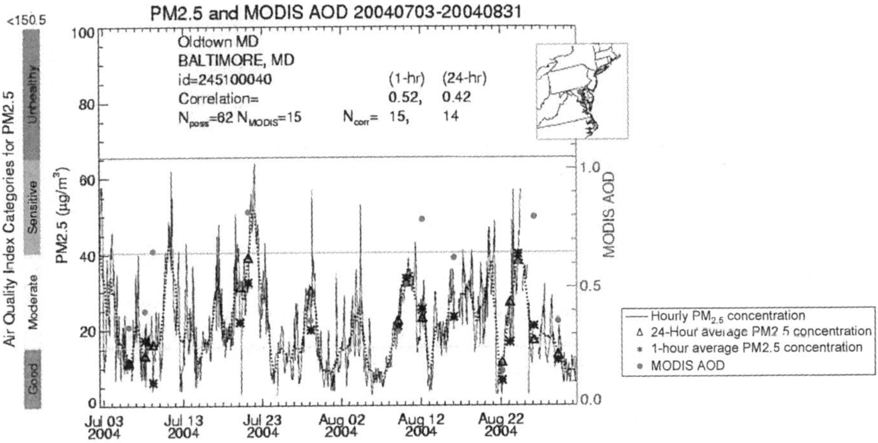


*Figure 5. Comparison of MODIS AOD and PM2.5 ground monitor for Baltimore, Maryland, USA. Chart courtesy of J. Engle-Cox, as presented at* Air & Waste Management Association Annual Meeting, *June 2006.*

In addition to the quantitative correlations of MODIS AOD with ground measurements, the visual, qualitative aspects of the satellite images provide significant opportunities to explain pollution events to the public and to support forecasts. The synoptic views that the near real-time satellite images, in combination with the other measurements, provide a strong means to communicate a pollution event to the public and the media. The introduction of air quality data in forecasts can build on the public's familiarity with overhead images of weather maps and satellite weather images. As data is available in near real-time, forecasters can support their analyses with visible, overhead images to explain the broader conditions leading to the pollution event.

Certainly, there are limitations to satellite observations and their routine use in air quality forecasts. Observations lack specificity about some pollutants, and the temporal, vertical, and spatial resolutions of current satellites may not satisfy those desired by air quality managers. The near real-time data and images do provide insights into contributing factors in pollution events as well as an ability to capture the present realities in atmospheric conditions. The prototype project provided data-fusion products to forecasters, who indicated that the forecast tools helped to identify the extent and frequency of the pollution events and regional transport. The visualizations of data fusion products provide information to distinguish local from regional pollution episodes, including transport from foreign sources. In the benchmark report accompanying the prototype project, the forecasters noted that the pseudo-synoptic view of aerosol loading allowed them to identify natural event influences and re-circulation influences to improve the context of their forecasts (NASA, 2003). In addition, the identification of an impending pollution event may allow state and local governments to take actions to mitigate the effects or request non-attainment waivers.

LONG-RANGE TRANSPORT

The synoptic view and global coverage of satellites have provided insights on the long-range transport of dust and pollutants. Figure 4 illustrated continental-scale transport of pollutants in a 2003 event down and across significant portions of North America. Satellites have observed numerous other large-scale transport events. MODIS and other satellites measured and tracked significant levels of smoke across North America from the 2004 fires in Alaska and Western Canada. In 2001, another satellite (Total Ozone Mapping Spectrometer (TOMS)) observed the transport of dust from East Asia across the Pacific Ocean, and several ground monitors across the United States measured significantly higher aerosol measurements (Mintz and Szykman, 2002). Figure 6 provides a visual display of this 2001 event. In the figure, higher index values and warmer colors indicate higher levels of aerosols. The images suggest that the satellite measurements provide significant detail over regions with no or few ground monitors, especially the oceans.

7-April-2001 9-April-2001

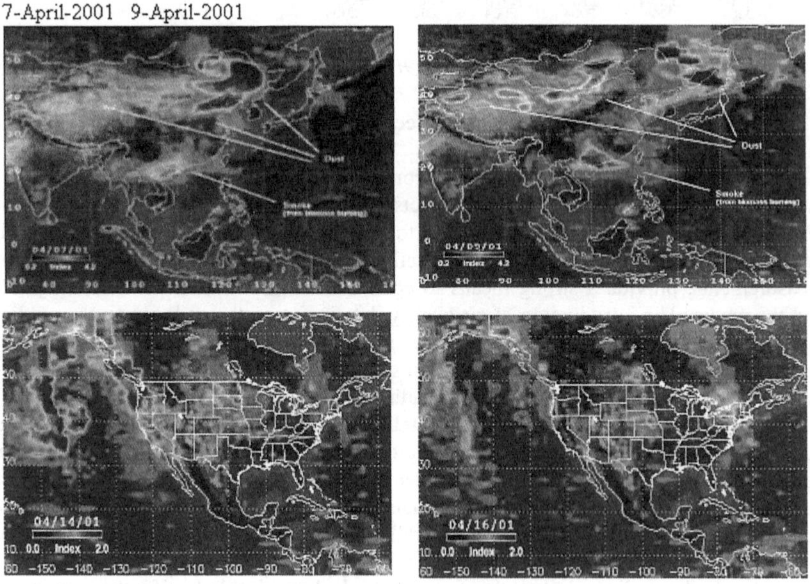


14-April-2001 16-April-2001

*Figure 6. Long-range transport observed by Total Ozone Monitoring Spectrometer (TOMS); aerosol optical index in April 7-16, 2001. Images courtesy of NASA/Langley. Higher index values and warmer colors indicate higher levels of aerosols.*

## INTERNATIONAL COORDINATION: GROUP ON EARTH OBSERVATIONS

The use of Earth science and environmental measurements has drawn significant attention recently on international fronts. In August 2003, ministerial representatives from over 30 nations gathered in Washington, DC, to initiate the Group on Earth Observations (GEO; *http://www.earthobservations.org*). By coordinating and raising awareness of Earth observations, GEO encourages the assimilation of Earth observations and model predictions to serve as inputs to nations and organizations decision-support tools. By August 2006, the GEO had grown to 65 countries and over 35 international organizations. In developing a 10-year implementation plan, the GEO decided to focus on integrating the scientific capacity of organizations and observing systems to support nine societal benefit areas:

- Natural & Human Induced Disasters
- Water Resources Ecosystems
- Oceans
- Sustainable Agriculture & Desertification
- Climate Variability & Change
- Weather Information, Forecasting
- Human Health & Well-Being (includes Air Quality)
- Energy Resources

The Group on Earth Observations (GEO; *http://www.earthobservations.org*) adopted the figure below (Figure 7) as a notional architecture to depict the contributions of Earth observations and models to organizations' decision support, management, and policy-making activities that provide value and benefit to society. The coordinated contribution and integration of Earth observations (satellite, airborne, ground, in situ) and Earth system models is referred to at the Global Earth Observing System of Systems (GEOSS).

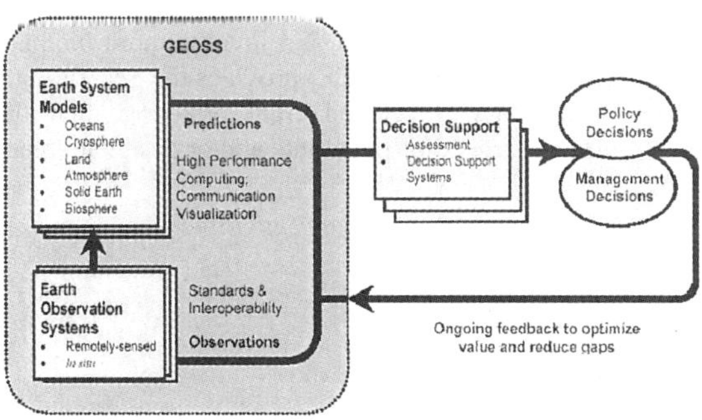


*Figure 7. Group on Earth Observation architecture depicting the contributions of Earth science data and models to support societal benefits.*

On the right side of the architecture, operational agencies own, develop, and operate decision-support tools to inform their decision-making processes. Decision-support systems serve different purposes, such as planning, forecasting, and early warning activities. Generally, decision-support systems are interactive, computer-involved systems that provide organizations with methods to retrieve and summarize information, analyze alternatives, and evaluate scenarios to gain insight on critical factors, sensitivities, risks, and consequences of potential decisions. Government agencies use decision-support systems to support their responsibilities to the public, such as resource management, security, regulations, public health, and economic development.

On the left side of the figure, research and operational agencies develop measurement techniques, collect measurements, produce new knowledge, and extend Earth observations, environmental and climate data records, and model predictions and forecasts. Where the Earth science products are determined to have potential value to a decision support system, organizations can collaborate to facilitate and streamline the flow of products to the tools, drawing on computational techniques and interoperability practices to support data sharing and system integration. Given the copious volumes of Earth observation data and the computationally demanding scientific models, decision support systems typically provide systematic mechanisms to incorporate data products and document the value derived from the inputs. The outcomes of the GEO approach are manifest in the organizations' enhanced policy and management decisions, and the impacts are the resulting socioeconomic benefits from the improved decisions.

## INTERNATIONAL POLICY: LONG-RANGE TRANSPORT OF AIR POLLUTION

In 1979, some countries adopted the Convention on Long-Range Transboundary Air Pollution (LRTAP), which calls on the parties to reduce transboundary air pollution using the best policies and strategies and best available technology which is economically feasible (Engle-Cox, 2005). 49 countries signed and ratified LRTAP, although the primary countries were Canada, USA, Former Soviet Republics, and several European countries. Countries from Asia, Middle East, northern Africa, and Central America are not currently included in LRTAP; Southern Hemisphere countries were not included due to different pollution types (e.g., pollution is less industrial and involve more biomass burning). The UN Economic Commission for Europe (UNECE) provides the Secretariat.

Under this convention, the parties have developed eight protocols, and the protocols have typically set specific emission targets for pollutants and/or designate types of actions:

- Sulfur (1985, 1994)
- Nitrogen Oxides (1988)
- Volatile Organic Compounds (1991)
- Persistent Organic Pollutants (1988)
- Heavy Metals (1998)
- Acidification, Eutrophication, and Ground-level Ozone (1999)

In December 2004, the UNECE established a new task force on Hemispheric Transport of Air Pollutants (HTAP) to address the intercontinental transport in the Northern Hemisphere. From 2005-2009, this technical task force will assess the scientific evidence concerning hemispheric transport for use in international policy discussions and reviews of LRTAP protocols. In addition, this task force will specifically involve Northern Hemisphere countries who are not signatories to LRTAP. The air pollutants of interest and assessment by the HTAP task force include Fine particles/PM, Ozone and precursors, Acidifying substances (NOx, SOx), Mercury, and Persistent organic pollutants.

The HTAP task force efforts involve model assessments, model intercomparisons, sensitivity studies, and use of satellite observations to constrain models and provide information over oceans and regions with minimal ground-based monitors. The task force issues an interim assessment on Ozone and PM in Spring 2007, and it will issue a final report in 2009.

## CONCLUSION

Satellite observations provide comprehensive information on aerosols and their global distributions and transport to advance fundamental knowledge about the Earth system and climatic radiative forcings. These measurements also provide unique information for operational use by public and private organizations. Combined with data from ground networks, the satellite measurements provide air quality managers and international policy makers with qualitative and quantitative methods to observe temporal and spatial relationships. Although significant attention is needed to continue the transition of research satellite measurements into operational observations and continuing development of Earth system research, modeling, and techniques.

Thus, in addition to the inherent value of increased knowledge, the Earth science research and measurements provides opportunities for scientists and scientific organizations to demonstrate the value and relevance of science to support direct societal benefits and decision making. This use and future developments of Earth science research, satellites, and models continue the roles and benefits that scientific research plays in providing societal value and benefits.

## REFERENCES

1. Al-Saadi, Jassim, J. Szykman, B. Pierce, C. Kittaka, D. Neil, D.A. Chu, L. Remer, L. Gumley, E. Prins, L. Weinstock, C. MacDonald, R. Wayland, F. Dimmick, and J. Fishman. "Improving National Air Quality Forecasts with Satellite Aerosol Observations." *Bulletin of the American Meteorological Society,* 86(9), 1249-1261, September 2005.
2. Chu, D.Allen, J. Al-Saadi, C. Kittake, B. Pierce, J.Szykman, L.A.Remer, and D. Neil. "Analysis of Relationship between MODIS Aerosol Optical Depth and PM2.5 over the Summertime U.S.," as submitted to *Atmospheric Environment,* 2005.
3. Engel-Cox, J., R. Hoff, and A. Haymet, "Recommendations on the Use of Satellite Remote-Sensing Data for Urban Air Quality," *Journal of Air and Waste Management,* 54, 1360-1371, November 2004.

4. Engle-Cox, J., R. Hoff, R. Rogers, F. Dimmick, A. Rush, J. Szykman, J. Al-Saadi, D.A. Chu, and E. Zell, "Integrating Lidar and Satellite Optical Depth with Ambient Monitoring for 3-Dimensional Particulate Characterization" Accepted for publication to *Atmospheric Environment* 2006.
5. Engel-Cox, J., C. Holloman, B. Coutant, and R. Hoff. "Qualitative And Quantitative Evaluation of MODIS Satellite Sensor Data For Regional And Urban Scale Air Quality," *Atmospheric Environment,* 38, 2495-2509, May 2004.
6. Engle-Cox, J., C. Holloman, M. Cupp, B. Coutant, and K. Swinton. "Satellite Data for Air Quality Analysis." Technical Report by Batelle for U.S. Environmental Protection Agency Office of Air Quality Planning and Standards (Contract No. 68-D-02-061, Work Assignment 1-05), September 30, 2003.
7. Engle-Cox, J., and E. Zell. "NASA Earth Science Research for International Air Quality Policy." Report for by Battelle for National Aeronautics and Space Administration (JCET sub-award CG0416). February 1, 2005. Available through website: *http://aiwg.gsfc.nasa.gov/dss.html*
8. Friedl, L., and NASA Air Quality Program Team. "Space-based Earth Science Support for Air Quality Management." *EM*, 28-32, September 2005.
9. Gupta, P., S.A. Christopher, J. Wang, R. Gehrig, Y.C. Lee, and N. Kumar, Satellite Remote Sensing of Particulate Matter and Air Quality over Global Cities, *Atmospheric Environment*, October, 2005.
10. National Aeronautics and Space Administration. Benchmark Report: "The Application of Satellite Data for Forecasting Particle Pollution." November 28, 2003. Available through website: *http://aiwg.gsfc.nasa.gov/dss.html*
11. Remer, L.A., Y J. Kaufman, D. Tanre, S. Mattoo, D.A. Chu, J.V. Martins, R-R. Li, C. Ichoku, R.C. Levy, R.G. Kleidman, T.F. Eck, E. Vermote, and B.N. Holben. "The MODIS Aerosol Algorithm, Products and Validation." *J. Atmos. Sci.,* 62, 947-973, April 2005.
12. Wang, J., and S.A. Christopher. "Intercomparison between Satellite-Derived Aerosol Optical Thickness and $PM_{2.5}$ Mass: Implications for Air Quality Studies," *Geophys. Res. Lett.*, 30(21), 2095, 2003.

*Websites of interest to this topic:*
NASA Air Quality Prototype project: *http://idea.ssec.wisc.edu/*
NASA MODIS Atmosphere, *http://modis-atmos.gsfc.nasa.gov/*
U.S. Air Quality (The Smog Blog), *http://alg.umbc.edu/usaq/*
Group on Earth Observations, *http://www.earthobservations.org*
Hemispheric Transport of Air Pollutants, *http://www.htap.org/*

# SATELLITE REMOTE SENSING OF AEROSOL CLIMATE EFFECTS—PROGRESS AND POTENTIAL

SUNDAR A. CHRISTOPHER
Department of Atmospheric Sciences
University of Alabama , Huntsville, USA

The importance of aerosols and their effects on regional and global climate have been addressed by numerous studies including comprehensive reports from the International Panel of Climate Change (IPCC, 2001). Aerosols also affect visibility and pose serious threats to health since fine particulate matter can cause respiratory illness. On a global average annual basis, the amount of incoming radiation from the sun into the earth-atmosphere system must equal to the amount of radiation leaving the system. Aerosols affect the radiation balance of the earth-atmosphere system primarily through two different mechanisms. Through the direct effect, atmospheric aerosols scatter and absorb the incoming solar radiation. The scattering of solar energy increases the earth's planetary albedo, thereby cooling the earth's surface. The absorption of solar energy by aerosols changes the atmospheric heating rate, thereby influencing atmospheric circulation. Through the indirect effect, aerosols modify the shortwave reflective properties of clouds by increasing their lifetime and suppressing drizzle formation, thereby altering precipitation processes. The study of aerosol radiative forcing is also critical because it counteracts the warming effect due to anthropogenic greenhouse gases such as carbon dioxide. The greenhouse gases have a positive forcing of around $+2.4 Wm^{-2}$ whereas the direct and indirect effect of aerosols are believed to have a negative forcing. This indicates that while GHG increases surface temperatures, aerosols tend to cool the surface. Therefore it is important to know the spatial distribution of aerosols, their properties, and how they counteract the GHG warming.

Based on previous IPCC reports in 2001, aerosols are one of the largest sources of uncertainty in climate change studies. The level of understanding for all types of aerosol forcing was considered to be very low and even the sign of the radiative forcing of aerosols is not well established. The lack of quantitative understanding is not only due to the large variations in aerosol physical and optical properties, but also due to the large variations in the spatial and temporal distributions of aerosols.

In 2000, the National Aeronautics and Space Administration (NASA) began to launch a series of well calibrated, high quality sensors on several satellites to study the earth-atmosphere system in an integrated fashion. Sensors such as the Moderate Resolution Imaging SpectroRadiometer (MODIS) and the Clouds and the Earth Radiant Energy System (CERES) scanners on board NASA's Terra and Aqua satellites were designed to study, among others, the role of aerosols and climate. Global aerosol concentrations can now be obtained on a daily basis from these polar orbiting satellites, which allows for the study of aerosol induced radiative energy changes.

The MODIS has 36 spectral channels and obtains aerosol properties globally with unprecedented accuracies. The CERES with improved spatial resolution and better instrument characteristics when compared to previous generation instruments can measure outgoing shortwave and longwave radiation from the earth with high accuracy.

Combining information from such instruments with radiative transfer models and ground-based instruments has helped us reduce the uncertainties in aerosol forcing studies.

In the IPCC report released in 2001, the uncertainties in the direct radiative forcing of aerosols were estimated to be greater than 80%. This assessment came largely from numerical modeling studies that had insufficient emission and aerosol data sets. During the last five years tremendous progress has been made in advancing our understanding of aerosol effects on climate. A recent study by Bellouin et al. 2005, notes that the uncertainties have been reduced to nearly 20% and that aerosol forcing is higher than what was previously estimated in the IPCC 2001 report.

A new paradigm is emerging in the scientific community for studying aerosols. To begin with, detailed insitu measurements are required over selected regions such as particle size, chemical composition, and other aerosol scattering and absorbing properties. These measurements are then used to refine and validate satellite algorithms. As part of this process, high quality ground-based instruments such as sunphotometers from the Aerosol Robotic Network (ARN) program are needed to routinely monitor aerosol properties from point locations. The knowledge gained from this combination of insitu-ground-satellite measurements coupled with radiative transfer modeling can be used to constrain global numerical modeling estimates of aerosol forcing since global models are best suited to predict the future role of aerosols on climate that satellite data sets cannot do.

While significant progress has been made in understanding the direct radiative forcing of aerosols, the problem is far from being solved. We need to understand how aerosols interact with clouds and how aerosol absorptive properties change globally. Another critical need is to obtain data on how the aerosol and cloud co-exists in a vertical column. Questions such as what type of aerosols and what fraction exist above or below clouds? Newly launched satellites such as CloudSAT and CALIPSO with vertical probing capabilities will be able to provide vital information to solve some of these problems.

In summary, I conclude with the following statements:

- High quality satellite data sets are now available to answer key questions related to the effect of aerosols on climate.
- The uncertainties in the direct radiative effect of aerosols have been reduced tremendously over the last few years largely due to better satellite data and improved understanding of aerosol properties and processes in numerical modeling simulations.
- The effect of aerosols on clouds remains a challenging problem and recently launched satellites such as CALIPSO and CloudSAT will play a major role in answering some key questions.

## REFERENCES

1. Bellouin, N., O. Boucher, J.Haywood and M. S. Reddy. "Global estimate of aerosol direct radiative forcing from satellite measurements." *Nature*, **438**:1138-1141, doi:10.1038/nature04348, 2005.

2. IPCC, 2001: Climate Change 2001: The Scientific Basis. Contribution of Working Group I to the Third Assessment Report of the Intergovernmental Panel on Climate Change, Cambridge University Press, United Kingdom and New York, NY, USA, 881pp, 2001.

# LINKING AEROSOLS SOURCES TO CLIMATE CHANGE AND AIR POLLUTION IMPACTS—HOW GOOD ARE OUR GLOBAL AND REGIONAL MODELS?

GREGORY R. CARMICHAEL
Department of Chemical and Biochemical Engineering
University of Iowa, Iowa City, USA

## INTRODUCTION

The largest uncertainty in the radiative forcing of climate change over the industrial era is that due to aerosols (cf., the IPCC assessments). Aerosols influence climate by scattering and absorption of radiation (called the direct effect) and by modifying the cloud properties (referred to as indirect effects). Aerosols, which act as cloud condensation nuclei increase the number of droplets in clouds and tend to decrease the mean droplet size and increase the cloud albedo and cloud lifetime. While this has a net cooling effect on climate, the absorption of shortwave radiation by aerosols results in heating of the atmosphere and can evaporate clouds. The largest uncertainty in the climate effects of aerosols are due to the indirect effects. However, a substantial fraction of the uncertainty is also associated with scattering and absorption of shortwave (solar) radiation by anthropogenic aerosols in cloud-free conditions. Quantifying and reducing the uncertainty in aerosol influences on climate is critical to understanding climate change over the industrial period and to improving predictions of future climate change for assumed emission scenarios.

Aerosols are also air pollutants and contribute to reduced visibility and to adverse ecosystem and human health effects. Health effects due to aerosols are evidenced by toxicological and epidemiological studies which are increasing the certainty that exposures to small combustion particles of combustion origin pose significant health risk globally. Assessments, e.g., the Comparative Risk Assessment (CRA) study (Ezzati et al. 2002) of the World Health Organization (WHO), are presently made using particle mass as a measure of health effects. Close to 3 million premature deaths are now attributed to aerosol exposures in occupational, indoor, and outdoor environments from combustion sources, mainly fossil and biomass fuels and tobacco (passive smoking). Combustion aerosol, thus, is by far the largest environmental source of ill-health in the world, far exceeding that of poor water and sanitation for example. Most of the impact occurs in developing countries, a significant fraction in young children, to quantify the burden of disease (premature death and illness) from various risk factors. Health impacts analysis also requires the spatial and temporal distribution of ambient aerosols: information that presently must be supplied by model simulations.

Through ambient aerosols the issues of radiative forcing of climate change, air quality, and health impacts are linked. Linking emissions to aerosol distributions is essential to attribute aerosol health and radiative effects to specific aerosol components and to provide policy makers with the information needed for adaptive management of atmospheric composition. Global and regional chemical transport models play a critical role in both science and policy by providing a predictive means to estimate the ambient aerosol distributions, based on specified emission distributions (as shown in Figure 1).

These aerosol distributions are in turn used by radiative transfer models to estimate radiative forcing due to aerosols, and are used in population exposure models to assess their impacts on health.

In this paper, the use of aerosol models in air quality and climate applications are illustrated and capabilities (strengths and weaknesses) demonstrated by comparison with observational data. Our ability to monitor atmospheric aerosols continues to increase, and these measurements during the past decade are contributing to an enhanced understanding of atmospheric aerosols and their effects on climate. Further improvements in our quantitative understanding of the linkages between emissions and the resulting pollutant distributions require a closer integration of observational data with chemical transport models. Such comparisons with observations and resultant reductions in uncertainties are essential for improving and developing confidence in climate model calculations incorporating aerosol forcing. Current approaches to the better integration of observations and models are also discussed.

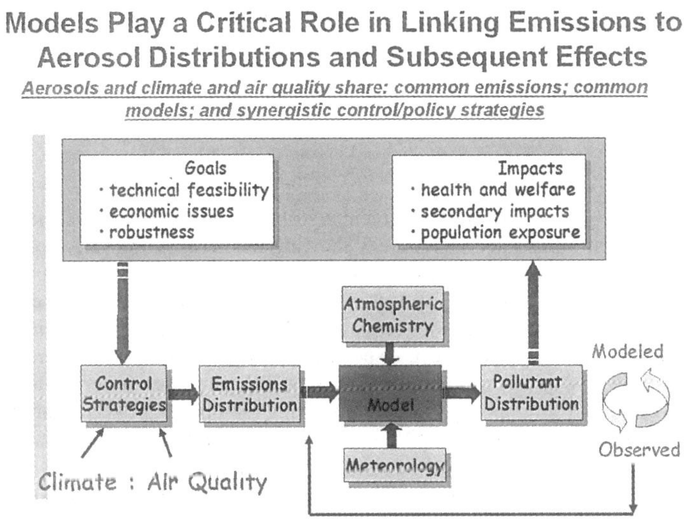


*Figure 1. Schematic of the analysis framework linking emissions to aerosol distributions. Also shown are the linkages between air quality and climate impacts and controls.*

## CHEMICAL TRANSPORT MODEL (CTM) CALCULATIONS OF THE DISTRIBUTIONS OF NATURAL AND ANTHROPOGENIC AEROSOLS

Chemical transport models provide a means to estimate 4-dimensional aerosol distributions, based on an emissions distribution. The aerosol mass and composition distributions, in turn can be used in radiative transfer models to estimate their impact on climate forcing, and in health outcomes studies to supply estimates of access mortality and morbidity. Linking emissions to aerosol distributions is essential to attribute aerosol effects to specific aerosol components and to provide policy makers with the information needed for adaptive management of atmospheric composition. It is necessary to estimate the size and chemical composition of the aerosol, as the effects depend on these. The composition of ambient aerosols in South Asia is shown in Figure 2. From a radiative forcing perspective, black carbon particles absorb radiation and act like $CO_2$ to warm the atmosphere, while sulfate particles are "white" and scatter incoming radiation and tend to cool the atmosphere.

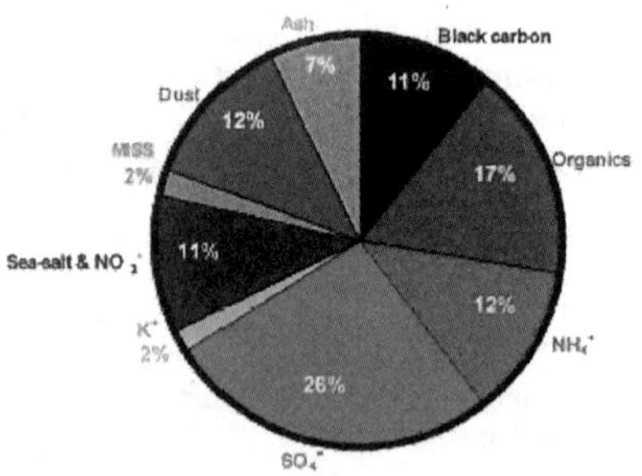


*Figure 2. Representative composition of the ambient aerosol in South Asia (data from the Atmospheric Brown Cloud (ABC) program (http://www-abc-asia.ucsd.edu/ )*

The processes that affect the ambient distribution of aerosols and their subsequent radiative effects are illustrated in Figure 3 showing the aerosol lifecycle. Some particles are emitted directly (e.g., dust, black carbon), while others are formed as a result of atmospheric chemistry (sulfate). Some emissions are due to natural process (e.g., sea salt), others due to anthropogenic activities (e.g., black carbon from coal combustion). Once in the atmosphere these particles are transported by the prevailing winds, gravitationally settle, and form and grow by chemical and physical processes (referred to as aging). These processes play an important role in modifying the optical properties of the aerosol. For example, black carbon when emitted is not effective in taking up ambient water, but after aging become hydrophilic, which in turn alters their impact on direct and indirect effects. The small particles that deposit deep into the lungs are important from health perspectives.

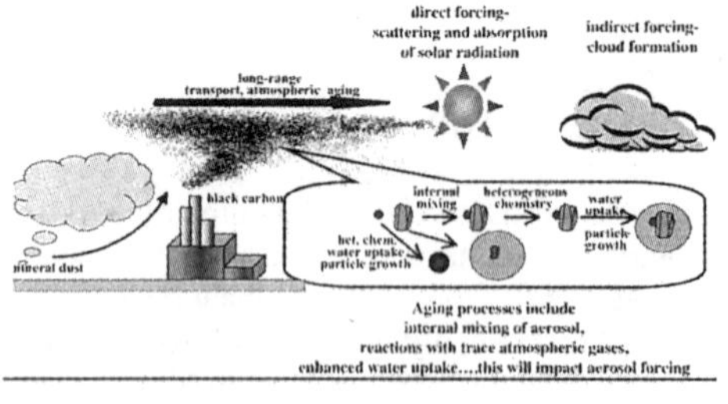


*Figure 3. Schematic of the life cycle of dust and black carbon aerosol and their interaction with climate forcing.*

Models of aerosols attempt to represent these processes. The analysis chain represented by aerosol models is illustrated in Figure 4. The analysis starts with the emission estimates, which are large sources of uncertainty. Major sources of emission uncertainties are listed in the figure. The chemical transport models calculate the 4-dimensional aerosol distributions, taking into account various transport, transformation and removal processes. These processes are an additional source of uncertainty. The mass loadings for sub and super micrometer sizes of non sea salt (nss) sulfate, black carbon, organic carbon, sea salt, and mineral dust, along with their spatial and temporal variation are calculated and then used in subsequent analysis of climate and health impacts.

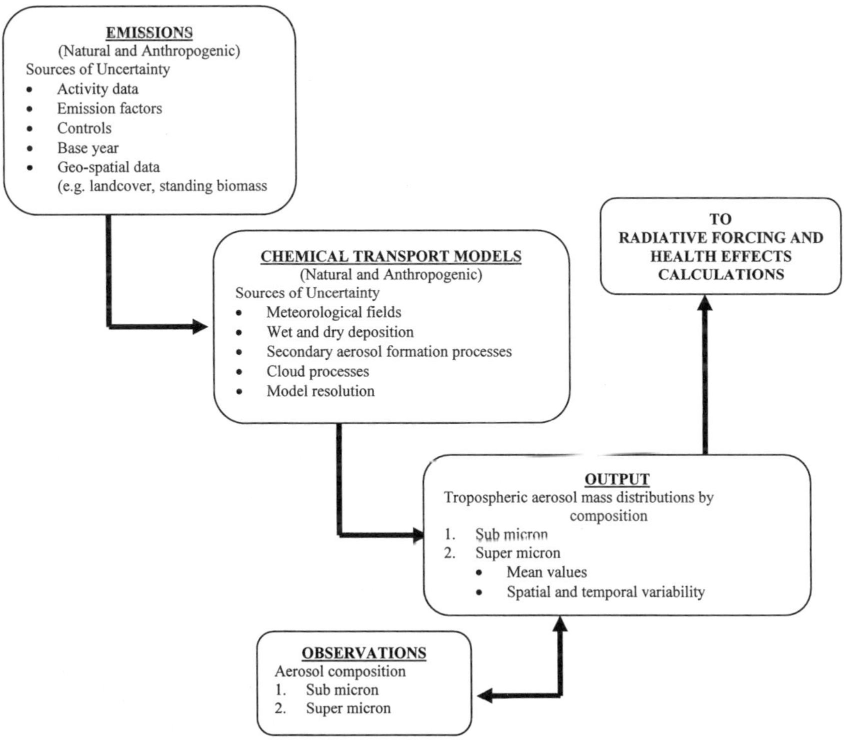


*Figure 4. Schematic of the calculation chain linking emissions to aerosol distributions discussed in this section.*

Each step along the analysis chain adds uncertainty to the estimates. A summary of the factor uncertainties (here a mixture of uncertainty types; i.e., parameter as well a scientific knowledge are included together) of the various processes to the calculated aerosol amounts of selected species is presented in Table 1 (the uncertainties associated with predicting the aerosol composition and size at a specific time and location are higher than those for the column quantities when time and space integrated). These results allow for a qualitative comparison of the sources of uncertainty in the analysis chain. While the relative sources of uncertainty vary from species to species, in general the uncertainties are ranked as follows: emissions > wet removal > chemical formation > vertical transport. Further information regarding emissions is shown in Figure 5, where the global spatial distribution of black carbon are presented, along with estimated uncertainties of various emissions for Asia. It is important to point out that the uncertainties in the aerosol emissions are very much greater than those for greenhouse gases. Furthermore the lifetime of an aerosol particle in the atmosphere is typically a few days, while that for $CO_2$ is many decades. This has important implications for policy.

**Table 1. Summary of estimated factor uncertainties in column amounts based on model intercomparisons, sensitivity studies, and expert opinion.**

| | Emissions | Wet removal | Vertical Transport | Chemical Formation | Total Uncertainty |
|---|---|---|---|---|---|
| Nss- SO4 | 1.3 | 1.3 | 1.5 | 1.3 | 1.8 |
| BC | 3 | 2 | 1.5 | -- | 3.9 |
| OC | 3.5 | 2 | 1.5 | 3 | 6.4 |
| Dust | 5 | 2 | 1.5 | -- | 6.0 |
| Sea Salt | 5 | 1.3 | 1.5 | -- | 5.4 |

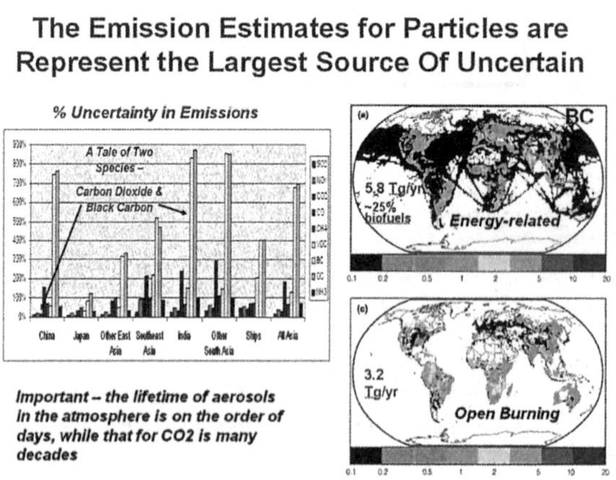


*Figure 5. Estimated uncertainties in the emissions of various pollutants in Asia. The right panels show the spatial distributions of black carbon emissions from energy related and open burning activities. Streets et al. 2004; and Bond et al. 2004.*

Given the large uncertainties in aerosol models it is important to assess: How good are the models? What is the consistency of model-derived information? How model-dependent are the results? There are important on-going studies that are designed to provide systematic and comprehensive aerosol model intercomparisons. For example, the AEROCOM (http://nansen.ipsl.jussieu.fr/AEROCOM/) study is comparing several global aerosol models and regional models are being compared for East Asia applications in the MICS study *Carmichael et al.* (2002). Despite these large uncertainties, models show a significant level of consistency in their predictions, and show appreciable predictive capability as shown in Figure 6.

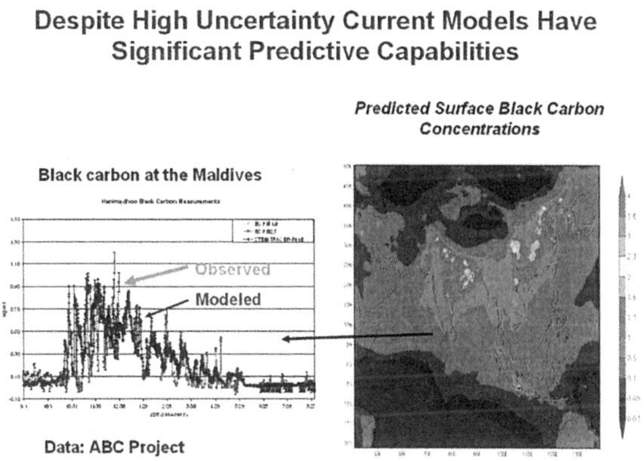


*Figure 6. Right panel: calculated mean near surface black carbon distribution over Asia for the month of February, 2005. Left panel: comparison of observed and predicted BC for a measurement site in the Maldives. This data and analysis was conducted under the Atmospheric Brown Cloud (ABC) study (http://www-abc-asia.ucsd.edu/).*

## AEROSOL EFFECTS ON CLIMATE FORCING

Of particular interest for climate models representing climate change over the industrial period are the top-of-atmosphere (TOA) and surface direct climate forcing, defined here as the changes in the respective net fluxes due to scattering and absorption of shortwave (solar) radiation by aerosols of anthropogenic origin in cloud-free conditions. TOA forcing is important to local and global radiation budgets; surface forcing is important to surface heating and water evaporation. Here direct climate forcing by aerosols (DCF) is defined as a change in a given radiative flux due to anthropogenic aerosols; this anthropogenic flux is additive to the direct radiative effect of natural aerosols (DRE). Both quantities are commonly expressed in units watts per square meter ($Wm^{-2}$).

Local instantaneous changes in shortwave radiative flux due to scattering and absorption of solar radiation by atmospheric aerosols in cloud-free conditions depend on the vertical integrals of the pertinent aerosol optical properties, the vertical distributions of these properties, the solar zenith angle, the surface reflectance and its angular distribution function, and water vapor amount and vertical distribution. The optical properties of the aerosol depend on its chemical composition and microphysical properties (size distribution, size distributed composition, and shape), which in many instances are strongly influenced by relative humidity (RH). In principle these properties can be calculated from Mie theory (or extensions thereof for nonspherical particles) for specified size dependent concentration, composition, shape, and mixing state. Calculations of DCF require the aerosol to be apportioned into natural and anthropogenic components. Because aerosol concentrations and compositions are spatially inhomogeneous, even the most intensive measurements are not able to represent the quantities needed to calculate DRE. Therefore, the requisite information must be approximated with the help of models.

To estimate aerosol radiative effects, the aerosol distributions discussed above are then used as inputs into radiative transfer models (in the case of estimates of direct effects). This analysis is shown in Figure 7. These calculations add additional uncertainties, and require further information on the optical properties as a function of aerosol size and composition. Over the last few years many laboratory and field experiments have focused on the optical properties of aerosols, and this information is being used to improve the model predictions. The effect of informing the models through the use of this new information on the estimates of direct radiative effects have recently been assessed and found to be large (Bates et al. 2006). The results are shown in Figure 8, and the uncertainties have been reduced by a factor of three. Again the largest uncertainty remains due to aerosols burdens (and the underlying emissions). The same has recently been found for the estimates of the aerosol indirect effects (Penner et al. 2006).

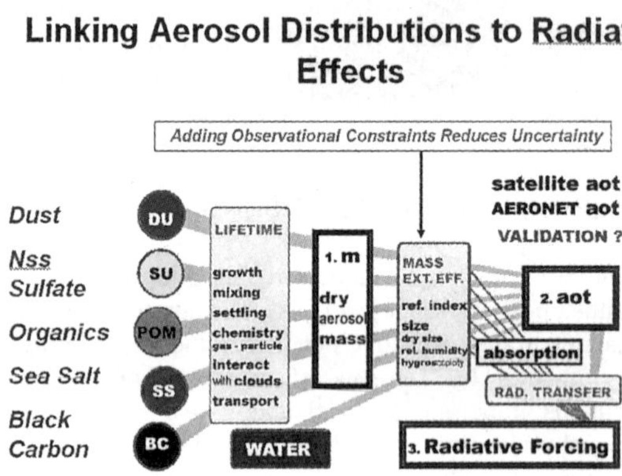


*Figure 7. Schematic of the linkages between aerosol distributions and radiative effects (from Kinne et al. 2006).*

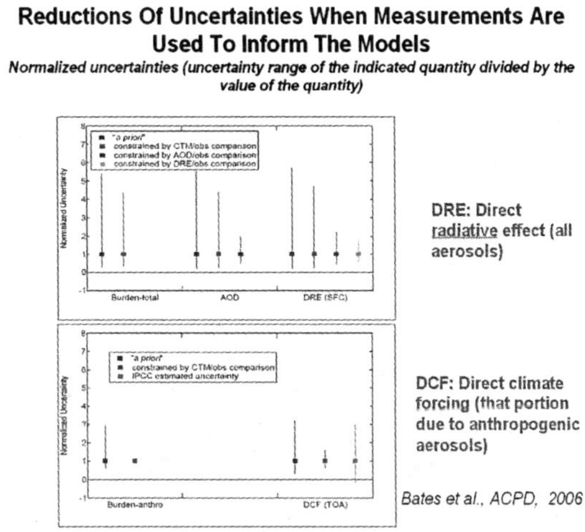


*Figure 8. Summary of the associated uncertainties with various components of the analysis linking aerosols to direct climate forcing. The Burden is the aerosol column resulting from emissions and subsequent transport. Aerosol optical depths and direct radiative effect (DRE) and direct climate forcing (DCF) are also shown. The reductions of uncertainty when new observational information is used to inform the model are shown to be large. The continued progress can be seen by comparing the current estimates of uncertainty to those estimated for the last IPCC. (Bates et al. 2006)*

A further reduction in uncertainty in the analysis of aerosol effects on health and climate will come from the utilization of the aerosol information contained in satellite observations. Sensors on satellites are providing important information on the aerosol optical depth, which as we discussed earlier contains information on the aerosol composition and spatial distributions. These observations can be used to evaluate and refine the model predictions. As shown in Figure 9, there is a difference between what is observed and the model predicted values. The model parameters (e.g., the emissions) can be adjusted to provide a best fit to the observational constraints. However using just the AOD values, there is no unique solution to the adjustment (i.e., no way to preferentially change one aerosol type differently than another without making additional assumptions). Satellites are providing additional useful information, such as the fine fraction of the AOD. This allows the adjustment of the fine mode aerosol components (i.e., sulfate, black and organic carbon) in a manner differently than the coarse mode (i.e., dust and sea salt). Additional information that would distinguish between the absorbing and scattering aerosol (e.g., single scattering albedo) would allow further refinement of the adjustment procedures. Such information is presently not available from satellites.

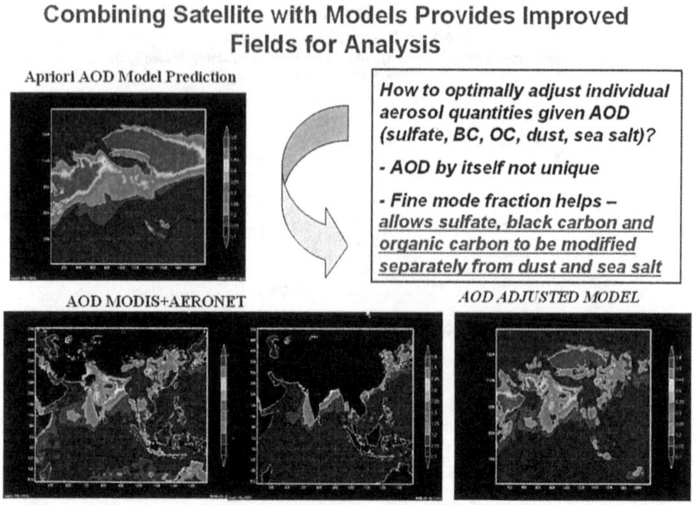


*Figure 9. Satellite data can be used to constrain model predictions to provide more accurate aerosol fields for subsequent analysis of climate and health impacts.*

SUMMARY

Models play a critical role in linking emissions to aerosol distributions and subsequent effects. Models have improved substantially over the past few years. Further improvements will require reductions in key uncertainties (e.g., emissions, better basic understanding of some processes). A closer integration of observations will also be necessary, and satellites are providing a wealth of new information that needs to be integrated (assimilated) with models. Finally, air quality, health and climate linkage offers synergistic policy/actions that require further cultivation and consideration.

REFERENCES

1. Bates, T.S., T.L. Anderson, T. Baynard, T. Bond, O. Boucher, G. Carmichael, A. Clarke, C. Erlick, H. Guo, L. Horowitz, S. Howell, S. Kulkarni, H. Maring, A. McComiskey, A. Middleborook, K. Noone, C.D. O'Dowd, J. Ogren, J. Penner, P.K. Quinn, A.R. Ravishankara, D.L. Svaoie, S.E. Schwartz, Y. Shinozuka, Y. Tang, R.J. Weber, and Y. Wu. (2006) "Aerosol Direct Radiative Effects over the Northwest Atlantic, Northwest Pacific, and North Indian Oceans: Estimates Based on In-Situ Chemical and Optical Measurements and Chemical Transortt Modeling." *Atmospheric Chemistry and Physics*, in press.
2. Bond, T.C., D.G. Streets, K.F. Yarber, S.M. Nelson, J.H. Woo, and Z. Klimont. (2004) "A technology based inventory of black and organic carbon emissions from combustion." *J. Geophys. Res.*, 109, (doi:10.1029/2003JD003697), D14203.

3. Carmichael, G.R., G. Calori, H. Hayami, I. Uno, S.Y. Cho, M. Engardt, S.B. Kim, Y. Ichikawa, Y. Ikeda, J.H. Woo, H. Ueda and M. Amann. (2002) "The Mics-Asia study: Model intercomparison of long-range transport and sulfur deposition in East Asia." *Atmos. Environ.* **36**(2), 175-199.
4. Carmichael, G.R., Y. Tang, G. Kurata, I. Uno, D. Streets, J.H. Woo, H. Huang, J. Yienger, B. Lefer, R. Shetter, D. Blake, E. Atlas, A. Fried, E. Apel, F. Eisele, C. Cantrell, M. Avery, J. Barrick, G. Sachse, W. Brune, S. Sandholm, Y. Kondo, H. Singh, R. Talbot, A. Bandy, D. Thorton, A. Clarke and B. Heikes. (2003a) "Regional-scale chemical transport modeling in support of the analysis of observations obtained during the TRACE-P Experiment," *J. Geophys. Res.*, 108, (D21), 8823, doi:10.1029/2002JD003117.
5. Carmichael, G.R., Y. Tang, G. Kurata, I. Uno, D.G. Streets, N. Thongboonchoo, J.H. Woo, S. Guttikunda, A. White, T. Wang, D.R. Blake, E. Atlas, A. Fried, B. Potter, M. A. Avery, G.W. Sachse, S.T. Sandholm, Y. Kondo, R.W. Talbot, A. Bandy, D. Thorton and A.D. Clarke, (2003b) " Evaluating regional emission estimates using the TRACE-P observations," *J. Geophys. Res.*, 108, (D21), 8810, doi:10.1029/2002JD003116.
6. Climate Change Science Program (CCSP), (2004) Our Changing Planet. The U.S. Climate Change Science Program for FY 2004 and 2005, 150 pages.
7. Kinne, S., et al. (2006) "An AeroCom initial assessment—optical properties in aerosol component modules of global models." *Atmos. Chem. Phys.* **6**:1815–1834.
8. Ezzati, Lopez, Rodgers, Vander Hoorn, Murray, and the Comparative Risk Assessment Collaborating Group. (2002) "Selected major risk factors and global and regional burden of disease". *Lancet.* **360**:1347-60.
9. Penner, J., J. Quaas, T. Storelvmo, T. Takemura, O. Boucher, H. Guo, A. Kirkevag, J. E. Kristj´ansson, and Ø. Seland., (2006) "Model intercomparison of indirect aerosol effects." *Atmos. Chem. Phys.* **6**:3391–3405.
10. Streets, D.G., T.C. Bond, G.R. Carmichael, S.D. Fernandes, Q. Fu, D. He, Z. Klimont, S.M. Nelson, N.Y. Tsai, M.Q. Wang, J.H. Woo, and K.F. Yarber. (2003) "An inventory of gaseous and primary aerosol emissions in Asia in the year 2000." *J. Geophys. Res.*, *108*, (D21), GTE 30-1-GTE 30-23, 8809, doi: 10.1029/ 2002JD003093.

# FUNDAMENTAL SCIENCE IN CLIMATE FORECASTING WITH MODELS

CHRISTOPHER ESSEX
Department of Applied Mathematics, University of Western Ontario
London, Ontario, Canada

Geophysical forecasting is an exceptional topic. There are few, if any, fields where the deep theory and the practice are so diametrically at odds with each other. The solutions to the basic dynamical equations used in weather forecasting are part of our daily weather reports, even though, paradoxically, no one knows how to actually solve those equations.

On one hand, fluid dynamics in long-term geophysical forecasting is considered a routine engineering application of Newtonian mechanics, while on the other hand, fluid dynamics also made it onto the list of the unsolved (Clay millennium) problems of mathematics. That list includes profound mathematical problems like the Riemann Hypothesis and the Hodge conjecture and mathematical issues concerning high-energy physics. Fluid dynamics is considered a deep scientific problem!

This isn't just empty theorist's talk either; money has been put on it. There is a million dollar prize for solving any of those problems. So if you feel like you could use some extra cash, I suggest that you dust off your Navier-Stokes theory and have a go at it. The good news is that you don't even need to solve the equations. All you need to do for the prize is to prove that solutions exist. You can even restrict yourself to the incompressible case! Easy money.

Of course anyone who has seriously tangled with the problem of turbulence knows that fluid dynamics remains a major scientific problem and that lottery tickets might be a better bet than winning this Clay prize. Nonetheless we are still interested in forecasting the future and, in the geophysical fluids case, the long-term future.

The classical problem of long-term forecasting or prediction is celestial mechanics, especially the study of the motions of bodies in the solar system. Success with this gave 18$^{th}$ century scientists, such as Laplace, the confidence to suggest that the future could be fully known with Newtonian mechanics and full knowledge of initial conditions.

Confidence in the Laplacian vision of forecasting began to fall apart with Poincaré's work on the three-body problem of celestial mechanics at the end of the 19$^{th}$ century. Today we have a whole century of uncertainty behind us: first there was the intrinsic uncertainty of modern physics, then in the latter half of the twentieth century, classical physics proved to be uncertain too with the emergence of what has come to be called the "butterfly effect," although Lorenz's original articles cited a flapping seagull rather than a delicate butterfly.

But things take time to sink in. As a child, I recall older meteorologists reciting 18$^{th}$ century Laplacian thinking: more accurate data plus bigger computers equals problem solved. Of course everything is different now. The butterfly effect (chaos, or natural variability as many meteorologists also call it) is now widely accepted. Sensitivity to initial conditions has even become a cliché in the movies.

But the cliché typically leaves off the second part of the story, wherein the rapid growth of error from those initially erroneous conditions is bounded. That latter part is what makes the butterfly effect something different. Sensitivity itself was not actually new. We always had sensitivity to initial conditions.

The butterfly effect allowed sensitivity, while preserving things like energy and momentum of the system, so the outcome could never be physically unreasonable. You just find a different, but still physically legitimate, state than the one you might have expected at the beginning. Thus you could never tell from the observations whether the butterfly stirred things up at the beginning or not.

Based on that "bounded sensitivity" that geophysical fluids are now believed to have, a figure of two weeks is widely circulated as a practical limit for forecasting by direct integration of the dynamical equations. Contrast this with the billion-year timescales possible for forecasting planetary movements. The extreme difference underscores how extraordinary attempts at long-term atmosphere-ocean forecasting are. This would seem to be the end of Laplacian style forecasting for geophysical fluids, if we care about what is going to happen months or years in advance.

Is there any way out of the dilemma that this situation represents? The only way is to average. Nothing has ruled out knowing an *average* weather, years in advance, even if the specific weather may not be known. Of course this is precisely what Rayleigh tried to do with turbulent fluids: average the dynamical equations and solve the equations in terms of the new averaged variables. This would work easily if the equations were linear, but it is precisely because they are not linear that we want to try to average them in the first place.

Rayleigh's approach to the fluid problem leads to the classical (unsolved) closure problem of turbulence, which many generations of scientists have attempted (unsuccessfully) to solve or to get around. In averaging the equations, you end up with too many averaged variables for the number of averaged equations. There are empirical treatments, of course, to fix this, but a solution from first principles is not known.

An alternative to this analytical approach is not to average the equations, but to use a computer instead. Repeatedly run the numerical integration over the same interval of time, but running each with slightly different initial conditions. Then average the ensemble of solutions that results. Large climate models do precisely this. Forecasts are always of ensemble behaviours, not of individual solutions, contrary to the 18th century prescription.

As inescapable as the demise of the Laplacian ideal is, it is minor compared to what else must be done: the known equations are not actually useable in practice. Large-scale geophysical computer models are not actually direct implementations of known laws. They are not even computational approximations of known laws. Instead they are substantially empirical in nature.

There is no known way around this because classical computation, as one learns in school, is not possible. Of course one can never actually put the basic equations directly onto a computer anyway. Instead one works with approximations set onto a computational grid, which thus introduces an artificial length scale associated with the computation. However this level of approximation is not the issue here. I must talk about the computational grid instead.

Classical computation calls for the computational grid to be finer than all natural length scales of the problem. That is because anything finer than the grid is "invisible" to the calculation. Looking for things finer than the grid is like looking at a computer or television screen for the picture details in between individual pixels! To avoid this in the atmosphere and oceans, this would require the grid to be very small indeed compared to the global scale of the computation. We might like to just dismiss small scales, but because omissions can all be amplified, in principle, by the butterfly effect, this cannot be done casually.

The smallest length scale for the purpose of this idea, which is not by any means the smallest length scale in reality, might be the Kolmogorov microscale. It is the length scale at which turbulent eddies get broken up by viscous dissipation. For air it is about 1 millimeter. Clearly this is already much larger than length scales associated with gradients in fluids and much larger than all of the aerosol microphysics that is the basis for much of the discussion of this session. But 1 millimeter is nonetheless quite small enough from a computational standpoint. If you compute everything on a 1 millimeter grid, with 1 second time steps, at about 10 gigaflops, it will take the computer at least $10^{20}$ years to compute the ten year forecast.

I note that about a year ago *Scientific American* was breathlessly anticipating an increase of a factor of 100 in processing power due to a parallelization on the Internet. OK, let's accommodate that, and throw in an extra factor of a thousand, just to be nice. We get $10^{15}$ years—100,000 times the age of the Universe. You wouldn't be able to do the computation once, let alone many times to generate an ensemble.

So the length scales that they have to work with in actual computer models for long-term integration are very much larger than 1 millimeter: typically hundreds of kilometers. Such a coarse grid is certainly too coarse for butterflies (or even whole thunderstorms) to show up in. All of those things and more are known as subgridscale phenomena.

The collective effects of subgridscale processes are so large that if they were left out, the model would behave in very unphysical ways. So a substitute is put in instead. A kind of pseudo physics is inserted, which is made up of computationally simpler quasi-empirical relationships called *parameterizations*. These parameterizations replace the more complex true dynamics below the resolution of the computation.

Subgridscale phenomena are well precedented in engineering. Parameterizations work in engineering because they can be tuned to make the computation conform to known and expected conditions. Such a circumstance emerges if one is computing turbulent flow over an aircraft wing, for example. We can put an actual aircraft wing, or a model wing, into a wind tunnel and tune the parameterizations of the computational model with reproducible, controlled experiments until the computation works to the desired level. Even so, when the aircraft is finally finished, it is not put into service at once but flown in repeated tests.

In the engineering case we have an experimentally validated, quasi-empirical model, which is absolutely legitimate and can make valid and accurate predictions for the regimes over which it has been developed. Outside of those regimes, insurance premiums will be high.

However, in long-term forecasting of geophysical fluids things are very different. There are no reproducible, controlled experiments. While it makes sense to try to do some kind of hindcasting (i.e., "predicting" the past) to improve things, hindcasting is not really the same thing as a controlled experiment and it is surely not reproducible as such.

In an unknown, uncontrolled dynamical environment the past cannot be presumed to be the same as the future. Hindcasting is not forecasting.

Moreover, in terms of hindcasting, we just do not have enough experience with the dynamics on the very long-term to have any but the most fleeting experience with the details, in terms of the time and space resolution necessary to tune models at the very low frequencies corresponding to climate. Dynamics at extremely low frequencies is simply not known. The faux physics on which such models are based simply cannot reliably give us insight into very long-term unforced natural variability, which is essential to understanding climate change.

If the empirical parameterizations only pertained to the physics of laboratory scales we could empirically validate them separately, and we might expect that they would remain unchanged in a future different climate regime. However this is not the case. They are not laboratory scale parameterizations. They entail nearly every aspect of global vertical energy transport from cloud formation, convection, water transport, aerosols and how they interact through dynamics and radiative transport.

All of these are below the grid scale resolution of the global dynamics. Some are less empirical than others, but they are all potentially climate regime dependent. They do not need to hold for all time scales and into new climate regimes. Thus they have no fundamental predictive power for new climate regimes.

Unfortunately, that does not mean that they do not produce good-looking results. Good-looking results are an intuitive trap for us in the 21st century. Hollywood's computers have brought them to us in modern special-effects films. Although puny compared to the demands of nature, we have enormous computational power compared to what we have had in the past. Hollywood can make visually interesting, believable and even quasi-physical visual effects entirely from computers.

Such visual effects easily seduce the eye into believing that fluids have been fully conquered by modern computation. But all Hollywood needs to do is to produce something that looks believable and visually interesting. That is very different than producing an accurate forecast. Visual effects are not unlike the engineering experiments wherein parameterizations get adjusted until desirable behaviours are reproducible in known regimes. As long as one stays within the known regime all is well, but, almost by definition, this is precisely what climate forecasting cannot do.

Consider the sketch of a Fourier power spectrum (Figure 1) of some model-produced time series.

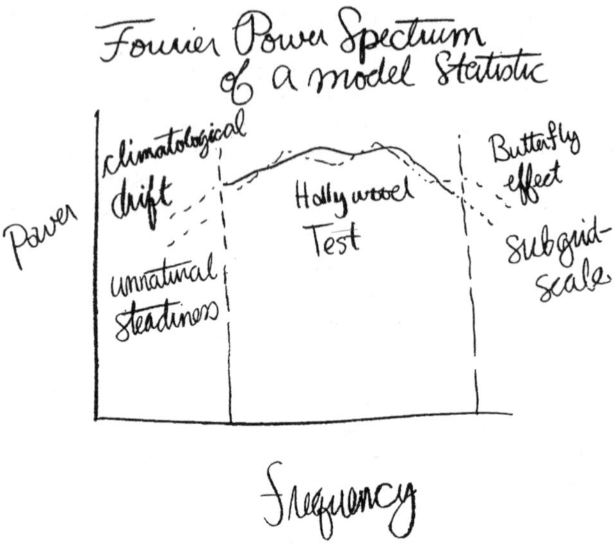


*Figure 1.*

What is important here is that there are three regions in the sketch. On the right side (high frequencies) is the subgridscale region. Models cannot forecast in this region because it is below model resolution. This regime is where parameterizations are imposed.

In the middle region (middle frequencies) climate model output can be compared to historical data. Parameterizations can be adjusted to some extent to make the model match the corresponding observed values in the middle. The sketch depicts dashed and solid curves to denote the comparison between observation and calculations. This criterion is denoted accordingly in the figure as the "Hollywood test" of model validation, because validation is ultimately limited to an observable region and has no validity outside of it.

The left hand region (low and very low frequencies) is the regime where climate and climate change happen. We need this part of the diagram to be correct for accurate forecasts. However, there is no observational data detailed enough to adjust models effectively in this invisible, low frequency, "infrared" region. The region remains unobserved because it corresponds to timescales that are longer than our data sets cover. In that sense it really is like infrared light: an invisible, low frequency limit.

There are special problems that models have in this region. Some models have stability problems there. This is endemic for large-scale simulation of complex systems involving nonlinear partial differential equations generally, not just for geophysical forecasting. The model climate drifts, and strange things like mass loss have been known to happen, particularly with a class of models called "spectral models."

Artificial fixes are imposed to stop the models from drifting and doing non-physical things. It is completely unknown whether or not these fixes eliminate true natural infrared dynamics in the attempt to eliminate computational instability. If they do, then the fix can also create an unnatural steadiness. There is no way to know for sure, as we really have little idea what the natural variability is like in the infrared. We have no independent way to determine whether we throw the "baby out with the bath water" when such fixes are made. It is however a very good bet that in many models the power spectrum has values that are unnaturally low in the infrared because of the fight against typical computational instabilities.

At this stage I would like to take a moment to prevent leaving the wrong impression. Much of what I have been saying so far might seem, in these times, to suggest that models are a waste of time and that they are hopeless, or that I would do things differently. If I were building models, I would be doing the same things, and maybe not as well. Modeling is done as well as it is possible to do. First-rate people are doing the work at a high level. But they face an extraordinary problem.

We need models. There is much to learn from them. They represent a legitimate academic exercise. It is one thing to compute from one classical physical theory alone. It is a far more difficult thing to link them all together in a way that coherently reflects reality. If nothing else, the discipline of mapping out details of the entire system and coherently linking them together to make the whole thing go at all has been an enormous accomplishment in itself.

But most importantly, climate models remain the best that we have.

That said, one point that is important to take from this talk is that models used in long-term forecasting are not forecasting the future in a Laplacian sense. Many educated people do not know this. It is also important to be aware of what a marvelous unsolved scientific opportunity the long-term geophysical forecasting problem is. There are so many fascinating things to understand and perhaps discover.

I have a couple of interesting ideas to tell you about regarding the limits of long-term prediction in light of Figure 1. Both of them relate to the nature of the dynamical equations to amplify errors. Because of tuning to reduce error in the middle region, long-term forecast error in models has to come from the right hand region. That is where the butterfly effect originates and where the parameterizations are installed too. So curiously, it is the region corresponding to small scales that can determine whether the large scale, low frequency end on the left is going to be accurate or not.

Clearly if we are going to average, random errors will tend to cancel each other out. One might presume that what one butterfly is doing will cancel out what another does on the average. More optimistically, one might also hope that averaging will eliminate errors produced in replacing the small-scale physics with parameterizations.

So how can there be any error in the long term if the error sources on the small-scale succumb to averaging? It is easy to make a vague claim that deviations from true physics by parameterizations create long-term error in the average, but it becomes rather difficult to envision specific error mechanisms. What mechanisms will be small enough to exist in the region at the right, while not appearing significantly in the middle, but which significantly alter average behaviour on the left side of the diagram?

Clearly purely random small-scale errors will be averaged out, but what if the error is not random? What if it is systematic instead? I have two potential mechanisms to present today, one is physical and one is computational. Both are meant to tie into the theme of basic science for this talk.

The first of these has to do with higher order fluid dynamics. It has been established that the standard dynamics can amplify small deviations, no matter the cause of the deviations. Clearly if some of the true dynamics is omitted, the true behaviour of the system will deviate from the forecasted behaviour.

If the forecasting dynamics and the true dynamics both amplify errors the deviation can also be expected to be amplified as if it where an error in initial conditions. That is, it can behave just like the butterfly effect, even if the error does not originate from errors in initial values but from neglected dynamics instead. This is called the *virtual butterfly effect,*[1] because it is just like the butterfly effect but it does not require a butterfly to set it in motion. Someone once joked that it could be called the "dead butterfly effect."

In the case of fluid dynamics, it is well known that Navier Stokes theory does not work for extremes of gradients or densities. It also does not work well for sound dispersion. Of course climate models work with the incompressible approximation of fluids, so they will not be able to handle sound at all. Certainly on scales of 100's of kilometers sound waves surely do not matter, at least for the Hollywood test.

The question remains though, is there a long-term average effect from neglected dynamics? And, from the point of view of this talk, can one envisage a physical mechanism that connects to basic science issues? In that light, the case of higher order fluid dynamics, Burnett stresses, are of interest. They come from kinetic theory and are presumed to be a general improvement over Navier Stokes theory for a so-called "dilute gas,"[2] which, in the minds of kinetic theorists, the atmosphere qualifies as.

While the deviation between Burnett and Navier Stokes stresses is small, it is certainly not random. And that is precisely what we are looking for. The question is whether or not it is so small that it cannot show up on any reasonable long-term climate time scale. Straightforward calculations show that its systematic effects would have a time scale of at most a few years.[1] This suggests that the fascinating idea that kinetic effects, negligible in a laboratory, could have an effect on averaged dynamics on a long term.

This is not to say that this is proven to happen, but rather we have a potential mechanism that could lead to erroneous behaviour at low and very low frequencies. That is, there is the possibility that certain very small systematic physical effects can use the mechanism of the butterfly effect to affect average behaviours on the long term. Clearly this issue is a basic rather than an empirical one. I can see no way to use a quasi-empirical computer climate model to resolve this issue.

Another idea of a mechanism has to do with the fact that relations used on a computer are different than the actual dynamical equations. I touched on this issue above, but set it aside to pursue the issue of subgridscale processes instead. However let's revisit it now in the new light of small-scale systematic differences playing a role on large scales averages.

It is a good thing that the computer relations are not the same as the equations they approximate. We are, for example, not able to solve the fluid equations directly, but a slightly different approximate *discrete map* used on computers can be used to generate an approximate solution.

The errors produced by differences between a discrete map and the differential equation it approximates are small by design. As in the case of neglected dynamics, small errors are actually desirable for the mechanisms we are attempting to envisage. Although they are sometimes modeled as such, these errors are certainly known not to be random.

How do we characterize the differences between the map and the equation? How do we know how accurate the calculation is for very long times if we do not have an independent standard such as observational data to determine the computational error? Recall that we are attempting to look at dynamics in a domain (i.e., left region of Figure 1) where there is only the weakest observational information, if any at all. Therefore we need some other basis to be satisfied that computation is accurate.

The degree to which a computational scheme conserves energy is used as a natural criterion for the case of very long time scale (~$10^9$ years) computationally approximate integrations for planetary motions[3]. This is a very compelling way to look at accuracy for long-term forecasting. In a properly formulated treatment, a physically conserved quantity tomorrow must remain the same value as today, and all the days afterward. Constant, or invariant, quantities are at the heart of all physical forecasting, because all change must ultimately be defined against them, no matter what the time scale.

The idea is easy to understand if we use the harmonic oscillator as an example. Consider the Euler map to approximate it. That map is only one possible computer scheme for approximating the true solutions. The nice thing about this example, is that we know the exact answer and thus can see fully the nature of computational error from the long-term point of view. We also know that the Euler map has many undesirable properties for accurate computation, which we want, and we can extract complete solutions for the map too.

In this example, energy should be conserved, but the computational scheme does not conserve it, as you can see from the left hand plot in Figure 2. That is, the energy (Hamiltonian), in scaled units, is plotted against the number of oscillations (or orbits). It grows instead of remaining fixed.

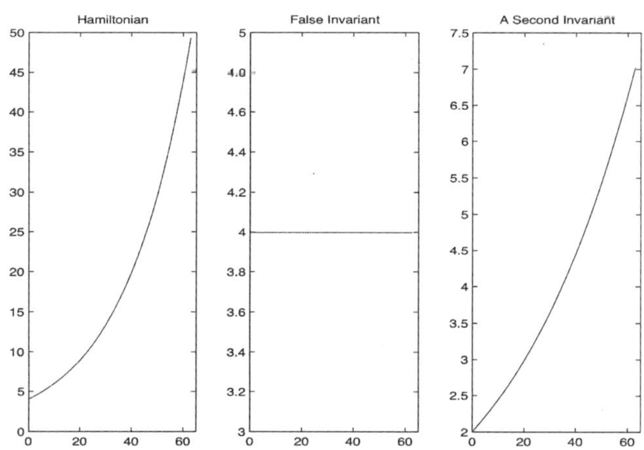


*Figure 2.*

Most practical computational schemes do not conserve energy for conservative systems either, but they are calculated over short enough time scales that this does not usually matter. As long as the integration is not carried out for so long that the error grows unmanageable, this is not an issue. But in this example it is an issue. The growth of energy corresponds to a rapid increase in amplitude, which is entirely non-physical.

The middle plot is the Hamiltonian (i.e., energy), scaled by a function of the computational step size and step number. The remarkable point is that this quantity is *not* conserved by the original equations, even though it is conserved, as you can see, by the Euler map! This is what is known as a *false invariant*.

I warned of false invariants some years ago in empirical or quasi-empirical models.[4] Parameterizations naturally conserve quantities that are not invariant in the actual physics. Inevitably it restricts dynamics to solutions that are unnatural. This concern clearly holds for computational schemes too.

Conservation properties, or symmetries as they are also called, clearly provide a natural language to compare completely unrelated equations and dynamics. In fact, the conserved quantities connect so closely with the identity of the equations, that if I could find an alternative computational scheme (there are an infinite number of possibilities) that not only conserved the energy (some computational methods do conserve energy) but a second conserved quantity too, such as that depicted on the right hand plot of Figure 2, the computational scheme would amount to an *exact* solution of the full equations. That is to say that the "forward error," as it is known, is strictly zero if the conservation properties are preserved.

But what do simple conservative systems have to do with the left hand region of Figure 1 and the physical problem at hand, which is *not* considered "conservative"? Ironically even non-conservative systems have conserved quantities. Computational schemes also fail to preserve those quantities too. The paradoxical terminology is historical. It reflects an old, energy-centric view.

In fact for the sorts of equations necessary for long-term geophysical forecasting, it can be shown that there are not just two independent conserved quantities, as in the case of the harmonic oscillator mentioned, but there are an *infinite* number of them instead.[5] Moreover my colleagues and I believe that it can be shown that the identity between accurately conserved quantities and error carries over generally to non-conservative systems.

This latter point is very important to the theme of this talk. It means that if, say, I could find a computational scheme that conserved all of the infinite number of invariants expected for, say, the Navier Stokes equation, the computational scheme would produce a solution with no error. That would be tantamount to the direct solution. The computational scheme with no error would amount to a solution of the Clay Millennium problem and I could collect the one million dollar prize!

Thus to eliminate the prospect of small systematic error affecting the low frequencies in this scenario, either observational data, currently unavailable, is required to tune and confirm the quasi-empirical models or a basic problem in Mathematics must be solved.

Both this example and the physical one preceding it illustrate how naturally the question of long-term forecasting emerges as a basic science problem.

REFERENCES


1. M. Davison and C. Essex (1998) *Open Sys. & Info.Dyn.* **5**:125-138: M. Davison, C. Essex and J.S. Shiner (2003) *Open Sys. & Information Dyn.* **10**:311-320.
2. e.g., J. Foch & G.E. Uhlenbeck (1967) *Phys. Rev. Let.* **19**:1025-1027.
3. e.g., K.R. Grazier, W.I. Newman, J.M. Hyman, P.W. Sharp, (2005) *ANZIAM J.* **46**:C1086-C1103.
4. C. Essex (1991) *Pure and Applied Geophysics* **135:**125-133.
5. e.g., (1986) J.A. Cavalcante and K. Tenenblat, *J. Math. Phys.* **29**:1044-1049.

# ON THE CONNECTIONS BETWEEN AEROSOL, ATMOSPHERIC RADIATION AND HYDROLOGICAL PROCESSES IN CLIMATE CHANGE

GRAEME L. STEPHENS
Department of Atmospheric Sciences
Colorado State University, Fort Collins, USA

## INTRODUCTION

Water is one of the distinguishing features of this planet. It is a precious commodity that is essential for life in general and for the sustainable development of human society. Although vast reservoirs of water exist on this planet primarily in the oceans and the frozen ice masses, the fraction of this water that is fresh and available to support life is less than one percent. Most of the freshwater we use is taken from the water found underground in aquifers (e.g., Clarke and King, 2004). These reserves of water are ultimately replenished by the rains and snow: water from the sky. The crisis looming for future generations is that current and projected future demands of freshwater exceed the anticipated supply. Confounding matters further is the uncertainty surrounding the influence of climate change and the effect of such change on the fresh water supply. A more quantitative understanding how fresh water is created in the sky and what factors influence where and how much is produced and how climatic change will alter this supply is essential for any strategy aimed in part at managing our precious reservoirs of freshwater.

Freshwater is created as part of the planet's water cycle—a cycle that sees water cycling continuously throughout the Earth system, rising from oceans, lakes, and the land surfaces to the atmosphere, forming into clouds, and eventually falling from clouds back to Earth's surface as rain and snow. The water in the sky is a critical component of this cycle, and processes that influence this water content fundamentally govern the supply of freshwater upon which we depend.

In this context, clouds play an obvious and important role in the cycling of water over the planet. They are the manifestation of the processes that convert a fraction of the invisible water vapor contained in the atmosphere into liquid and solid water that falls back to Earth, replenishing our reservoirs of freshwater. If our world had no clouds, there would be no way to replenish these reservoirs of freshwater and there would be no life.

The amount of water in the atmosphere that is involved in producing clouds and precipitation is tiny compared to the water found in the other reservoirs on this planet. For example, if all of the water in the world oceans was wrapped around the Earth in a uniform layer, the Earth would be engulfed by a layer of water approximately 1 km thick. The atmosphere, on the other hand, contains an amount of water (mostly in vapor form) corresponding to a layer about 3 cm thick. The water condensed in clouds is an even tinier amount filling a hypothetical layer less than 0.1 mm thick. It is this tiniest and perhaps most variable of components that control the supply of all freshwater.

There is still a great deal we don't know about clouds and their contribution to the water cycle. We don't know how much of the water in clouds falls as rain or snow, and we can't predict with any certainty how clouds—and thus our reservoirs of freshwater—

might change as our climate changes. Although the role of aerosol in influencing this cycle remains highly uncertain, there are many reasons to expect that such influences may be significant. This article considers some of the ways aerosols are thought to influence the planets water cycle.

PARTICLES IN THE SKY

The Earth's atmosphere is largely gaseous being composed of 99.9% oxygen, nitrogen and argon, the remaining <0.1% contains trace gases instrumental in supporting the climate and all life forms on Earth. Through the absorption of UV radiation in the stratosphere, ozone provides protection from the sun. Carbon dioxide and methane are greenhouse gases whose changing abundances in time perturb the climate. Water vapor is a highly variable gas in the atmosphere. Being the principle greenhouse gas, it exerts a fundamental control on the observed temperature distribution of the atmosphere and serves as an essential ingredient of the Earth's hydrological cycle.

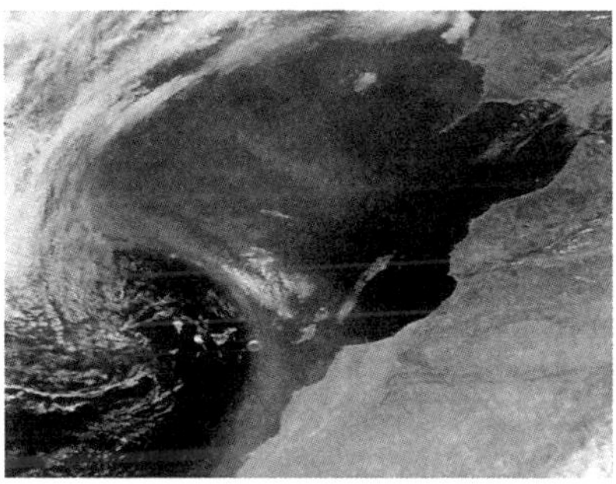

*Figure 1. A dramatic example of the different particles in the sky hinting at interactions amongst one another. Shown is a false color image of a sandstorm swept up by winds and carried into clouds. There has been much recent discussion about possible effects of such sandstorms on formation of thunderstorms and tropical cyclones (Dunion and Veldon, 2004; Stephens et al. 2004).*

The atmosphere also supports particulate matter that remains suspended in air on a vast variety of time scales. These particles vary enormously in composition (implied in Figure 1) and size ranging from micron size particles to particles exceeding a millimeter (Figure 2). These suspended particles can be grouped into two general categories: aerosol particles of varying composition, either of solid or liquid form, and cloud and precipitation particles composed primarily of water and formed initially via condensation of water on small aerosol particles.

Like the so-called greenhouse (trace) gases, suspended particles exist only in minute concentrations compared to the other gases yet exert a profound influence on our climate, on our weather and on life on Earth. Among the most important effects are:

Direct aerosol effects on climate

Evidence is mounting that aerosols produced as result of human activity are altering climate through effects on the planet's energy balance: a topic introduced in previous lectures.

Cloud effects on climate

Clouds and related processes not only control the pace of the water cycle as mentioned but also exert a dominant influence on the energy balance of the Earth. Clouds introduce their own greenhouse effect that is substantially larger than that associated with build up of carbon dioxide, for example. Clouds also grossly effect the amount of sunlight that enters the planet (the amount of sunlight reflected from Earth to space is termed the planets albedo). The eventual response of the climate system to imposed aerosol and greenhouse gas changes depends on how clouds change in response to factors that force climate change and how these cloud changes affect the earths greenhouse and albedo. These responses, referred to as cloud feedback, remain the principal source of uncertainty in climate model predictions of global warming.

Cloud effects on weather

Latent heat is released through the process of water condensing on small aerosol particles and the subsequent production of precipitation. This heat, when organized into large weather systems, fuels the severe storms of our planet's weather and is an important source of energy that drives the larger scale atmospheric circulation.

Indirect aerosol effects on climate

Changing aerosols also affect the properties of clouds, and are possibly affecting weather systems. The so-called aerosol indirect effect is complex, not fully understood, nor easily quantified and includes effects on cloud microphysical properties (such as cloud particle sizes and concentrations) as well as through direct effects on radiation that contribute to heating in clouds thus inducing motions that in turn change clouds (the so-called semi-indirect effect, Hansen et al. 1997). This paper considers only the first of these effects for discussion.

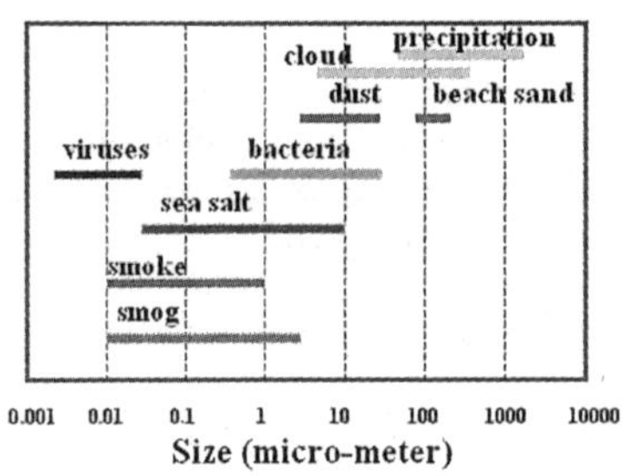


*Figure 2. The size range of typical small particles found in the Earth's atmosphere, whose sizes range from a fraction of a micron to millimeters are also suspended in the air.*

A MICROSCOPIC PERSPECTIVE ON THE FIRST INDIRECT EFFECT OF AEROSOL ON CLOUDS

The profound influence of suspended particles on climate and on life on Earth stems in part from processes that take place on the microscopic level. It is on this scale that aerosol influences are most easily understood:

Cloud particle activation

Clouds form on aerosol particles via a process referred to as activation. The particles that produce cloud (liquid) droplets or ice crystals are referred to as cloud condensation nuclei (CCN) and ice nuclei (IN), respectively. Not all aerosol particles serve as the nucleus for cloud particle growth. The ability to activate growth depends on:

- The amount of water vapor in the environment. Activation is generally initiated in air that is supersaturated with respect to water or ice. The availability of water vapor is also influenced by the removal of the vapor from the environment as the particles grow and the replenishment of water vapor through advection by cloud scale vertical motions.
- The chemistry of the nuclei, their size and number concentrations.
- Activation of ice is much more complex than activation of water droplets involving other complicated processes like secondary ice production by splinter, among others.

The effects discussed imply that higher concentrations of aerosol, such as occur in heavily polluted regions of the planet, imply higher concentrations of CCN. If all other conditions remain the same, specifically the supply of water to clouds, then higher concentrations of CCN lead to clouds composed of larger concentrations of smaller droplets (Figure 3). The consequences of such changes are considered below.

Although it does not necessarily follow that more aerosol particles correlate with more CCN or IN, we do generally find that more highly polluted air contains more CCN (less is really known about IN and its variability). In fact, those climate models that attempt to include aerosol-cloud effects do so by simply correlating CCN (related to the total number of cloud droplets in a given volume of air as shown in Figure 4) to aerosol amount. Such relationships as shown on Figure 4 are highly variable and perhaps too simplistic overlooking (by necessity) key factors in the activation process.

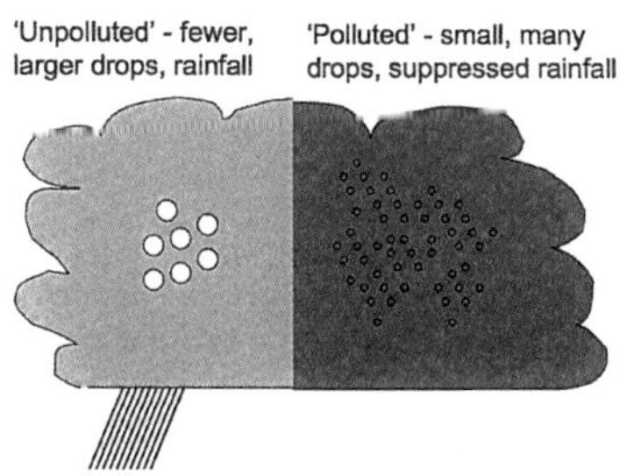


*Figure 3. A simple schematic of possible effects of increased pollution (CCN) on cloud 'microphysical' properties for two clouds of the same water content. The consequences of such changes are referred to as aerosol indirect effects.*

Making rain

A typical cloud drop must increase its size by about two orders of magnitude to become a raindrop. This growth occurs in two general ways:

- *Via warm rain processes*: Collisions with neighboring droplets are followed by coalescence into a larger particle. As these larger particles fall, they further collect the smaller droplets, growing yet larger thus producing a cascading effect. We understand the general nature of the coalescence process and observe this process both in the real atmosphere and in the laboratory. Coalescence is initiated in water clouds when particles grow to about 20 microns in radius. However, coalescence theories are at odds with general observations, especially for shallow clouds. Theory predicts that growth to a size large enough to trigger coalescence takes several hours yet rain is typically observed soon after formation of short-lived cumulus clouds. One explanation for this discrepancy hypothesizes the existence of a few giant nuclei on which large cloud droplets are activated. A second explanation argues for local effects of turbulence bringing droplets together thereby enhancing the collection process. Fundamentally, the triggering mechanism for warm rain remains uncertain.
- *Via cold rain processes*: Ice crystals also grow at the expense of liquid droplets in the clouds. This is sometimes referred to as the ice crystal process or the Wegener-Bergeron-Findeison process. Ice crystals grow more rapidly under generally large super-saturation (with respect to ice), fall, coalesce with other crystals and collide with super-cooled droplets freezing them on impact.

Thus it is to be expected that changing the microphysical properties of clouds via changes in aerosol affect the collision-coalescence process and the amount of rain that falls from warm clouds (Albrecht, 1988) such that that polluted clouds would be expected will rain less than non-polluted clouds (Figure 3). Such changes would then lead to a chain reaction such that many other properties of clouds including their lifecycle will be altered. Aerosol influences on cold rain processes are much less 'obvious' and not at all understood at this time although many recent model studies suggest the effects may be large.

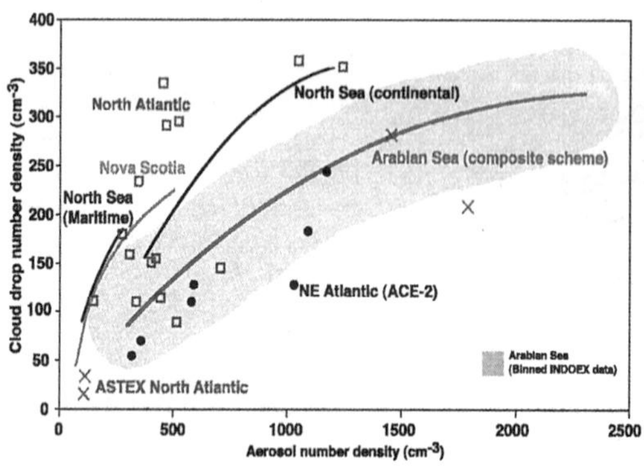


*Figure 4. Models that attempt to treat the cloud drop activation process do so via simple relationships between aerosol and cloud number densities. Observations suggestion such relationships are too simple, glossing over other important influences.*

POSSIBLE CONSEQUENCES

The types of changes to cloud microphysics simply illustrated in Figure 3 can lead to a number of consequences that are important to climate:

(i) Clouds composed of higher concentrations of smaller droplets reflect more solar radiation than clouds of equivalent water content but composed of a few, larger droplets. This is referred to as the Twomey effect (Twomey, 1977). It has been observed that mostly shallow clouds, when formed in locally polluted air-masses, reflect more radiation than surrounding clouds formed in cleaner air (Figure 5). The example shown in Figure 5 is a special case where it is clear that aerosols are affecting the clouds reflection of solar radiation. On the broader global scale, however, we are unable to establish the water budget of clouds so we cannot yet separate out changes in cloud albedo by aerosol from other processes that affect the water budget. Thus the global magnitude of the 'Twomey' effect is highly uncertain. Climate model studies (e.g., Lohmann and Feichter, 2002) suggest that the effect may at least as large as the direct effect of aerosol on solar radiation.

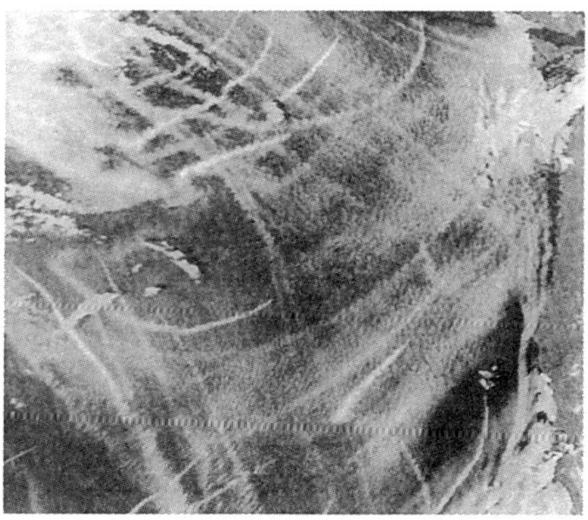

*Figure 5. Clouds composed of higher concentrations of small droplets are referred to as colloidally stable. For the same total water content, more colloidally stable clouds reflect more solar radiation than do clouds composed of larger droplets. This enhanced reflectivity is observed in this satellite image of clouds off the west coast of California. Shown is an extreme example of the effects of aerosol emitted from the chimney stacks of ships. This higher aerosol air enters the cloud creating local areas of higher CCN, smaller droplets and higher albedo.*

(ii) The Twomey effect described above assumes that the water content does not differ between polluted and unpolluted clouds. As suggested earlier this is a dubious assumption and is one that at this time is impossible to test. Shifting the cloud droplet populations to smaller sizes also inhibits the coalescence rain-forming process as mentioned, implying less precipitation from clouds formed in polluted air masses (e.g., Rosenfield, 1999) and thus very different water budget. Thus for the same total mass of water, less precipitation is produced implying a reduction of the precipitation efficiency of clouds and further implying different life-cycle characteristic. As with the Twomey effect, identifying the possible suppression of rainfall by aerosol is also confounded by other meteorological influences. For example, there has been much recent speculation of the effects of dust storms like that shown in Figure 1 on the development of convection and tropical cyclones (Dunon and

Veldon, 2004). Further study, however, reveals that it is the meteorological conditions associated with the dust air mass (i.e., excessively dry in middle levels) that is the primary cause of suppression of convection (Stephens et al. 2004).

## AEROSOL-CLOUD EFFECTS ON A BROADER SCALE—THE A-TRAIN

The simple microphysical arguments offered previously point to an obvious pathway along which aerosols exert an influence on clouds and precipitation (Figure 3). However, identifying such effects in the real world is much more complex for reasons already hinted at. In reality, the microphysical processes that connect aerosol to clouds and precipitation are shaped by many other factors that operate on a much larger scale more typical of atmospheric weather patterns. For example, the process of particle activation is strongly dependent on the availability of water and the amount of lift provided to air parcels. These properties are typically dictated by the circulation flows associated with weather patterns so any strategy aimed at measuring aerosol effects has to be able to separate the direct microphysical influences from the large-scale effects of weather.

Thus the challenge in observing aerosol effects on clouds is to devise ways of observing both of the intricate microscopic processes that determine particle properties and the larger scales typical of the weather patterns that control and organize these processes. Satellite observations have been used in a number of studies to study aerosol effects on clouds. For the most part, however, these studies are ambiguous as they have not adequately separated effects of aerosol on clouds from parallel effects of meteorology on clouds. What is required is more quantitative information about the water budgets of clouds and precipitation and information about aerosol and their vertical distributions.

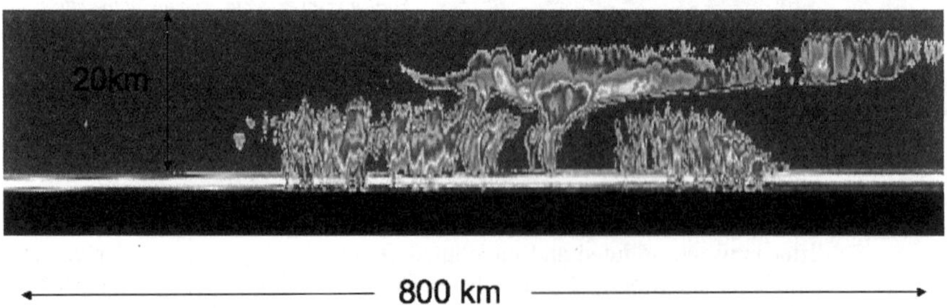


*Figure 6. A spectacular early CloudSat view of the vertical structure tropical convection over the equatorial mid-pacific on June 10th, 2006. This image shows the power of CloudSat view of both clouds and precipitation. A thick layer of cirrus cloud is observed above about 10 km overlying shallow convection. This cirrus is generated as outflow from a deeper thunderstorm. The color refers to the echo power of the radar from which the water, ice and precipitation contents all can be inferred. These quantities are crucial for understanding aerosol indirect effects.*

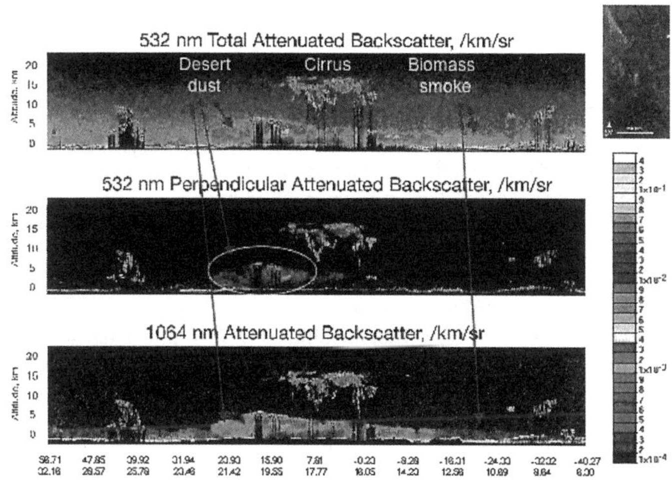


*Figure 7. A first light image of part of an orbit of CALIPSO showing the vertical distribution of different aerosol types and thin cirrus clouds as the space craft passed over central Africa (courtesy, D. Winker, NASA LaRC). The images are derived from the 3 channels of the laser system.*

Recently two new satellites were jointly launched on April, 28th 2006. Each satellite carries an active sensor designed to study clouds and aerosol and the interactions between them. One is the CloudSat satellite that provides observations necessary to advance our understanding how clouds affect the radiation balance of the atmosphere, the water budgets of clouds and the relation between cloud properties and the precipitation they produce. CloudSat flies the first space-borne millimeter wavelength radar. The unique feature of this radar lies in its ability to observe jointly most of the clouds and precipitation within its nadir field of view (Figure 6). The second satellite is the CALIPSO satellite that carries a laser radar (lidar) system as its primary payload. This instrument provides a unique view of the vertical structure of aerosol (Figure 7). Knowing where the layers of aerosol exist is critical for understanding how aerosols and cloud interact.

CloudSat and CALIPSO also fly as part of a constellation of satellites that includes EOS Aqua and EOS Aura at each end of the constellation and another small satellite, PARASOL, carrying the POLDER polarimeter (Deschamps et al. 1994) inserted in the formation between the larger EOS spacecraft. This constellation is referred to as the A-train. CloudSat and CALIPSO fly just seconds apart which means these we obtain coincident views of the atmosphere from these two different sensors. This combination of observations together with other sensors of the A-train provide an unprecedented and revolutionary view of the atmosphere and its particulate properties from which we expect to advance our understanding of aerosol indirect effects (Stephens et al. 2002).

REFERENCES


1. Albrecht, B. A. 1989. "Aerosols, cloud microphysics, and fractional cloudiness." *Science* **245**:1227-1230.
2. Clarke, R. and J. King, 2004. The Water Atlas. The New Press, 127 pp.
3. Deschamps, P.Y., F.M. Breon, M. Leroy, A. Podaire, A. Bricaud, J.C. Buriez and G. Seze. 1994. "The POLDER mission: Instrument characteristics and scientific objectives." *IEEE Trans. Geosci Remote Sensing.*, **32**:598-615.
4. Dunion, J. and C. Veldon. 2004. "The impact of the Saharan air layer on Atlantic tropical cyclone activity." *Bull.Amer.Met. Soc.*, 353-365.
5. Hansen, J., M. Sato, and R. Ruedy. 1997. "Radiative forcing and climate response." *J. Geophys. Res.*, **102**:6831-6864.
6. Lohmann, U., and J. Feichter. 2001. "Can the direct and semi-direct aerosol effect compete with the indirect effect on a global scale?" *Geophys. Res. Lett* **28:**159-161.
7. Rosenfield, D. 1999. "TRMM Observed First Direct Evidence for smoke from forest fires inhibiting precipitation." *Geophys.Res.Lett.* **26**:3105-3108.
8. Stephens, G.L., N.B. Wood and L.A. Pakula, 2004. "Radiative effects of dust on tropical convection." *Geophys. Res. Lett.* 31, doi:10.1029/2004GL021342.
9. Stephens, G.L., D.G. Vane, R J. Boain, G.G. Mace, K. Sassen, Z. Wang, A.J. Illingworth, E.J. O'Connor, W.B. Rossow, S.L. Durden, S.D. Miller, R.T. Austin, A. Benedetti, C. Mitrescu, and the CloudSat Science Team. 2002. "The CloudSat mission and the A-TRAIN: A new dimension to space-based observations of clouds and precipitation." *Bull. Am. Met. Soc.*, **83**:1771-1790.
10. Twomey, S., 1977. "The influence of pollution on the shortwave albedo of clouds," *J. Atmos. Sci.*, **34**:1149-1152.

# 6. POLLUTION

# FOCUS: PLASTIC CONTAMINANTS IN WATER

# SYNTHETIC POLYMERS IN THE MARINE ENVIRONMENT: WHAT WE KNOW. WHAT WE NEED TO KNOW. WHAT CAN BE DONE?

CHARLES MOORE
Algalita Marine Research Foundation
Long Beach, California, USA

ABSTRACT

Synthetic Polymers, commonly known as plastics, have been entering the marine environment in quantities paralleling their level of production over the last half century (Thompson). However, during the last decade of the 20$^{th}$ Century, the deposition rate accelerated exponentially (Copello, Ogi). Thirty years ago the prevailing attitude of the industry was that "plastic litter is a very small proportion of all litter and causes no harm to the environment except as an eyesore" (Derraik). Plastics became the fastest growing segment of the U.S. municipal waste stream between 1970 and 2003, increasing nine-fold (U.S. EPA), and marine litter is now 60-80% plastic, reaching 90-95% in some areas (Derraik). While undoubtedly still an eyesore, plastic debris today is having significant harmful effects on marine biota. Albatross, fulmars, shearwaters and petrels mistake floating plastics for food and few, individuals of these species remain unaffected; in fact, 44% of all seabird species ingest plastic. Sea turtles ingest plastic bags, fishing line and other plastics, as do 26 species of cetaceans. In all, 267 species worldwide are known to have been affected (Derraik). The numbers of fish, birds, and mammals that succumb each year to derelict fishing nets and lines in which they become entangled cannot be reliably known, but estimates in the millions have been made (Moore). Marine plastic debris can be divided into two categories; macro, >5mm and micro, <5mm. While macro debris may sometimes be traced to its origin by object identification or markings, micro debris, consisting of particles of two main varieties, degraded pieces broken from larger objects, and resin pellets and powders, the basic thermoplastic industry feedstocks, are difficult to trace. Ingestion of small plastics by filter feeders at the base of the food pyramid is known to occur (Moore, Thompson), but has not been quantified. Ready ingestion of degraded plastic pellets and fragments (U.S. EPA) raises toxicity concerns since they are known to sorb hydrophobic pollutants (Moore, Takada). The potential bioavailability of compounds added to plastics at the time of manufacture, as well as those sorbed from the environment is a complex issue that merits more widespread investigation (Andrady). The physiological effects of any bioavailable compounds desorbed from plastics by marine biota have not been directly investigated, but Ryan et al. found that the mass of ingested plastic in Great shearwaters was positively correlated with PCBs in their fat and eggs. Field and laboratory studies of the physiological effects on seabirds that ingest plastic resin pellets are in progress (Takada), and a fish study to examine possible xenoestrogenic activity of ingested plastics has been designed by Michael Baker of the UC San Diego Department of Medicine. "Studies by Gregory, Zitko and Hanlon have drawn attention to … small fragments of plastic …derived from hand cleaners, cosmetic preparations and airblast cleaning media" (Derraik). The quantities and effects of these contaminants on the marine environment have yet to be determined, but in a study conducted on the Los Angeles and San Gabriel Rivers in 2004-

2005, 2 billion plastic particles of all types, <5mm, were found to flow toward the ocean in three days of sampling (Moore). Colonization of all plastic marine debris by alien species poses one of the greatest threats to global marine biodiversity (Barnes). "There is also potential danger to marine ecosystems from the accumulation of plastic debris on the sea floor...The accumulation of such debris can inhibit the gas exchange between the overlying waters and the pore waters of the sediments..."(Derraik). The extent of this problem and its effects, have yet to be investigated, but based on resin sales in the United States, a little more than half of all thermoplastics will sink in seawater (U.S. EPA).

## INTRODUCTION

A major unforeseen consequence of the "Plastic Age" is the material's ability to proliferate in infinite configurations throughout the marine environment worldwide (Moore). The modern trend is for nearly all consumer goods to contain and/or be contained by plastic, and recovery of the material does not provide readily realizable profits, or options for reuse (Unnithan), therefore plastics are the fastest growing component of waste. Some of this waste makes it to disposal sites, but much of it litters the landscape. Since the ocean is downhill from virtually everywhere humans live, and about half live within 50 miles of the ocean, lightweight plastic trash, lacking recovery infrastructure, blows and runs off into the sea. There, it moves to innumerable environmental niches, where it causes at least eight complex problems, none of which are well understood.

1) Plastic trash fouls beaches worldwide, devaluing the experience of beachgoers. Medical waste, plastic diapers and sanitary waste often found among this debris constitutes a public health hazard.

2) Plastic entangles marine life and kills through drowning, strangulation, and drag on feeding efficiency. So called "ghost nets" continue to fish after being lost or abandoned by their owners and kill untold numbers of commercial species.

3) Ingestion of plastic items that mimic natural food fails to provide nutrition proportionate to its weight or volume. It kills seabirds through starvation and false feelings of satiation, irritation of the stomach lining, and failure to put on fat stores necessary for migration and reproduction. Sea turtles and marine mammals with ingested plastic have been found washed up or floating dead, but linking mortality unequivocally to the ingested debris is difficult.

4) Petroleum-based plastic polymers do not biodegrade, and are long-lived and slow moving in the ocean. As such, they are hosts for "bryozoans, barnacles, polychaete worms, hydroids and mollusks (in order of abundance)," and may present a more effective invasive species dispersal mechanism than ship hulls or ballast water (Barnes).

5) Plastic resin pellets and fragments of plastic broken from larger objects are sources and sinks for xenoestrogens and persistent organic pollutants (POPs) in the marine and aquatic environments, and are readily ingested by invertebrates at the base of the food pyramid (Andrady).

6) Since the majority of consumer plastics are neutrally buoyant (within 0.1g/mL of seawater density), grains of sand caught in their seams or fouling matter make many objects sink to the sea floor. Much of this material consists of thin packaging film and has

the potential to inhibit gas exchange, possibly interfering with $CO_2$ sequestration. It also has the potential to interfere with or smother inhabitants of the benthos.

7) Marine litter threatens coastal species by filling up and destroying nursery habitat where new life would otherwise emerge (UNEP). 8) Marine plastic litter fouls vessel intake ports, keels and propellers, and puts crewman at risk while working to free the debris; incurring significant damage and economic costs for vessels.

Given the variety of problems caused by plastic debris, it is important to gauge its rate of change. From the 1960s to the 1990s evidence from archived plankton samples suggests that marine plastics increased at a rate approximating their steadily increasing production (Thompson). Then, during the decade of the 1990's, plastics in the U.S. municipal waste stream tripled (U.S. EPA) and researchers found accelerating levels in the marine environment. Moore found maximum neuston plastic levels three times greater in the North Pacific Gyre than Day had found a decade earlier. From 1994 to 1998, debris levels around the United Kingdom coastline doubled, "and in parts of the Southern Ocean it increased 100-fold during the early 1990s" (Barnes). Ogi found that neuston plastic increased 10-fold in coastal areas of Japan during the 1970s to 1980s, but that during the 1990s, densities increased 10-fold every two to three years. The most extreme rate of change is at polar latitudes, threatening to turn the pristine shores of Antarctica into a wasteland (Barnes).

Once it reaches the ocean, plastic debris is dispersed in various ways. Onshore winds force land-based debris entering the ocean from rivers and storm drains back to the shore, with greater effect on objects that have appendages above the sea surface, while offshore winds push debris towards major ocean current transport systems. In the deep ocean, large high-pressure systems known as gyres tend to accrete the debris, while low-pressure systems tend to disperse it (Ingraham). In the largest gyre, located in the central North Pacific, neuston trawls for plastic debris yielded the astounding figure of six kilos of plastic fragments for every kilo of zooplankton >0.333mm in size (Moore).

*Trawl Sample, Aug., 2005, AMRF survey: $40^0$ No.L., $140^0$ W.Lo. Photo: Captain Charles Moore.*

Effective solutions for plastic pollution have not yet been developed that yield measurable overall reductions to this rapidly increasing despoiler of marine and aquatic environments. Current measures, as yet unsuccessful at reducing marine plastics include:

1) Recycling: Plastic is lipophilic and hard to clean. It is also difficult to separate composites and mixed plastic waste into the many different plastic types that require different reprocessing technologies. Furthermore, many thermoplastics melt at temperatures not far above the boiling point of water. Therefore, oily contaminants are not driven off during remanufacture. The price of recycled plastic materials far exceeds the current price of virgin plastic resin. Because of contamination, recycled plastics can rarely be used in true "closed loop" recycling: a layer of virgin plastic must be added onto the recycled material for food contact applications. In spite of separation schemes for households, only about 5% of plastics in the U.S. are recycled in any way.

2) Structural controls: Devices to capture plastic debris before it reaches rivers and oceans are being installed at urban catch basins, storm drains and pumping stations, and debris booms are being placed across rivers draining urban areas. Containment structures cover only a small percentage of debris conduits, and during heavy storms, these devices overflow, and release debris.

3) Source reduction: Because plastic packaging extends the shelf life of products by providing an air and moisture barrier, it is increasingly used in global trade. In some applications, where space is at a premium, bulk, rather than individual containers are preferred, but the trend is for more individual packaging. Producers in the United States have little incentive to minimize the use of their products, or to design them for ease of

recycling, but European countries are responding to "green dot" initiatives with some packaging reductions. A few U.S. companies have adopted a "zero waste" policy which requires that their suppliers take back packaging, and they provide take back programs for their customers, but these companies remain a tiny part of industry as a whole.

4) Beach cleanups: While beach cleanups by civic groups raise awareness among the general public of the plastic debris problem, they are infrequent and do not stem the tide of debris. In municipalities that regularly groom their beaches mechanically, the amount of debris removed depends on amount of rainfall, not on frequency of cleaning.

5) Reef cleanups: In the Northwest Hawaiian Islands, the National Oceanic and Atmospheric Administration (NOAA) spends 2 million U.S. dollars per year to remove 60 tons of derelict fishing nets and gear in an effort to save the critically endangered Hawaiian Monk Seal, over 200 of which have become entangled since records were kept (Foley). The amount retrieved does not diminish, year to year, and efforts are currently being made to find accumulation zones where the nets can be retrieved at sea before they damage coral reef habitat.

6) Escape of Plastic Industry Feedstocks and Production Scrap: Evidence suggests that pre-production plastic resin pellets accidentally released from plastic processors contribute approximately 10% by count to the plastic debris problem (Moore, McDermid). In response, the American Plastics Council (APC) and the Society of the Plastics Industry (SPI) in the United States have adopted a voluntary program of Best Management Practices known as "Operation Clean Sweep." Measurements of industrial discharge before and after implementation of the program showed reductions of approximately 50% in pellet discharge (AMRF), but because of the voluntary nature of the program, only a small percentage of the industry participates.

7) Biodegradable plastics: Many polymers originate from non-petroleum sources. In general, these plastics biodegrade more rapidly than their petroleum-based counterparts. However, typical tests for biodegradability rely on hot, aerated composting media, rich with bacteria and fungi. The marine environment is cold and devoid of fungi by comparison, and many compostable "bioplastics" degrade very slowly, or hardly at all in the deep ocean (Wirsen). Currently, substitution for conventional plastics is limited by the cost of bioplastics, which are five to ten times greater than petroleum-based resins. A 1999 projection of the world biodegrables market was to grow from 30 to $250 \times 10^6$ pounds per year, while petroleum plastics consistently sell at the rate of $250 \times 10^9$ pounds annually (New York Times).

## DEBRIS IS A CONCERN

Plastic debris littering the world's oceans is becoming a major concern in three main areas:

1. Aesthetics
2. Entanglement
3. Ingestion

Aesthetics

According to the World Health Organization, a clean beach is one of the most important characteristics sought by visitors. The negative effects of debris, defined as solid materials of human origin, are: loss of tourist days; resultant damage to leisure/tourism infrastructure; damage to commercial activities dependent on tourism; damage to fishery activities; and damage to the local, national and international image of a resort. "Such effects were experienced in New Jersey, USA in 1987 and Long Island, USA in 1988 where the reporting of medical waste, such as syringes, vials and plastic catheters, along the coastline resulted in an estimated loss of between 121 and 327 million user days at the beach and between U.S. $\$1.3x10^9$ and U.S. $\$5.4x10^9$ in tourism related expenditure" (WHO, Bartram). Naturally clean beaches, free from debris, are a thing of the past. In the 20 years since the Ocean Conservancy organized the first annual International Coastal Cleanup Day, 6 million volunteers from 100 countries have removed 100 million pounds of litter from 170,000 miles of beaches and inland waterways. Reports of groups finding nothing to pick up do not exist. While the International Cleanup Day effort expands each year, so does the amount of debris recovered. Between 1996 and 2006, at Escondido Beach, California, 310 total debris items were removed, but 182 of those were found in 2005, representing 59% of the total recovered in the last year of the 10-year effort. At Torrey Pines State Beach, California, in all four quarters of 2005, 136 items were removed, but in the second quarter of 2006 alone, 189 items were found (Ocean Conservancy).

It must be remembered that beach cleanups focus on macro debris. Numerous studies have found micro debris in beaches worldwide, many of them remote from human activity (McDermid, S. Moore, Gregory, Thompson, Ng). A random 2'x 2'x 4" quadrant study of a beach, near an urban river mouth, found the sand to be 1% plastic by volume (Moore, unpublished data). Whether, or to what extent, mixing lighter plastics with heavier sediments contributes to beach erosion has not been determined. Mechanical raking and grooming of beaches to remove debris tills in plastic fragments and may contribute to erosion by removing plant roots and seaweed that anchor sand (U.S. EPA).

Floating debris is an aesthetic issue for swimmers, mariners, coastal and inland water body dwellers, and submerged debris is an aesthetic issue for divers.

Entanglement

In the 1980s, researchers estimated that there were 100,000 marine mammal deaths per year in the North Pacific related to entanglement in plastic nets and fishing line (Wallace). Currently NOAA is using digitally enhanced photos of wounds suffered by marine mammals to identify the type of line they were entangled in. Lost and abandoned nets, termed "ghost nets," continue to fish and destroy resources. A report by Canada's FAO estimates that 10% of all static fishing gear is lost and that this results in a loss of 10% of the target population. Efforts to remove this gear are growing, but are not widespread, and the great cost of removal of derelict gear is rarely, if ever, borne by those who manufacture it or lose it. Indeed, if it were, commercial fishing would be extremely uneconomical.

Documentation of entanglement of seabirds and other marine species in 6-pack rings used to hold cans and bottles, has resulted in changes to the plastic formula to speed up disintegration in the environment. The polymer can be changed chemically during manufacture so that it absorbs UV-B radiation from sunlight and breaks down into a very brittle material in a fairly short period of time (Andrady), however, the resulting particles are no more biodegradable than the untreated polymer. Such embrittlement accelerators are not used in nets and lines, however, and volunteer groups worldwide are regularly called on to free entangled cetaceans and other sea life.

Ingestion

The term "plastic" means "capable of being formed into any shape." The plastic objects that populate the marine and aquatic environments, with the exception of fishing lures, are not made to look like natural food to marine creatures, though thin plastic shopping bags balloon out in water to appear like jellyfish and are regularly consumed by sea turtles. It seems probable that the infinite ways in which the mega-tons of multi-colored plastic debris break down create mimics for virtually every natural food source. Andrady reported on feeding studies by Alldredge at UC Santa Barbara, using Ivlev's Electivity Index (designed to quantify prey-selection by predators, especially planktivores), showing that two common species, *Euphausia pacifica,* and *Calanus pacificus* had values of the index very close to zero and that ingestion of contaminant free, uncolonized plastic particles, versus natural prey, appeared to be non preferential. Most feeding that takes place in the ocean is accomplished by indiscriminate feeders with mucus bodies or appendages designed to adhere to and capture anything of an appropriate size with which the organism comes in contact. Collection of salps in the North Pacific Central Gyre by Algalita Marine Research Foundation (AMRF), using both plankton trawls and hand nets, found individuals with plastic particles and fishing line embedded in their tissue. The optimum size class of plastic for filter feeder ingestion appears to be <1mm diameter, although individuals with larger particles have been found. A 1999 AMRF study of 27,448 plastic particles trawled from the surface of the Gyre found 9,470 particles near 1mm in size, 4,646 near 0.5mm, and 2,626 near 0.3mm, suggesting that smaller particles are being removed, or are leaving the system by some unknown mechanism (Moore). Thompson kept intertidal invertebrates in aquaria with microscopic plastic <2mm in diameter. The microscopic plastics were ingested by polychaete worms, barnacles, and amphipods during these laboratory trials. Documentation of transmission of these types of particles up the food pyramid has been provided by Eriksson, who surveyed Southern Fur Seal Scat on Macquarie Island. She found that all scat contained plastic particles associated with the night-feeding myctophids (Lantern fish), active near the sea surface, and consumed by the seals.

When plastic debris hits the sea, the proportion that floats heads for surface accumulation zones. Modeling done by Ingraham using the Ocean Surface Current Simulator (OSCURS) seeded 113 drifters uniformly over the North Pacific from the U.S. Coast to China. After 12 years, winds and currents had gathered 75% of the drifters into an area of the Central Gyre equal to 28% of the total area seeded. The five enormous high-pressure gyres in the oceans comprise 40% of the sea surface, or 25% of the area of the entire earth. The mountains of air that create the highs, force the sea level lower near their centers and create accumulation zones described as "gentle maelstroms" (Moore). These areas are in the deep ocean and are oligotrophic, oceanic deserts. Thus, the ratio of plastic particles to plankton is highest here on average, although after heavy rains cause runoff of plastic particles from urban areas, higher ratios are found in the coastal zone

(Moore). Detritus feeders, like the Laysan albatross, have been demonstrated to feed primarily in the gyres (Henry), and the stomach contents of their chicks, receiving nutriment only by regurgitation from adult birds, contain alarming quantities of plastic (Auman). Sileo documented eighty species of seabirds to ingest plastic in 1990. Carpenter found plastic pellets in eight of fourteen species of fish and one chaetognath in 1972 off Southern New England. Steimle found pellet ingestion more common in lobster than winter flounder in the New York Bight in 1991 (U.S. EPA).

*Laysan albatross chick, Kure Atoll, 2002, photo: Cynthia Vanderlip, www.algalita.org*

Plastics as a means to transport pollutants in aquatic and marine ecosystems have become the focus of scientific research as levels of macro and micro plastics in these environments increase (Thompson, Moore, Ogi, Copello). Mato and Takada at the Tokyo University of Agriculture and Technology have studied how polypropylene (PP) pellets in the marine environment adsorb, (with adsorption coefficients of $10^5$-$10^6$ from ambient seawater), and transport PCBs, DDE and nonylphenols (NP). Moore found polycyclic aromatic hydrocarbons and phthalates in all marine and river samples of both pre-production pellets, and post consumer fragments, of the same size class. The extent to which these compounds desorb when ingested by different organisms has not been studied. Whether or to what extent estrogenic compounds in plastics are implicated in findings such as a high percentage of intersex in Mediterranean swordfish (De Metrio) has not been investigated, but the presence of micro plastics in the sea surface microlayer where xenoestrogens are known to accumulate, has been documented by Ng.

## COLLATERAL CONCERNS

Just as plastics are infinitely variable, so are the concerns raised by their ubiquitous presence as uncontrollable, marginally degradable waste. Foremost among these concerns is the recent exponential explosion in what may be termed "pelagic plastics." For most of their history, synthetic, petroleum-based polymers were used and discarded principally in Europe and the United States, and more recently, Japan. Levels of plastic pollution off these coasts paralleled the level of plastic production until recently. During the last decade of the 20th Century, and continuing to the present day, proliferation of plastic packaging and products accelerated worldwide. Sales of plastic water bottles alone rose from 3.3 billion in 1997, to 15 billion in 2002 (Container Recycling Institue). Many of these bottles are shipped around the world for disaster relief and other purposes, where no recycling infrastructure exists. Ebbesmeyer has estimated that a single, one litre plastic water bottle will photo degrade into enough pieces to put one on every mile of beach in the world. Studies (Ogi, Moore, Barnes) cited above show that the increase in marine plastic debris is now exponential, going up by a factor of ten every two to three years off Japan. There are now 15,000 plastic processors in India, necessitating importation of plastic resin. Exports of primary plastic resins from the Middle East are growing rapidly in every global market except North and South America (Al-Sheaibi) Consumer plastics are going global. Tracking their fate is difficult if not impossible. Based on statistics compiled in a 2003 California "Plastics White Paper," that included amounts of plastics made, disposed of, and recycled nationwide, approximately 25% of all disposable plastics remain unaccounted for. With total U.S. thermoplastic resin sales at $50x10^6$ tons, $25x10^6$ tons (50%) is disposed of as municipal waste, 5% is recycled and an estimated 20% is made into durable goods, leaving 12.5 million tons(25%) unaccounted for. Much of that 12.5 mt of unaccounted for plastic makes its way via rivers to the sea. In three days of sampling on the Los Angeles and San Gabriel Rivers, AMRF found 60 tons of plastic debris flowing towards the sea, representing 2.3 billion individual pieces of plastic trash of all size classes >1mm.

Many islands, which act as sieves for ocean borne plastics, have already been heavily impacted by plastic debris originating far from their shores. On the surface of one square foot of beach sand on Kamilo Beach, Hawaii, 2,500 plastic particles >1mm were found, and the fact that 500 of them were pre-production plastic pellets, with no processors located in Hawaii, lends credence to the concept that these particles are of foreign origin (Moore, unpublished data, 2003). McDermid collected 19,100 plastic particles from nine remote Hawaiian beaches separated by 1500 miles, and 11% were pre-production pellets by count. These pellets come in a variety of shapes, including rounded, flattened oval, and cylindrical, and are normally <5mm in diameter. Plastic producers make these pellets then ship them to plastic manufacturers or processors to be melted into consumer products. A 1998 study of Orange County Beaches in Southern California showed plastic pellets to be the most abundant items, with an estimated count of over 105 million, comprising 98% of the total debris (S. Moore). Southern California has the largest concentration of processors in the western United States. A 2005 study by AMRF of the two main rivers draining the Los Angeles, California basin found in one dry and two rainy days of sampling, over $2.3x10^9$ plastic objects and fragments being transported to the Pacific Ocean at San Pedro Bay. Macro debris accounted for ten per cent of the total. Of the identifiable objects, the largest single component was pre-production plastic pellets at $2.3x10^6$. Ignoring these inputs results in underestimates of the total pieces of litter entering the ocean worldwide on a daily basis, like the widely quoted figure of 8 million pieces per day (UNEP). In reality, 8 million is only one per cent of the total number of plastic pieces flowing to the sea from the Los Angeles area in a single day, based on AMRF's three-day totals. AMRF's figures do not include anthropogenic debris other than plastic.

Plastics form a stable substrate for colonization by marine organisms, with larger floating items generally having one side exposed to the sun, and one side ballasted with fouling organisms. Less than 10% of the micro debris in a 1999 North Pacific Central Gyre study, however, appeared to have fouling organisms at all. This may be due to their frequency of tumbling in wavelets and changing the side exposed to the sun (Moore). Barnes estimates "that rubbish of human origin in the sea has roughly doubled the propagation of fauna in the subtropics and more than tripled it at high (>50°) latitudes." Globally, the proportion of plastic among marine debris ranges from 60 to 80%, although it has reached over 90-95% in some areas (Derraik). Bartram points out certain exceptions to these generalizations found in United Kingdom beach surveys, and states that, "Litter sourcing seems to be highly site specific."

A report by the United Nations Environmental Programme titled, "Marine litter—trash that kills," states:

> *"Marine litter is found resting or drifting on the seabed at all depths. In the North Sea, it has been estimated that some 70 per cent of the marine litter ends up on the seabed... Assessments made in the Dutch sector of the North Sea have indicated an average of over 110 pieces of litter per $km^2$ of seabed. If this is characteristic of the North Sea at large, a volume of at least 600,000 $m^2$ of marine litter could be found on the seabed. During a survey in the Mediterranean, 300 million pieces of garbage were found at a depth of 2,500 metres between France and Corsica. Consequently, large quantities of the entire input of marine litter around the world could be sinking to the bottom and be found on the seabed, both in shallow coastal areas and in much deeper parts of seas and oceans."*

Plastics made up 80-85% of the seabed debris in Tokyo Bay (Kanehiro). The consequences of partially covering the seabed with materials resistant to gas and water transport have not been investigated, although Goldberg speculated that it may interfere with carbon cycling in the ocean. In an article entitled: "Trashed," in Natural History Magazine, Moore speculated that the weight of plastic debris in an area of the North Pacific Central Gyre known as the "Eastern Garbage Patch," an area 1000 miles in diameter, was three million tons.

## SOLUTIONS ELUSIVE

The prevailing attitude among manufacturers of consumer plastics in the United States is that they are responding to the demands of the market, and that it is the responsibility of individuals and governments to create infrastructure for dealing with the resultant waste. Rarely are U.S. processors required to subsidize the cost of landfilling or otherwise disposing of the plastic waste they manufacture. Ten of 52 U.S. states have implemented "bottle bills" which require a deposit on certain plastic bottles to aid in their recovery and recycling, and in 2005, only 17% of the over 50 billion polyethylene terephthalate (PET) plastic water bottles consumed in the U.S. were recycled. The number of plastic bottles as a percentage of total debris recovered in beach cleanups is rising (Container Recycling Institute). Thin high-density polyethylene (HDPE) and thicker LDPE shopping bags are recycled at a rate of around 1% in the U.S (U.S. EPA), with trillions being produced worldwide. Many become airborne and soar to waterways and seas on the wind. An effort to put a deposit on the bags in San Francisco, California, was met with resistance by industry and failed, although eleven countries have such fees, and thirteen countries have enacted complete or partial bans (ERF).

In December 1994, the European Union issued the "Directive on Packaging and Packaging Waste." This legislation places direct responsibility and specific packaging waste reduction targets on all manufacturers, importers and distributors of products on the EU market. To meet the requirements of this legislation, manufacturers, importers and distributors must either develop their own take-back scheme or join industry-driven non-profit organizations, such as the Green Dot Program, to collect, sort and recycle used packaging. Green Dot is currently the standard take-back program in 19 European countries and Canada. Such programs encourage product and packaging design that gives waste value when it is recycled as another product in a "cradle to cradle" system (McDonough). Such schemes may help to reduce plastic waste that ends up in the ocean, but they are far from universal.

Pre-production plastics (in the form of pellets or powders) are discharged to waterways during the transport, packaging, and processing of plastics when Best Management Practices (BMPs), i.e., proper housekeeping practices, are not adequately employed. For pellets transported by rail, cars are emptied via a valve that connects to a conveyance hose. The valve should be capped when not in use. Caps often are not replaced, causing pellet loss within the rail yard adjacent to a facility. A similar conveyance system exists for resins transported by hopper trucks. Pellets and powders escape when hoppers are emptied through pipes connected to valves at the bottom of the truck. When handled improperly, resin pellets and powders are also released from conveyance mechanisms on site. In addition to plastic resins, additives used for coloring or creating specific characteristics of processed plastics are also delivered in pellet and powder form. The discharges to local waterways include colorants and additives, not just plastic resins. Grindings, cuttings and fragments from the processing of plastics, known as production scrap, are often part of the mix of debris that is conveyed by wind, storm water, or runoff from plastics facilities to storm drains and nearby waterways.

Operation Clean Sweep (OCS) is a program of voluntary BMPs that was first developed in 1980 by SPI. It was recently revised and improved by a collaborative effort between AMRF, APC, and SPI. Monitoring done by AMRF indicates reductions of pellet loss of greater than 50% can result from getting processors to implement the voluntary program. But recruiting processors to the program has proved challenging, and less than one per cent of the industry participates in OCS (Moore).

Pellets, powders, and fragments are widely dispersed from their places of origin. The impacts of powders and plastic debris smaller than pellets are not known but ingestion by plankton and other small marine organisms does occur. The impacts of pelletized and powdered plastic additives, such as colorants and chemicals, in the marine environment are not well understood, as research is in the initial phases. Dr. Michael Baker at the UC San Diego School of Medicine has made progress in developing a cross-species microarray that can screen many genes at once in fish. This is more than measuring the estrogenic action of a chemical. It can also be determined if the chemical affects cortisol (stress steroid), progesterone and testosterone (important reproductive steroids), thyroid hormone, vitamin D, or other bioactive compounds. Thus, the effects on fish of interest of a variety of plastics and the chemicals that bind to the plastics on expression (up or down) of genes important in all aspects of steroid hormone action can be determined.

A proposal to feed plastics to male fish held in aquaria and subject them to such analyses has been developed, but funding for such research is not a current priority in the United States, and these trials have not been conducted.

REFERENCES


1. Al-Sheaibi, Fahad, President, Saudi Basic Industries Polymers Group, *Address to 5th Dubai Plast Pro Congress*, Riyadh, Saudi Arabia, April 15, 2002. www.sabic.com/sabic-www/index_press_speech_09.htm
2. Algalita Marine Research Foundation, "*Assess and reduce sources of plastic and trash in urban and coastal waters*," Report to the State of California Water Resources Control Board, March, 2006. www.swrcb.ca.gov , www.algalita.org
3. Andrady, T.L., *Plastics in Marine Environment*, 2005 In: Proceedings of the Plastic Debris Rivers to Sea Conference, 2005. www.plasticdebris.org
4. Michael E. Baker, Research Professor, *Personal Communication*, Department of Medicine, 0693, University of California, San Diego, 9500 Gilman Drive, La Jolla, CA 92093-0693. http://medicine.ucsd.edu/faculty/mbaker/
5. Barnes, David K.A., "Invasions by marine life on plastic debris." *Nature*, Vol. 416, 25 April, 2002, 808-9.
6. Barnes, David K.A., "Remote Islands Reveal Rapid Rise of Southern Hemisphere, Sea Debris." *The Scientific World,* (2005) **5**:915-921.
7. Bartram, Jamie, Coordinator, Department of Water, Sanitation and Health, World Health Organization, Editor, *A Practical Guide to the Design and Implementation of Assessments and Monitoring Programmes*, 2000 (email: bartramj@who.int)
8. California Integrated Waste Management Board, *Plastics White Paper*, 2003, www.ciwmb.ca.gov
9. Copello, Sofia, and Quintara, Favio, *Marine debris ingestion by Southern Giant Petrels and its potential relationships with fisheries in the Southern Atlantic Ocean*, Marine Pollution Bulletin 46 (2003) 1513-1515.
10. Container Recycling Institute, *Plastic Recycling Rates*, Pat Franklin, Executive Director http://www.container-recycling.org/plastic_rates.htm
11. De Metrio, G. et al, 2003. "Evidence of a high percentage of intersex in the Mediterranean swordfish (Xiphias gladius L.)." *Marine Pollution Bulletin* **46**: 358-361.
12. Derraik, J.G.B. 2002. "The pollution of the marine environment by plastic debris: a review." *Marine Pollution Bulletin* **44**(9):842-852.

13. Ebbesmeyer, Curtis, Beachcombers and Oceanographers International Association, *personal communication*, www.beachcombers.org
14. Earth Resource Foundation, *Campaign Against the Plastic Plague*, www.earthresource.org
15. Ericsson, Cecilia and Burton, Harry. "Origins and Biological Accumulation of Small Plastic Particles in Fur Seals from Macquarie Island." *AMBIO: A Journal of the Human Environm*ent. Vol. **32**, No. 6, pp. 380–384.
16. Foley, D. and Veenstra, T. *Characterizing and surveying oceanic sources and sinks of marine debris*, Proceedings of the Southern California Academy of Sciences Annual Meeting, 2006, recorded remarks.
17. Franklin Associates, *Characterization of Municipal Solid Waste in the United States: 2003 Update* Washington, D.C., United States Environmental Protection Agency.
18. Goldberg, E.D. 1997. "Plasticizing the seafloor: An overview." *Environmental Technology* **18**:195-201.
19. Gregory, M.R. 1977. "Plastic pellets on New Zealand beaches." *Marine Pollution Bulletin* **8**:82-84.
20. Gregory, M.R. 1978. "Accumulation and distribution of virgin plastic granules on New Zealand beaches." *New Zealand Journal of Marine and Freshwater Research* 12(4):399-414.
21. Gregory, M.R. 1983. "Virgin plastic granules on some beaches of eastern Canada and Bermuda." *Marine Environmental Research* **10**:73-92.
22. Gregory, M.R. 1991. "The hazards of persistent marine pollution: Drift plastics and conservation islands." *Journal of the Royal Society of New Zealand* **21**:83-100.
23. Gregory, M.R. 1996. "Plastic 'scrubbers' in hand cleansers: a further (and minor) source for marine pollution identified." *Marine Pollution Bulletin* **32**:867-871.
24. Gregory, M.R. 1999. "Plastics and South Pacific island shores: environmental implications." *Ocean and Coastal Management* **42**(6-7):603-615.
25. Henry, William, Long Marine Laboratory, University of California, 100 Shaffer Road, Santa Cruz, CA 95060, *Plastic and Seabirds, Regional Patterns in Plastic Accumulation in Laysan Albatrosses*, 2004, Report to the Will J. Reid Foundation, available from AMRF, www.algalita.org
26. Ingraham Jr., W.J. 2001. *Surface current concentration of floating marine debris in the North Pacific Ocean: 12-year OSCURS Model Experiment*. Honolulu Marine Debris Conference 2001.
27. Kanehiro, H. et al. 1995. "Marine litter composition and distribution on the seabed of Tokyo Bay." *Fisheries Engineering* **31**:195-199.
28. Lattin, G.L., Moore, C.J., Zellers, A.F., Moore, S.L., Weisberg, S.B. 2004. "A comparison of neustonic plastic and zooplankton at 3 different depths near the southern California shore." *Marine Pollution Bulletin.*
29. Mato, Y., et al. 2001. "Plastic resin pellets as a transport medium for toxic chemicals in the marine environment." *Environmental Science & Technology* **35**:318-324.

30. McDermid, K.J., McMullen, T.L. 2004. "Quantitative analysis of small plastic debris on beaches in the Hawaiian archipelago." *Marine Pollution Bulletin* **48**:790-794.
31. McDonough, William and Braungart, Michael, *Cradle to Cradle-The Way we Make Things*, 2002, North Point Press.
32. Moore, C.J., Moore, S.L., Leecaster, M.K., Weisberg, S.B. 2001. "A comparison of plastic and plankton in the North Pacific central gyre." *Marine Pollution Bulletin* **42**:1297-1300.
33. Moore, C.J., Moore, S.L., Weisberg, S.B., Lattin, G., Zellers, A. 2002. "A comparison of neustonic plastic and zooplankton abundance in southern California's coastal waters." *Marine Pollution Bulletin* **44**(10):1035-1038.
34. Moore, Charles J. et al, "A comparison of neustonic plastic and zooplankton at different depths near the southern California shore." *Marine Pollution Bulletin* **49**:(2004) 291-294.
35. Moore, Charles J., et al, *A Brief Analysis of Organic Pollutants Sorbed to Pre and Post-Production Plastic Particles from the Los Angeles and San Gabriel River Watersheds*, in Proceedings of the Plastic Debris Rivers to Sea Conference, 2005, www.plasticdebris.org
36. Moore, C.J., G.L. Lattin, A.F. Zellers, *Working our way upstream: a snapshot of land-based contributions of plastic and other trash to coastal waters and beaches of Southern California*, in Proceedings of the Plastic Debris Rivers to Sea Conference, 2005, www.plasticdebris.org
37. Moore, C.J., G.L. Lattin, A.F. Zellers, *Measuring the Effectiveness of Voluntary Plastic Industry Efforts: AMRF'S Analysis of Operation Clean Sweep*, in Proceedings of the Plastic Debris Rivers to Sea Conference, 2005, www.plasticdebris.org
38. Moore, Charles, 2003. "Trashed: Across the Pacific Ocean, plastics, plastics everywhere." *Natural History* Vol. 112, Number 9, Nov. 2003.
39. Moore, S.L., Gregorio, D., Carreon, M., Leecaster, M.K., Weisberg, S.B. 2001. "Composition and distribution of beach debris in Orange County, California." *Marine Pollution Bulletin* **42**(3):241-245.
40. National Oceanic and Atmospheric Administration, *Mapping of Marine Mammal Entanglement Wounds*, http://www.esri.com/news/arcnews/fall03articles/sources-of-mortality.html
41. New York Times, *Trying to be Green, Many Companies Invested in Biodegradable Plastics; They Ended Up in the Red*, February 6, 1999.
42. Ng, K.L., and Obbard, J.P., 2005. "Prevalence of microplastics in Singapore's coastal marine environment." *Marine Pollution Bulletin.*
43. Ocean Conservancy, International Coastal Cleanup Day, www.oceanconservancy.org
44. Ogi, H., Baba, N., Ishihara, S., Shibata, Y. 1999. "Sampling of plastic pellets by two types of neuston net and plastic pollution in the sea." *Bulletin of Faculty of Fisheries*, Hokkaido University **50**(2):77–91.
45. Ogi, H., Fukumoto, Y. 2000. "A sorting method for small plastic debris floating on the sea surface and stranded on sandy beaches." *Bulletin of the Faculty of Fisheries*, Hokkaido University **51**(2):71-93.

46. Ryan, P.G. et al. 1988. "Plastic Ingestion and PCBs in Seabirds: Is There a Relationship?" *Marine Pollution Bulletin*, **19**:174-176.
47. Thompson, Richard C., et al, 2004. "Lost at Sea: Where Is All the Plastic?" *Science*, Vol. 304, 2004.
48. United Nations Environment Program, Global Programme of Action Coordination Office; "*Marine Litter-trash that kills*", Global Marine Litter Information Gateway, http://marine-litter.gpa.unep.org/documents/documents.htm
49. United States Environmental Protection Agency, M*unicipal Solid Waste Generation in the United States:2003,* http://www.epa.gov/msw/msw99.htm
50. Unnithan, Sandeep, "Through thick, not thin, say ragpickers." *Indian Express* 23 Nov 1998 http://www.mindfully.org/Plastic/Ragpickers-Hate-Plastic.htm
51. U.S. EPA, *Plastic pellets in the aquatic environment: Sources and recommendations*. 1992, EPA Oceans and Coastal Protection Division Report 842-B-92-010. Washington, DC.
52. Wallace, N. 1985. *Debris entanglement in the marine environment: a review*. pp. 259-277. In: R.S. Shomura and H.O. Yoshida (eds.), Proceedings of the Workshop on the Fate and Impact of Marine Debris. November 27-29, 1984. Honolulu, Hawaii. U.S. Dep. of Comm., NOAA Tech. Memo. NMFS, NOAA-TM-NMFS-SWFC-54.
53. Wirsen, Carl. 1971. "*Microbial Degradation of Organic Matter in the Deep Sea.*" *Science* Vol. 171, No.3972, pp. 672-675.

# OCCURRENCE AND FATE OF PLASTIC ADDITIVES IN NATURAL AND ENGINEERING SYSTEMS

JEAN-FRANÇOIS DEBROUX, PH.D.
Kennedy/Jenks Consultants, San Francisco, USA

ANGELA YU-CHEN LIN, PH.D.
National Taiwan University, Taiwan

JESSICA HUYBREGTS
Kennedy/Jenks Consultants, San Francisco, USA

## INTRODUCTION

Since the start of chemical industrialization, there has been a consistent influx of chemicals into the environment due to waste discharges. Although mechanisms in the environment reduce their levels, pg to µg/l levels of many chemicals, or their metabolites, exist in waters impacted by human activity. Advances in analytical chemistry (e.g., GC-MS-MS, LC-MS-MS, etc.) allow environmental scientists to quantify the occurrence of trace levels of contaminants in the non-saline aqueous environment. Although the concentrations of civilization's chemicals may be decreasing in waste streams due to higher levels of treatment, the number of chemicals in waste streams is surely increasing as the introduction of new chemicals to the marketplace outpaces the discontinuation of problematic ones.

Plastic additives are widely used to give thermo-polymers the properties necessary for their numerous of applications. Typical additives include: plasticizers, pigments, light stabilizers, antimicrobials, flame retardants, impact modifiers and heat stabilizers. Non-covalent additives, as opposed to reactive additives that covalently bond to the polymer matrix, are especially susceptible to leaching when used.

This paper presents occurrence data and describes attenuation mechanisms for plastic additives in the environment. Three compounds/families of compounds have been identified in wastewater effluents, receiving waters and their associated sediments:

- Three commonly used phthalates
- Bisphenol A
- Three formulations of poly-brominated diphenyl ethers, or PBDEs

The three phthalates chosen are: DEHP (di (2-ethylhexyl) phthalate), DBP (dibutyl phthalate), and BBP (butylbenzyl phthalate). There are 209 congeners of PBDEs that differ in the number and location of bromine atoms on the molecule. The three formulations of PBDEs include numerous congeners. This paper will address the number of bromine atoms per molecule, which is reflected in the prefix of a family of congeners (e.g., deca-brominated diphenyl ether or DeBDE), but will not address location. PBDEs are commonly used in three formulations and contain various families of congeners, deca- (>97% DeBDEs), octa- (62% HxBDEs and 38% OcBDEs), and penta- (50–62% PeBDEs and 24–38% TeBDEs).

The three phthalates are common plasticizers and additives used in the manufacturer of goods such as PVC, resins, building materials, home furnishings, food packaging and insect repellents. Bisphenol A is primarily used as a monomer for the production of polycarbonate and epoxy resin and also as an ingredient in polyester-styrene resins and flame retardants. Bisphenol A also has numerous other industrial uses. PBDEs are flame retardants added to plastics, upholstery fabrics and foams used in a variety of products including computers, televisions, furniture and carpet pads. Global demands of these chemicals are presented in Table 1.

## OCCURRENCE AND PHYSICAL-CHEMICAL PROPERTIES

Plastic additives can enter the environment by point and non-point sources. Although difficult to quantify, it is acknowledged that significant loadings to surface and ground waters occur without passage through wastewater treatment plants. Nonetheless, industrial and domestic waste streams contain plastic additives and occurrence studies reflect those inputs.

Numerous studies have detected plastic additives in wastewater treatment plant solids and effluents, surface waters and sediments impacted by civilization's waste streams (Table 1). The selected phthalates and Bisphenol A are typically present in wastewater effluents and receiving waters in the ng to μg/L range, while PBDEs are typically found at levels three orders of magnitude less, or pg to ng/L range. The consistent flow of many receiving waters dilutes contaminant concentrations, although some discharges in arid areas can provide the majority of river flow during certain portions of the year. Typically, there is a decrease in contaminant concentration from wastewater effluent to receiving water.

**Table 1: Occurrence of Plastic Additives in Surface and Wastewaters.**

| Compound Family | Compound | Global Annual Production (tonnes)[a] | Surface Water (μg/l)[b] | Sewage Effluent (μg/l) | Sediment (mg/kg dw)[c] | Sewage Sludge (mg/kg dw) |
|---|---|---|---|---|---|---|
| Phthalates | DEHP[d] | $0.5X10^6$ | 0.33-97.8[e,m] | 1.74-182[e,f] | 0.21-8.44[e] | 27.9-154[e] |
| Phthalates | DBP[g] | $0.04X10^6$ | 0.12-8.8[e] | 0.2-10.4[e] | 0.06-2.08[e] | 0.2-1.7[e] |
| Phthalates | BBP[h] | $0.018X10^6$ | ND[i]-2.4[j] | [k] | ND-0.567[e,j] | ND[e] |
| BPA[l] | BPA | $2.5X10^6$ | 0.0005-12[e,m] | 0.018-0.702[e] | 0.01-0.19[e] | 0.004-1.36[e] |
| PBDEs[n] | Sum of all | $0.07 X10^6$ | 0.000003-0.000513[o] | 0.000004-0.029[p] | ND-0.212[p] | 0.061-1.44[p] |

Notes
(a) Actual or estimated 2004 production
(b) μg/l = micrograms per liter.
(c) mg/kg dw = milligrams per kilogram, dry weight.
(d) DEHP = Di (2-ethylhexyl) phthalate.
(e) Fromme *et al.* 2002.
(f) Marttinen *et al.* 2003.
(g) DBP = Dibutyl phthalate.
(h) BBP = Butylbenzyl phthalate.
(i) ND = non-detect concentration.
(j) Gledhill *et al.* 1980.
(k) Insufficient data in the literature.
(l) BPA = Bisphenol A.
(m) Koplin *et al.* 2002.
(n) PBDEs = Poly-brominated diphenyl ethers.
(o) Oros *et al.* 2005.
(p) North, 2004.

Aqueous contaminants, in general, can be attenuated during treatment and after discharge into the environment. This can occur through: 1) a transfer from the water phase to solid or vapor phases, 2) partial transformation from the parent contaminant to a potentially problematic daughter compound(s), or 3) a sustainable complete transformation to benign products. How contaminants behave in natural or engineered environments depends not only on the characteristics of the environment but also on the physical-chemical properties of the contaminants (Table 2).

## ATTENUATION IN NATURAL SYSTEMS

Primary attenuation mechanisms for aqueous contaminants in natural environments include: 1) adsorption, 2) biodegradation, 3) photolysis, 4) volatilization, and 5) hydrolysis. As hydrolysis does not appear to be a significant attenuation mechanism for the plastic additives discussed in this paper, only the first four mechanisms are discussed below.

### Adsorption

Hydrophobic compounds adsorb on to solid surfaces. The octanol-water coefficient ($K_{OW}$) for a specific compound is a measure of hydrophobicity and is a fair indicator of its propensity for adsorption onto solids. The $K_{OW}$ is an experimentally determined ratio of the fraction of contaminant that resides in octanol, a relatively non-polar solvent, versus water, a polar solvent, presented in logarithmic form. A higher $K_{OW}$ indicates a greater likelihood for adsorption.

Bisphenol A possesses the lowest and PBDEs the highest $K_{OW}$ values (Table 2). Studies reflect this trend as less than 50% of Bisphenol A was estimated to be absorbed onto solids (Staples et al. 1998) while 96% of PBDEs were determined to reside in sludge at a wastewater treatment plant (North, 2004). Occurrence data in wastewater treatment plant sludge and surface water sediments are presented in Table 1.

### Biodegradation

Biodegradation pathways are highly dependent on the acclimation of bacteria and the presence or absence of oxygen. Microbial populations typically need to acclimate to a certain contaminant. Aerobic pathways typically possess greater degradation rates than anaerobic pathways. Biodegradation can be complete but often it results in somewhat stable metabolites that can persist in the environment.

Trace levels (pg to μg/l) of contaminants may not be sufficient to promote acclimation of bacteria in water or soil. Wild and Reinhard (1999) observed that although a common biodegradable detergent metabolite quickly degraded to 0.2 μg/l, degradation stalled from this point forward. Their explanation included a minimum threshold substrate concentration, or $S_{MIN}$, value. This may be an explanation for the presence, albeit very low, of certain contaminants in groundwater recharged by wastewater effluents.

The three phthalates and Bisphenol A are fairly to readily degradable in microbially-rich environments. PBDEs do not degrade well either aerobically and limited DeBDE degradation to NoBDE and OcBDE have been observed anaerobically.

**Table 2: Physical-Chemical Properties of Selected Plastic Additives.**

| Compound Family | Compound | Molecular Weight (daltons) | Water Solubility (mg/l) | Log $K_{ow}$ (a) (unitless) | Henry's Law Constant (atm-m$^3$/mol) | Biodegradability |
|---|---|---|---|---|---|---|
| Phthalates | DEHP | 391 | 0.29-1.2 (b) | 4.2-8.4 (b) | $1.7 \times 10^{-5}$ (b) | Fairly readily, following acclimation |
| Phthalates | DBP | 278 | 11.2 (b) | 3.7-5.2 (b) | $8.8 \times 10^{-7}$ (b) | Readily, especially under anaerobic conditions |
| Phthalates | BBP | 312 | 2.7 (b) | 3.6-4.9 (b) | $7.6 \times 10^{-7}$ (b) | Readily |
| BPA | BPA | 228 | 120 (c) | 3.4 (c) | $1.0 \times 10^{-10}$ (c) | Fairly to readily |
| PBDEs | Various congeners | 565-959 | ~1 (d) | 4.3-9.9 (d) | $2.1 \times 10^{-4}$ to $7.3 \times 10^{-8}$ (d) | Low, anaerobically or aerobically |

Notes:
(a) $K_{ow}$ = Octanol-water equilibrium constant.
(b) Staples *et al.* 1997
(c) Staples *et al.* 1998
(d) UNEP chemicals 2002

Photolysis

Photolysis, bond cleavage or molecular rearrangement, of trace level contaminants can occur at significant rates in surface waters (Lin et al. 2005). Photolysis rates are highly dependent on contaminant structure, depth of photic zone, and strength of sunlight. Direct photolysis, or photolysis directly from light energy, occurs only when the contaminant can absorb the sun's ultra-violet (UV) energy. Indirect photolysis involves light absorption by other organic or inorganic constituents, which further react with the contaminant. In the aqueous matrix, indirect photolysis is highly dependent on water quality.

The three phthalates and Bisphenol A photolysis rates are too low to impact concentrations in surface waters. PBDEs are much more photosensitive; 95% DeBDE photolysis was documented to occur in less than one hour (Bezares-Cruz et al. 2004). In this study, DeBDE was observed to photodegrade to over 40 different lower-brominated PBDE congeners, some of which are more toxic and/or more bioaccumlative.

Volatilization

Henry's Law constant is a good measure of a contaminants propensity to volatilize in waters exposed to atmosphere. Contaminants with a Henry's Law constant less than that of water, $1\text{x}10^{-7}$ atm-$m^3$/mol, are considered not very volatile.

The lesser brominated PBDEs are more volatile than the greater brominated PBDEs, the selected phthalates are moderately volatile and Bisphenol A is not very volatile.

## ATTENUATION IN ENGINEERED SYSTEMS

Water quality goals of wastewater and water treatment are not based on trace level contaminant removal. That said, mechanisms present in treatment plants can remove contaminants to very low levels. Generally, engineered systems capitalize on the same physical chemical characteristics of contaminants; removal mechanisms are maximized by design but residence times are typically much shorter. Greater levels of treatment typically leads to greater trace contaminant removal. All things equal, concentrations of trace contaminants are higher in secondary treated wastewaters than tertiary treated waters. Furthermore, advanced treatment (i.e., membrane filtration) has the capability to decrease trace contaminant concentrations by two to four orders of magnitude (Debroux, 2006).

Within the water industry, engineered systems comprise of physical, biological, and chemical treatment. Physical treatments (e.g., screening, settling, clarification, coagulation, filtration, etc.) remove solids from the aqueous matrix. Hydrophobic contaminants, those with high $K_{OW}$ values can be significantly removed from the product water. Transferring contaminants to a solid waste stream does not guarantee their isolation from the environment. The concern of less biodegradable contaminants in biosolids continues to grow.

**Table 3: Behavior of Plastic Additives in Natural Systems.**

| Compound Family | Compound | Adsorption | Biodegradation | | | Aqueous Photodegradation |
|---|---|---|---|---|---|---|
| | | | Surface water | Sediments | Wastewater | |
| Phthalates | DEHP | 17-35% of DEHP in sewage removed by sedimentation[a] | Mean half life of 78 days. Some samples up to >99% degradation in 20 days. [a] | Average primary degradation half life of 3 months. [a] | Primary aerobic degradation up to 95% in 1 week. [a] | 0.12-1.5 years [a] |
| Phthalates | DBP | | Exceeds 90% in 1 week. [a] | Primary aerobic degradation of 30-100% over 1-6 weeks. [a] | Primary aerobic degradation of 70-100% over 1 week. Anaerobic longer at 84% over 80 days. [a] | 2.4-12 years [a] |
| Phthalates | BBP | | Exceeds 90% in 1 week. [a] | Typically exceeds 90% degradation in 1 week. [a] | Typically exceeds 90% degradation in 1 week. [a] | >100 days [a] <5% degradation over 1 month [b] |
| BPA | BPA | <50% of mass associated with sediments[c] | 99% degradation in 4 days. [d] | | | Little direct photolysis |
| PBDEs | Various congeners | 96% of mass associated with sediments[e] | Very little to none | Very little to none | Very little to none | 99% reduction in 30 minutes; >40 less-brominated PBDE congeners were found to be degradation products. [f] |

Notes:
(a) Staples *et al.* 1997.
(b) Gledhill *et al.* 1980.
(c) Staples *et al.* 1998
(d) Dorn *et al.* 1987.
(e) North, 2004.
(f) Bezares-Cruz *at al.*, 2004.

Adsorption is often the first step to biodegradation. As contaminants concentrate on a solid surface, often rich with bacteria, conditions are often more favorable for acclimation and degradation. This is often the case with wastewater treatment (e.g., activated sludge, trickling filters) and water treatment (e.g., activated carbon filtration progressing to biologically active activated carbon). Biodegradation can be a sustainable mechanism for contaminant attenuation, but often only the parent contaminant degrades leaving numerous benign and/or equally problematic daughter metabolites.

Chemical wastewater and water treatment typically consists of adding oxidants or disinfectants to water. In addition, advanced oxidation processes (AOPs), which encourage the generation of hydroxyl radicals, are used to treat waters. Chlorine, ozone, UV light and AOPs have been shown to be effective ways of decreasing trace contaminant concentrations (Snyder et al. 2003). Chlorine can oxidize organic compounds but can also substitute chlorine atoms for hydrogen atoms resulting in chlorinated analogues that may or may not be more toxic than the original contaminant

(Alum et al. 2004). Ozone can completely oxidize contaminants to $CO_2$ but rarely does as ozone doses are targeted for pathogen removal rather than complete contaminant oxidation. UV light is typically used for pathogen inactivation; significantly greater doses and a larger spectrum of light energy are needed to effectively remove trace contaminants. Hydroxyl radicals, generated by AOPs, are powerful and less selective oxidants. However, generation is again rarely focused on complete contaminant oxidation.

Table 4 presents estimated removal of select plastic additives through engineered treatment processes.

**Table 4: Estimated Removal of Plastic Additives in Engineered Systems.**

| Compound Family | Compound | Coag /floc[a] | AC[b] | RO[c] | $Cl_2$[d] | UV[e] | $O_3$[f] |
|---|---|---|---|---|---|---|---|
| Phthalates | DEHP | Poor | Fair to Good | Excellent | Good[g] | Fair to good[h] | Good to Excellent |
| Phthalates | DBP | Poor | Fair to Good | Excellent | Good[g] | Fair to good[h] | Good to Excellent |
| Phthalates | BBP | Poor | Fair to Good | Excellent | Good[g] | Fair to good[h] | Good to Excellent |
| BPA | BPA | Poor | Fair to Good | Excellent | Good[g] | Fair to good[h] | Good to Excellent |
| PBDEs | Various congeners | Fair | Good | Excellent | Fair | Good[i] | Good to Excellent |

Notes:
(a) Coagulation/flocculation
(b) AC = Activated carbon
(c) RO = Reverse osmosis membrane filtration
(d) $Cl_2$ = Chlorination
(e) UV = Ultra-Violet light radiation
(f) $O_3$ = Ozone
(g) Potentially problematic chlorinated analogues may form
(h) UV dose and wavelength must be optimized
(i) DeBDE is very photosensitive; other lower-brominated PBDEs are formed in the process

## PLASTIC ADDITIVES LEVELS IN RECEIVING AND DRINKING WATERS

The combination of wastewater treatment, natural attenuation, and water treatment does a good job in reducing trace contaminants to sub-ng/L levels in drinking water. In addition, drinking water sources are chosen to minimize impacts to waste, where that option is available. Regardless, some compounds persist. Recently, DEHP and other trace contaminants were found in drinking waters in southern California, USA, where source waters are impacted by human activity (Loraine and Pettigrove, 2006). Similar findings will surely follow.

Unfortunately, aquatic wildlife does not have the benefit of all of these removal mechanisms as they can reside in receiving waters that possess many trace contaminants at higher concentrations than those found in drinking water.

## REINTRODUCTION OF PLASTIC ADDITIVES PRIOR TO CONSUMPTION

This paper's review of trace contaminants in water ends at drinking water leaving the treatment plant. It does not discuss the reintroduction of plastic additives due to contact with plastic in the distribution system, in plastic storage containers, and in conjunction with plastic use in the home which can be significant.

## REFERENCES

1. Alum, A., Yoon, Y., Westerhoff, P. and Abbaszadegan M., 2004. "Oxidation of bisphenol A, 17β-estradiol, and 17α-ethynyl estradiol and byproduct estrogenicity." *Environmental Toxicology*, **19**:257-264.
2. Bezares-Cruz, J., Jafvert, C.T. and Hua, I., 2004. "Solar Decomposition of Decabromodiphenyl Ether: Products and Quantum Yield." *Environmental Science & Technology*, **38**:15:4149-4156.
3. Debroux, 2006, unpublished data.
4. Dorn P., Chou C.-S., Gentempo J.J., 1987. "Degradation of Bisphenol A in Natural Waters." *Chemosphere*, **16**:7:1501-1507.
5. Fromme, H., Küchler, T., Otto, T., Pilz, K., Müller, and Wenzel A., 2002. "Occurrence of phthalates and bisphenol A and F in the environment." *Water Research*, **36**:1429-1438.
6. Gledhill, W.E., Kaley, R.G., Adams, W.J., Hicks, O., Michael, P.R. and Saeger, V.W., 1980. "An environmental safety assessment of Butyl Benzyl Phthalate." *Environmental Science & Technology*, **14**:3:301-305
7. Ikonomou, M.G., Rayne, S., Fischer, M., Fernandez, M.P. and Cretney, W., 2002. "Occurrence and congener profiles of polybrominated diphenyl ethers (PBDEs) in environmental samples from coastal British Columbia, Canada." *Chemosphere*, **46**:649-663.
8. Kolpin, D.W., Furlong, E.T., Meyer, M.T., Thurman, E.M., Zaugg, S.D., Barber, L.B. and Buxton, H.T., 2002. "Pharmaceuticals, Hormones, and Other Organic Wastewater Contaminants in U.S. Streams, 1999-2000: A National Reconnaissance." *Environmental Science & Technology*, **36**:6:1202-1211.
9. Lin, Angela Y-C. and Reinhard, Martin, 2005. "Photodegradation of common environmental pharmaceuticals and estrogens in river water." *Environmental Toxicology and Chemistry*, **24**:6:1303-1309.
10. Loraine, G.A. and Pettigrove, M.E., 2006. "Seasonal Variations in Concentrations of Pharmaceuticals and Personal Care Products in Drinking Water and Reclaimed Wastewater in Southern California." *Environmental Science & Technology*, **40**:687-695.
11. Marttinen, S.K., Kettunen, R.H. and Rintala, J.A, 2003. "Occurrence and removal of organic pollutants in sewages and landfill leachates." *The Science of the Total Environment*, **301**:1-12.
12. North, K., 2004. "Tracking Polybrominated Diphenyl Ether Releases in Wastewater Treatment Plant Effluent, Palo Alto, California." *Environmental Science & Technology*, **38**:17:4484-4488.

13. Oros, D.R., Hoover, D., Rodigari, F., Crane, D. and Sericano, J., 2005. "Levels and Distribution of Polybrominated Diphenyl Ethers in Water, Surface Sediments, and Bivalves from the San Francisco Estuary." *Environmental Science & Technology*, **39**:1:33-41.
14. Snyder, S.A., Westerhoff, P., Yoon, Y. and Sedlak, D.L., 2003. "Pharmaceuticals, Personal Care Products, and Endocrine Disruptors in Water: Implications for the Water Industry/" *Environmental Engineering Science*, **20**:5:449-469.
15. Staples, C.A., Peterson, D.R., Parkerton, T.F. and Adams, W.J., 1997. "The Environmental Fate of Phthalate Esters: A Literature Review." *Chemosphere*, **35**:4:667-749.
16. Staples, C.A., Dorn, P.B., Klecks, G.M., O'Block, S.T. and Harris, L.R., 1998. A "Review of the Environmental Fate, Effects, and Exposures of Bisphenol A." *Chemosphere*, **36**:10:2149-2173.
17. UNEP Chemicals, 2002, Regional Reports of the Regionally Based Assessment of Persistent Toxic Substances Program, http://www.chem.unep.ch/pts
18. Washing State Department of Ecology, 2006, Washington State Polybrominated Diphenyl Ether (PBDE) Chemical Action Plan, http://www.ecy.wa.gov/biblio/0507048.html
19. Wild, D. and Reinhard, M., 1999. "Biodegradation Residual of 4-Octylphenoxyacetic Acid in Laboratory Columns under Groundwater Recharge Conditions." *Environmental Science & Technology*, **33**:24:4422-4426.

# LEACHING OF BISPHENOL A FROM POLYCARBONATE PLASTIC DISRUPTS DEVELOPMENT VIA EPIGENETIC MECHANISMS

FREDERICK S. VOM SAAL, JULIA A. TAYLOR, BENJAMIN L. COE, JAMES R. KIRKPATRICK, MAREN E. BELL, JIUDE MAO
Division of Biological Sciences, University of Missouri
Columbia, MO, USA

WADE V. WELSHONS
Department of Biomedical Sciences, University of Missouri
Columbia, MO, USA

STEFANO PARMIGIANI
Department of Evolutionary and Functional Biology
University of Parma, Parma, Italy

## ABSTRACT

Bisphenol A (BPA) is the monomer used to manufacture polycarbonate plastic and is produced in excess of 6-billion pounds per year. This is an unstable polymer, and significant leaching of BPA into the environment occurs. Animal research has shown a myriad of adverse health effects due to exposure to BPA, and while only a few epidemiological studies have been conducted, they confirm the animal findings. BPA is referred to as an endocrine disrupting chemical, since BPA mimics the activity of the endogenous hormone estradiol. Similar to other hormones, hormonally active drugs and other endocrine disrupting chemicals, BPA can permanently alter gene activity when exposure occurs during "critical periods" in cell differentiation. The mechanism by which this "programming" of genes occurs is via "epigenetic" modification of the proteins associated with genes, as well as covalent addition of methyl groups to specific sites in the promoter region of genes. The mechanisms by which BPA and other chemicals in plastic cause adverse effects are thus known in considerable detail.

## INTRODUCTION

Bisphenol A (BPA) is one of the highest volume chemicals in worldwide production, with annual production capacity exceeding 6-billion pounds in 2003 (Burridge 2003). BPA is not a plastic additive and, instead, is the monomer used to manufacture polycarbonate plastic, the resin that lines metal cans, and dental sealants. In addition, BPA is used as an additive (plasticizer) in other types of plastic. All human fetuses that have been examined have measurable blood levels of BPA (Welshons et al. 2006). Mean and median levels of BPA found in human fetuses are higher than levels in fetal mice due to maternal exposure to very low doses of BPA (Zalko et al. 2002) that disrupt development of the brain and behavior, reproductive system, immune function, and body weight homeostasis; these effects, as well as prostate cancer, are often not detected until later in adult life (Ho et al. 2006; vom Saal and Welshons 2006).

EVIDENCE FOR LEACHING OF BPA FROM POLYCARBONATE PLASTIC AND CANS

The amount of leaching of BPA from products has to be considered in relation to its potency as an endocrine disrupting chemical. BPA is a chemical that can bind to and activate estrogen receptors. The "classical" receptors for estrogen are found in the cell nucleus associated with specific genes, and, in addition, receptors for estrogen that are associated with the cell membrane have more recently been discovered (Welshons et al. 2006). In a variety of tissues that express the receptors associated with the cell membrane, BPA not only has the efficacy of estradiol, but is also equally potent, with changes in cell function being observed at the extremely low dose of 0.23 pg/ml (0.23 parts per trillion) culture medium (Wozniak et al. 2005).

Exposure of humans and wildlife occurs due to the fact that when BPA molecules are polymerized to make polycarbonate plastic, they are linked by ester bonds that are subject to hydrolysis, which is accelerated as temperature increases and in response to contact with acidic or basic substances (Figure 1). The consequence is that as polycarbonate products are repeatedly washed, or polycarbonate plastic or metal cans are exposed to heat and/or acidic or basic conditions, significant leaching of BPA due to hydrolysis of the ester bond occurs (Welshons et al. 2006). In numerous studies virtually everyone examined has measurable and significant levels of bioactive (unconjugated) BPA in the low part per billion range (Ikezuki et al. 2002; Schonfelder et al. 2002; Calafat et al. 2005). There are many sources of human exposure due to leaching of BPA from products: plastic food and beverage containers, other polycarbonate products that leach BPA into landfill and subsequently into ground water, dental sealants, and the lining of metal cans (Welshons et al. 2006). A consequence of the leaching of BPA from products and the massive amounts produced each year is that BPA is detected in rivers and streams (Kolpin et al. 2002), in drinking water (Kuch and Ballschmiter 2001), and in indoor air (Rudel et al. 2001).

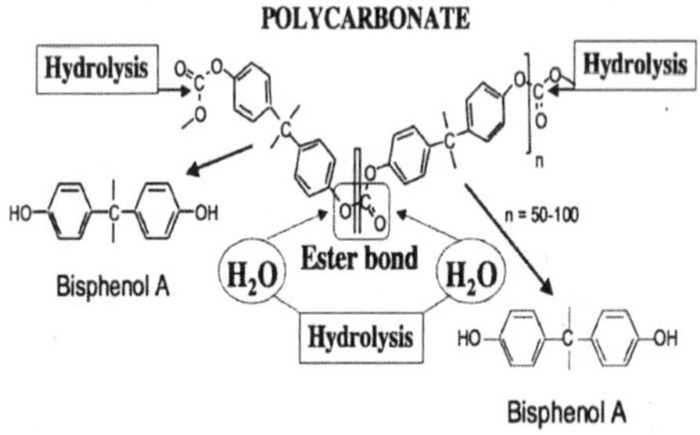


*Figure 1. Schematic diagram showing the ester bond linking bisphenol A (BPA) molecules to form polycarbonate plastic. The rate of hydrolysis of the ester bond, resulting in leaching of free BPA molecules from polycarbonate, increases as a function of temperature and either acidic or basic conditions.*

## HEALTH EFFECTS DUE TO FETAL EXPOSURE TO VERY LOW DOSES OF BPA

Exposure during gestation and lactation to BPA has been shown to result in a wide range of effects observed during postnatal life in mice and rats, and exposure during development also has adverse effects in other vertebrate and invertebrate species (vom Saal and Hughes 2005; vom Saal and Welshons 2006). For example, we initially reported (Howdeshell et al. 1999), and other studies have confirmed (Takai et al. 2000; Rubin et al. 2001; Markey et al. 2003; Akingbemi et al. 2004; Nikaido et al. 2004), that prenatal exposure to very low doses of BPA increases the rate of postnatal growth in mice and rats. Similarly, neonatal exposure to a low dose (1 μg/kg/day) of the estrogenic drug diethylstilbestrol (DES) also stimulated a subsequent increase in body weight and an increase in body fat in mice (Newbold et al. 2004). Related to these findings is the report that in vitro, a 2 μg/ml dose of BPA accelerated the conversion of mouse 3T3-L1 fibroblast cells into adipocytes, and also increased lipoprotein lipase activity and triacylglycerol accumulation; BPA resulted in the presence of larger lipid droplets in the differentiated cells (Masuno et al. 2002). Insulin and BPA interacted synergistically to further accelerate these processes. In a related study BPA stimulated an increase in the glucose transporter GLUT4 and glucose uptake into 3T3-F442A adiposities in cell culture (Sakurai et al. 2004). In a separate study up-regulation of GLUT4 increased basal and insulin-induced glucose uptake into adipocytes (Deems et al. 1994). In addition, very low doses of BPA stimulated rapid secretion of insulin in mouse pancreatic ß cells in primary culture through the cell membrane-associated estrogen receptor. In contrast, prolonged exposure to a low oral dose of BPA (10 μg/kg/day) resulted in stimulation of insulin secretion in adult mice that was mediated by the classical nuclear estrogen receptors; the prolonged hypersecretion of insulin was followed by insulin resistance (Alonso-Magdalena et al. 2006). Taken together, these findings suggest that developmental exposure to BPA is contributing to the obesity epidemic that has occurred during the last two decades in the developed world, associated with the dramatic increase in the amount of plastic being produced each year.

BPA also causes effects on the reproductive system. Exposure during fetal development (via the mother) to very low doses of BPA results in a permanent decrease in testicular sperm production (vom Saal et al. 1998) as well as enlargement of the prostate in male mice, which is due to an increase in the rate of the proliferation of basal epithelial cells; these are the progenitor cells thought to be the source of prostate cancer as males age (Nagel et al. 1997; Gupta 2000; Timms et al. 2005). Consistent with these findings, male rats administered a very low dose of BPA during early postnatal life all developed prostate cancer in adulthood (Ho et al. 2006). Thus, in addition to causing reproductive abnormalities, BPA is an animal carcinogen (Huff 2001). BPA exposure during development also dramatically stimulates mammary gland duct growth and causes changes detected during later adulthood; these changes are precancerous and suggest that BPA could contribute to breast cancer (Munoz-de-Toro et al. 2005). BPA also disrupts brain development (Belcher et al. 2005) and causes hyperactivity (Ishido et al. 2004) and changes in learning (Kubo et al. 2001).

## EPIGENETIC CHANGES MEDIATE SOME DEVELOPMENTAL EFFECTS OF BPA AND OTHER ENDOCRINE DISRUPTING CHEMICALS

Estrogens and other sex hormones regulate the functioning of tissues in adults, and these are termed "activational" effects that only occur when the hormone is present. However, when exposure occurs during the critical periods in development when cells are differentiating estrogens and other hormones as well as hormone-mimicking chemicals cause permanent changes termed "organizational" effects. Extensive research is currently directed at elucidating the mechanisms by which genes are "programmed" during cell differentiation under the influence of hormones such as estradiol, as well as endocrine disrupting chemicals such as BPA. The mechanisms that determine which genes in a cell are able to be transcribed, as well as the level at which transcription occurs, involves "epigenetic" modifications of DNA as well as the associated histone proteins. A schematic depicting "imprinting" mechanisms that determine whether specific genes are silenced or activated is shown in Figure 1 (Weinhold 2006). Specifically, activation of methyltransferase and addition of methyl groups within the promoter region of genes results in the inability of an activated transcription factor (such as the estrogen receptor bound to BPA) to initiate transcription of the gene. This is also typically associated with modifications of the histone proteins that are involved in determining whether DNA can be transcribed, with the combination of loss of acetyl groups and DNA methylation being associated with "gene silencing." The consequence is equivalent to a mutation that renders a gene product non-functional, but standard analyses that only examine whether a chemical has altered the sequence of bases that make up the "genetic code" will not reveal these "epigenetic" chemical modifications. Instead, analytical techniques that reveal whether these chemical modifications have occurred are required. There are data indicating that it is via these epigenetic changes that BPA "programs" gene activity during critical periods in development and determines whether male rats develop prostate cancer later in life (Ho et al. 2006).

Figure 2. Schematic diagram showing the "epigenetic" chemical modification of histone proteins by removal of acetyl groups as well as modification of cytosine bases by addition of methyl groups that result in repression of gene transcription. Epigenetic changes that occur early in development are transmitted to daughter cells during mitosis and thus permanently alter gene transcription in tissues (Modified from: Weinhold 2006).

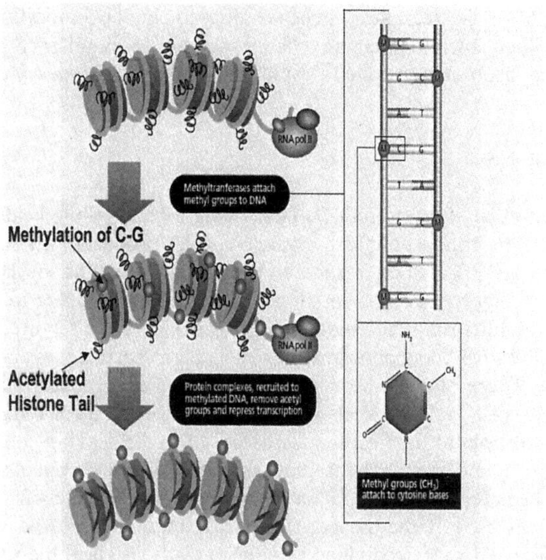


EVOLUTIONARY IMPLICATIONS OF HERITABLE EPIDENETIC CHANGES CAUSED BY ENVIRONMENTAL CHEMICALS

In addition to the clear importance of developmental exposure to environmental endocrine disrupting chemicals for health and disease of individual, recent discoveries about the mechanisms of action of these chemicals are challenging the traditionally accepted neo-Darwinian theory of evolution. Neo-Darwinian theory, also known as the "modern synthesis," states that a change in the base sequence of DNA (a point mutation, which is very rare) is the basis of the variability upon which natural selection operates very gradually over very long periods of time. Up until recently this view was generally accepted. In this view the basis of macroevolutionary processes is the gradual accumulation of mutations, which are selected by the particular environment in which the organism lives. Thus, the environment is not actively creating "adaptations" but simply selecting traits most advantageous for that particular environment (Darwin's concept of "survival of the fittest"). However, in the 1970s, the "gradualist" view of evolution was challenged in a series of papers by the paleontologists Eldridge and Gould. Evidence against "gradualism" was the absence from the fossil record of "intermediate" forms. Rather, there appeared to be periods of evolutionary "stasis" followed by very rapid shifts in traits and the origin of entirely new taxa (macrovolution). This model of evolution is called "punctuated equilibrium." Until recently, no molecular mechanism existed to explain a basis for the hypothesis of punctuated equilibrium, even though the very low frequency of point mutations could not possibly account for what appeared to be marked rapid change leading to macrovolutionary events. Now, as described above there is evidence that environmental contaminants can act via epigenetic mechanisms to actively "silence" or "activate" major regulatory genes leading to chemical modifications in genes (not mutations). When such epigenetic changes occur in regulatory genes and are heritable (which has been demonstrated), rapid and dramatic changes in phenotype can occur.

## CONCLUSIONS

The leaching of BPA from products poses a significant public health threat due to the massive amount (>6-billion pounds per year) of BPA produced, the instability of the polymers made from BPA, which results in significant leaching of BPA into food, beverages, the water supply and into air. Significant human exposure to BPA has been documented, and a number of small epidemiological studies have reported a relationship between blood levels of BPA and abnormalities such as miscarriage, ovarian disease, and obesity in humans. These studies were all conducted after similar findings had been reported in animals (vom Saal and Welshons 2006). However, there remains significant confusion by the public about the health hazards posed by BPA. This confusion is based on a concerted effort by the chemical corporations that produce chemicals used in plastic to create what has been termed "manufactured uncertainty" (Michaels 2005). This is the same strategy that was used by the tobacco industry in their attempt to discredit research showing a relationship between secondhand smoke and adverse health effects. In fact, the same groups who managed this campaign for the tobacco industry are now employed by the plastic industry to disseminate misinformation (vom Saal 2005).

The other strategy used by the chemical industry is to fund research that always leads to the conclusion that BPA is safe at virtually any dose. As shown in Table 1, in sharp contrast, the majority (>90%) of studies funded by government agencies find BPA to cause a wide range of adverse health effects at human exposure levels. This campaign to "manufacture uncertainty" has been successful in that regulatory agencies in both the EU and USA have ignored the massive literature by independent scientists showing that BPA causes adverse health effects. These agencies, such as the U.S. EPA and U.S. FDA, which are under intense political pressure to ignore health hazards posed by chemicals in commerce, will only recognize BPA and other plastics as a health problem when the public demands action.

**Table 1. Biased distribution of outcomes in research conducted with low doses of BPA in animals.**

| | **Study Outcome** | | |
|---|---|---|---|
| **Source of Funding** | **Harm** | **No Harm** | **Total** |
| Government | 138 (93%) | 11 (7%) | 149 |
| Chemical Corporations | 0 (0%) | 12 (100%) | 12 |
| | 138 | 23 | 161 |

The published studies and abstracts used in this analysis are available in a document online at: http://endocrinedisruptors.missouri.edu/vomsaal/vomsaal.html.

ACKNOWLEDGEMENTS

Funding during the preparation of this manuscript was provided to FvS by NIEHS (ES11283).

REFERENCES

1. Akingbemi, B.T., Sottas, C.M., Koulova, A.I., Klinefelter, G.R. and Hardy, M.P. (2004). "Inhibition of testicular steroidogenesis by the xenoestrogen bisphenol A is associated with reduced pituitary luteinizing hormone secretion and decreased steroidogenic enzyme gene expression in rat Leydig cells." *Endocrinol.* **145**:592-603.
2. Alonso-Magdalena, P., Morimoto, S., Ripoll, C., Fuentes, E. and Nadal, A. (2006). "The estrogenic effect of bisphenol A disrupts pancreatic beta-cell function in vivo and induces insulin resistance." *Environ. Health Perspect.* 114: 106-112.
3. Belcher, S.M., Le, H.H., Spurling, L. and Wong, J.K. (2005). "Rapid estrogenic regulation of extracellular signal- regulated kinase 1/2 signaling in cerebellar granule cells involves a G protein- and protein kinase A-dependent mechanism and intracellular activation of protein phosphatase 2A." *Endocrinol.* **146**:5397-S406.
4. Burridge, E. (2003). "Bisphenol A: Product Profile." *European Chemical News* April 14-20: 17.
5. Calafat, A.M., Kuklenyik, Z., Reidy, J.A., Caudill, S.P., Ekong, J. and Needham, L.L. (2005). "Urinary concentrations of bisphenol A and 4-nonyl phenol in a human reference population." *Environ. Health Perspect.* **113**:391-395.
6. Deems, R.O., Evans, J.L., Deacon, R.W., Honer, C.M., Chu, D.T., Burki, K., Fillers, W.S., Cohen, D.K. and Young, D.A. (1994). "Expression of human GLUT4 in mice results in increased insulin action". *Diabetologia* **37**:1097-1104.
7. Gupta, C. (2000). "Reproductive malformation of the male offspring following maternal exposure to estrogenic chemicals." *Proc. Soc. Exp. Biol. Med.* **224**:61-68.
8. Ho, S.M., Tang, W.Y., Belmonte de Frausto, J. and Prins, G.S. (2006). "Developmental exposure to estradiol and bisphenol a increases susceptibility to prostate carcinogenesis and epigenetically regulates phosphodiesterase type 4 variant 4." *Cancer Res.* **66**:5624-5632.
9. Howdeshell, K.L., Hotchkiss, A.K., Thayer, K.A., Vandenbergh, J.G. and vom Saal, F.S. (1999). "Exposure to bisphenol A advances puberty." *Nature* **401**:763-764.
10. Huff, J. (2001). "Carcinogenicity of bisphenol-A in Fischer rats and B6C3F1 mice." *Odontology* **89**:12-20.
11. Ikezuki, Y., Tsutsumi, O., Takai, Y., Kamei, Y. and Taketani, Y. (2002). "Determination of bisphenol A concentrations in human biological fluids reveals significant early prenatal exposure." *Human Reprod.* **17**:2839-2841.

12. Ishido, M., Masuo, Y., Kunimoto, M., Oka, S. and Morita, M. (2004). "Bisphenol A causes hyperactivity in the rat concomitantly with impairment of tyrosine hydroxylase immunoreactivity." *J. Neurosci. Res.* **76**:423-33.

13. Kolpin, D.W., Furlong, E.T., Meyer, M.T., Thurman, E.M., Zaugg, S.D., Barber, L.B. and Buxton, H.T. (2002). "Pharmaceuticals, hormones, and other organic wastewater contaminants in U.S. streams, 1999-2000: A national survey." *Environ. Sci. Technol.* **36**:1202-1211.

14. Kubo, K., Arai, O., Ogata, R., Omura, M., Hori, T. and Aou, S. (2001). "Exposure to Bisphenol A during the fetal and suckling periods disrupts sexual differentiation of the locus coeruleus and of behaviour in the rat." *Neurosci. Lett.* **304**:73-76.

15. Kuch, H.M. and Ballschmiter, K. (2001). "Determination of endocrine-disrupting phenolic compounds and estrogens in surface and drinking water by HRGC-(NCI)-MS in the picogram per liter range." *Environ. Sci. Technol.* **36**:3201-3206.

16. Markey, C.M., Coombs, M.A., Sonnenschein, C. and Soto, A.M. (2003). "Mammalian development in a changing environment: exposure to endocrine disruptors reveals the developmental plasticity of steroid-hormone target organs." *Evol. Dev.* **5**:67-75.

17. Masuno, H., Kidani, T., Sekiya, K., Sakayama, K., Shiosaka, T., Yamamoto, H. and Honda, K. (2002). "Bisphenol A in combination with insulin can accelerate the conversion of 3T3-L1 fibroblasts to adipocytes." *J. Lipid Res.* **43**:676-684.

18. Michaels, D. (2005)."Doubt is their product." *Sci. Amer.* **292**:96-101.

19. Munoz-de-Toro, M., Markey, C.M., Wadia, P.R., Luque, E.H., Rubin, B.S., Sonnenschein, C. and Soto, A.M. (2005). "Perinatal exposure to bisphenol-A alters peripubertal mammary gland development in mice." *Endocrinol.* **146**:4138-4147.

20. Nagel, S.C., vom Saal, F.S., Thayer, K.A., Dhar, M.G., Boechler, M. and Welshons, W.V. (1997). "Relative binding affinity-serum modified access (RBA-SMA) assay predicts the relative in vivo bioactivity of the xenoestrogens bisphenol A and octylphenol." *Environ. Health Perspect.* **105**:70-6.

21. Newbold, R.R., Jefferson, W.N., Padilla-Banks, E. and Haseman, J. (2004). "Developmental exposure to diethylstilbestrol (DES) alters uterine response to estrogens in prepubescent mice: low versus high dose effects." *Reprod. Toxicol.* **18**:399-406.

22. Nikaido, Y., Yoshizawa, K., Danbara, N., Tsujita-Kyutoku, M., Yuri, T., Uehara, N. and Tsubura, A. (2004). "Effects of maternal xenoestrogen exposure on development of the reproductive tract and mammary gland in female CD-1 mouse offspring." *Reprod. Toxicol.* **18**:803-811.

23. Rubin, B.S., Murray, M.K., Bamassa, D.A., King, J.C. and Soto, A.M. (2001). "Perinatal exposure to low doses of bisphenol A affects body weight, patterns of estrous cyclicity, and plasma LH levels." *Environ. Health Perspect.* **109**:657-680.

24. Rudel, R.A., Brody, J.G., Spengler, J.D., Vallarino, J., Geno, P.W., Sun, G. and Yau, A. (2001). "Identification of selected hormonally active agents and animal mammary carcinogens in commercial and residential air and dust samples." *J. Air Waste Manage. Assoc.* **51**:499-513.

25. Sakurai, K., Kawazuma, M., Adachi, T., Harigaya, T., Saito, Y., Hashimoto, N. and Mori, C. (2004). "Bisphenol A affects glucose transport in mouse 3T3-F442A adipocytes." *Br. J. Pharmacol.* **141**:209-14.
26. Schonfelder, G., Wittfoht, W., Hopp, H., Talsness, C.E., Paul, M. and Chahoud, I. (2002). "Parent bisphenol A accumulation in human maternal-fetal-placental unit." *Environ. Health Perspect.*. **110**:A703-A707.
27. Takai, Y., Tsutsumi, O., Ikezuki, Y., Kamei, Y., Osuga, Y., Yano, T. and Taketan, Y. (2000). "Preimplantation exposure to bisphenol A advances postnatal development." *Reprod. Toxicol.* **15**:71-74.
28. Timms, B.G., Howdeshell, K.L., Barton, L., Bradley, S., Richter, C.A. and vom Saal, F.S. (2005). "Estrogenic chemicals in plastic and oral contraceptives disrupt development of the mouse prostate and urethra." *Proc. Natl. Acad. Sci.* **102**:7014-7019.
29. vom Saal, F.S. (2005)."Low-dose BPA: confirmed by extensive literature." *Chem. Ind.* **7**:14-15.
30. vom Saal, F.S., Cooke, P.S., Buchanan, D.L., Palanza, P., Thayer, K.A., Nagel, S.C., Parmigiani, S. and Welshons, W.V. (1998). "A physiologically based approach to the study of bisphenol A and other estrogenic chemicals on the size of reproductive organs, daily sperm production, and behavior." *Toxicol. Ind. Health* **14**:239-60.
31. vom Saal, F.S. and Hughes, C. (2005). "An extensive new literature concerning low-dose effects of bisphenol A shows the need for a new risk assessment." *Environ. Health Perspect.* **113**:926-933.
32. vom Saal, F.S. and Welshons, W.V. (2006). 'Large effects from small exposures. II. The importance of positive controls in low-dose research on bisphenol A." *Environ. Res.* **100**:50-76.
33. Weinhold, B. (2006). "Epigenetics: The science of change." *Environ. Health Perspect.* **114**:A160-A167.
34. Welshons, W.V., Nagel, S.C. and vom Saal, F.S. (2006). "Large effects from small exposures. III. Endocrine mechanisms mediating effects of bisphenol A at levels of human exposure." *Endocrinol.* **147**:S56-S69.
35. Wozniak, A.L., Bulayeva, N.N. and Watson, C.S. (2005). "Xenoestrogens at picomolar to nanomolar concentrations trigger membrane estrogen receptor-a mediated Ca++ fluxes and prolactin release in GH3/B6 pituitary tumor cells." *Environ. Health Perspect.* **113**:431-439.
36. Zalko, D., Soto, A.M., Dolo, L., Dorio, C., Rathahao, E., Debrauwer, L., Faure, R. and Cravedi, J.P. (2002). "Biotransformations of bisphenol A in a mammalian model: answers and new questions raised by low-dose metabolic fate studies in pregnant CD-1 mice." *Environ. Health Perspect.* **111**:309-319.

# HUMAN EXPOSURE TO PHTHALATES AND THEIR HEALTH EFFECTS

SHANNA H. SWAN, PH.D.
Center for Reproductive Epidemiology
University of Rochester, Rochester, USA

## INTRODUCTION

In this brief overview I will first present recent data on human exposure to phthalates—diesters of 1,2-benzenedicarboxylic acid (phthalic acid)—and then review what is known about their impacts on human health. I will draw on the rapidly growing literature on the toxicity of these ubiquitous compounds, as well as data from our own study of pregnant women and their children. There are dozens of phthalates many have not yet been examined toxicologically. This discussion is limited to the seven phthalates, and their nine urinary metabolites, that are currently monitored by the U.S. Centers for Disease Control and Prevention (CDC) (Centers for Disease Control and Prevention 2003; Centers for Disease Control and Prevention 2005).

## EXPOSURE TO PHTHALATES

### Uses of phthalates

This group of man-made chemicals has a wide spectrum of industrial applications and they are also widely used in personal care and other consumer products (Wormuth, M. et al. 2006). Phthalates have been measured in residential indoor environments in both house dust and indoor air (Rudel, R. et al. 2003). They have also been measured in foods, milk and drinking water. However, the relative contribution from the various sources and routes of exposure to phthalates is unknown (Wormuth, M. et al. 2006). Di(2-ethylhexyl) phthalate (DEHP), di-n-butyl phthalate (DBP), and butylbenzyl phthalate (BBzP) are considered to be the most reproductively toxic of the phthalates, with toxicity ranking: DHEP>DBP>BBzP.

High molecular weight phthalates, such as DEHP, are primarily used as plasticizers in the manufacture of polyvinyl chloride (PVC), which is used extensively in consumer products, flooring and wall coverings, as well as food contact applications, and medical devices (Agency for Toxic Substances and Disease, 2002). Exposure to DEHP can occur in the workplace (in the manufacture of DEHP or DEHP containing products, or workers using these products), during consumer use of these products, or through environmental media (food, air, water, dust). The inclusion of DEHP in children's toys, particularly those used by young children, and medical devices such as tubing and blood and nutrient bags are of particular concern because of the vulnerability of the exposed population.

Lower molecular weight phthalates (for example, DBP, BBzP and diethyl phthalate [DEP]) are used as solvents and plasticizers for cellulose acetate, in making lacquers, varnishes, personal-care products (e.g., perfumes, lotions, and cosmetics), and coatings, including those used in making timed-release pharmaceuticals (Agency for Toxic Substances and Disease 1995; Agency for Toxic Substances and Disease 2001).

DEP, whose urinary metabolite MEP is found in population samples at levels that are often an order of magnitude higher than those of DEHP and DBP (Centers for Disease Control and Prevention 2003; Centers for Disease Control and Prevention 2005), is commonly found in cosmetics and personal care products.

Routes of exposure to phthalates

Humans are exposed by all possible routes; orally (phthalate-contaminated food, water and other liquids and children through mouthing of toys and teethers), dermally (particularly cosmetics and other personal care products, but also in occupational settings), via inhalation (phthalates volatilizing from PVC, nail polish, hair spray, and other phthalate containing products and is found in indoor air (Rudel, R. et al. 2003)), and subdermally (medical tubing (FDA 2001)). In contrast, in almost all rodent studies to date, exposure is oral. Therefore, these rodent studies may not reflect toxicity of phthalates to humans who are exposed via other routes. For example, human exposure to DEP is primarily dermal and via inhalation. While most animal studies on DEP and MEP do not find reproductive toxicity (Agency for Toxic Substances and Disease 1995), several of the human studies that have examined this phthalate have found adverse effects (Duty, S.M. et al. 2004; Main et al. 2005; Swan et al. 2005), and this apparent inconsistency may be the result of differing exposure routes.

Human exposure to phthalates

Phthalates are rapidly metabolized in the body, with half-lives of metabolites of only days (Koch, H.M. et al. 2004). These chemicals have been measured in all body fluids and matrices, including; urine, serum, saliva, seminal fluid, breast milk, amniotic fluid, meconium and even placenta. However, because levels are higher in urine than in other samples, and because urinary metabolites can be measured free of the introduction of phthalate contamination during collection, storage and analysis, urine is the preferred matrix for phthalate determination in humans.

The CDC has published data on levels of nine phthalate metabolites in a large population-based sample of the U.S. population (Centers for Disease Control and Prevention 2003; Centers for Disease Control and Prevention 2005). These data demonstrate the ubiquitous nature of these chemicals and demonstrate variation in concentration by ethnicity, sex and age. However, these data do not include children less than six years of age. A European group (Wormuth et al. 2006) (Wormuth, M. et al. 2006) has modeled estimated phthalate levels in a European population, including infants, and estimates of exposure to DEHP and DBP for infants are about one order of magnitude higher than those in adults.

## THE STUDY OF PHTHALATES IN PREGNANT WOMEN AND CHILDREN

Description of the study

Women participating in the *Study of Phthalates in Pregnant Women and Children (PPWC)* were originally recruited at prenatal clinics in Los Angeles, CA (Harbor-UCLA and Cedars-Sinai), Minneapolis, MN (University of Minnesota Health Center) and Columbia, MO (University Physicians), between September 1999 and August 2002 (Swan et al. 2003). All couples whose pregnancy was not medically assisted were eligible, unless the woman or her partner was less than 18 years, either partner did not read and speak Spanish or English, or the father was unavailable or unknown. Women provided a urine sample at entry to the study (which took place, on average, when they were 28.6 weeks pregnant). When the baby was at least three months old, the mother and baby were invited to participate in PPWC. At the first study visit, the mother gave a urine sample and a sample of the baby's urine was collected.

After obtaining standard anthropometric measurements (height, weight, head circumference and skin-fold thickness), a detailed examination of the breast and genitals, developed specifically for this study, was conducted on both boys and girls, under the supervision of pediatric physicians who were trained in its administration (Swan et al. 2005). Boys' examinations included a description of the testes and scrotum, location of testes, and measurement of the penis. We also obtained two measures of anogenital distance (AGD, discussed below). The CDC measured phthalate metabolite concentrations in all urine samples using a sensitive method that involves the enzymatic deconjugation of the phthalate metabolites from their glucuronidated form, automated on-line solid-phase extraction, separation with high performance liquid chromatography, and detection by isotope-dilution tandem mass spectrometry (Silva et al. 2004). CDC had no access to subject data, and neither the pediatric physicians nor the support staff had any knowledge of the mother's or child's phthalate concentrations.

Exposure in mothers and babies

At the first study visit, each woman completed a questionnaire that included questions about her use of phthalate-containing products on herself or her child. Detailed questions regarding baby product use were phrased as, "We would like to know if you or anyone else has used any of the following products on your baby in the 24 hours prior to the time we collected his/her urine sample today." – yes/no. Product categories were: baby powder/talc/cornstarch, desitin/diaper creams, baby wipes, baby shampoo, and baby lotion. We used linear regression to explore the relationships between individual log phthalate metabolite concentration and mother's report of use of individual baby products. We categorized product use into low, moderate (any 2-3 products) and high (any 4-5 products) to obtain an analysis that examined the combined contribution of product use.

At least one phthalate metabolite was detectable in all 163 infant urine samples and seven or more urinary phthalate metabolites were above the limit of detection in over 80%, with MEP having the highest concentration (mean of 178.2 mg/L). Mothers' report of baby lotion use was predictive of MEP concentrations (R =1.7, 1.1–2.5) and MBZP concentrations (RR=1.4, 1.03, 2.0). When categorizing personal care product use by moderate (2-3 products), and high use (4-5 products), moderate (1.6, 1.2–2.2) and high personal care product use (1.9, 1.1–3.3) was related to MEP concentration in the baby's urine.

We examined urinary phthalate metabolites in the mother's sample given the same day as the baby's, to see how well the concentration of phthalate metabolites in the mother's urine predicted those in the baby's urine (Table 1).

**Table 1: Summary statistics for concentration (ng/mL) of selected phthalate metabolites in mother's prenatal, postnatal and baby's urine (mean (SD)) (N=235).**

| Phthalate | Metabolite | Mom's prenatal | Mom's postnatal | Baby's |
|---|---|---|---|---|
| DEHP | MEHP | 15.7 (105.3) | 12.8 (19.8) | 6.1 (14.9) |
| | MEOHP | 40.2 (217.1) | 33.2 (74.9) | 33.3 (143.5) |
| | MEHHP | 51.7 (322.2) | 45.7 (107.3) | 48.2 (42.1) |
| DEP | MEP | 807 (2872) | 378 (1057) | 155 (412) |
| DBP | MBP | 24.9 (29.9) | 28.1 (40.3) | 31.8 (45.2) |
| BBzP | MBzP | 21.1 (38.8) | 26.0 (33.7) | 29.6 (46.7) |

We modeled the baby's phthalate metabolite concentration as a function of the mother's at the post-natal visit, controlling for baby's age and creatinine in both samples. Concentration in the baby's urine was strongly predicted by that in the mother's sample for the two oxidative metabolites of DEHP (MEHHP, adjusted $R^2 = 0.63$ (P=0.0004) and MEOHP: $R^2 = 0.61$ (P=0.0004)) as well as somewhat predictive for MBP: $R^2$=0.53 (P=0.040), MEP: $R^2 = 0.34$ (P=0.001). This suggests that the mother and child share a common exposure to DEHP, and, to a lesser extent, DBP and DEP.

Women were also asked about their personal product use in the 24 hours prior to sampling. As has been shown in men (Duty, S.M. et al. 2005), women's MEP levels were significantly related to the number of personal products used in the 24-hour period preceding urine collection.

## REPRODUCTIVE OUTCOMES ASSOCIATED WITH PRENATAL EXPOSURE TO PHTHALATES

### The phthalate syndrome

Newborn male rodents have no scrotum, and the external genitalia are undeveloped only a genital tubercle is apparent for both sexes. The distance from the anus to the insertion of this tubercle, the AGD, is androgen dependent and about twice as long in males as in females (Moore et al. 2001). The AGD has been shown to be a sensitive measure of prenatal anti-androgen exposure (Rhees et al. 1997) and prenatal phthalate exposure impairs testicular function and shortens anogenital distance (AGD) in male rodents. Prior to our study (Swan et al. 2005), only a single study had evaluated AGD in males (Salazar-Martinez et al. 2004), and two other studies had evaluated AGD in female infants (Callegari et al. 1987; Phillip et al. 1996): ours was the first study to examine AGD in relation to *in utero* phthalate exposure though none examined AGD in relation to any prenatal exposure.

A standardized measure of AGD was obtained in children in PPWC at 2-30 months of age and adjusted by expected weight for age, as determined by national standard curves, to estimate the expected AGD. On average the male: female ratio in AGD was 1.6. AGD was examined in males in regression and categorical analyses. In these mother-son pairs urinary concentrations of MEP, MBP, and the three DEHP metabolites were significantly and inversely related to AGD.

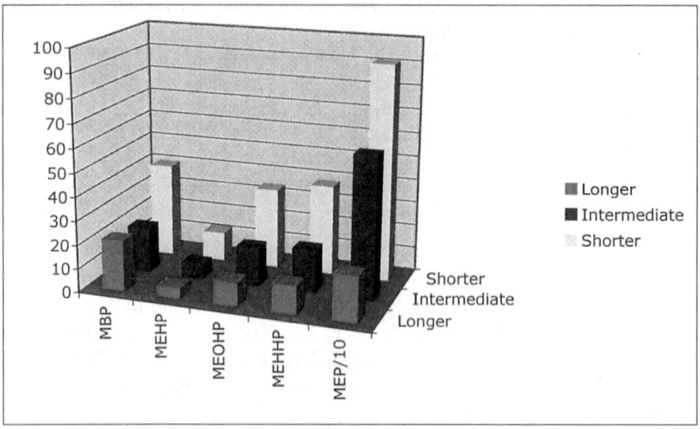


Adjusted AGD also decreased significantly with increases in a score that reflects joint exposure to these five phthalate metabolites, and the data suggested that the association with multiple phthalates is dose additive, as has been shown in rodents (Gray, L.E. Jr. et al. 2006). Shorter AGD was also significantly associated with smaller penile size and degree of testicular descent. These results are consistent with recent data linking hormone levels in male infants to phthalate metabolites in breast milk (Main et al. 2005). Together, these studies suggest that the phthalate syndrome originally identified in rodents (Gray, L.E. Jr. et al. 2006) may also be occurring in male babies whose mothers had higher levels of one or more phthalates while pregnant.

Phthalates and gestational age at delivery

Latini and coworkers measured serum DEHP and MEHP in the cord blood of 84 newborns born in Brindisi, Italy. Using logistic regression, they showed a correlation between the presence of MEHP in cord blood and earlier gestational age at delivery (odds ratio = 1.50, 95% CI, 1.013–2.21; p=0.043).

## OTHER HEALTH OUTCOMES ASSOCIATED WITH PHTHALATE EXPOSURE

A number of adverse outcomes have been associated with phthalate exposure. These health effects as well as sources of exposure to these phthalates are summarized in Table 2 and discussed in the text.

**Table 2: Sources of exposure and reported health effects of selected phthalates.**

| Phthalate/metabolites | Sources of exposure | Human health effects reported in one or more studies |
|---|---|---|
| Diethyl phthalate (DEP)/MEP | Personal care products, dyes, pharmaceuticals | Shortened AGD, sperm DNA damage, respiratory disease (adult males) |
| Di-n-butyl phthalate (DBP)/MBP | Cellulose acetate, nail polish, lacquers, medical coatings | Decreased sperm motility and concentration, respiratory disease (adult males) |
| Butylbenzyl phthalate (BBzP)/MBzP | Vinyl flooring, adhesives and sealants, industrial solvents | Decreased sperm concentration, respiratory disease (children) |
| Di(2-ethylhexyl) phthalate (DEHP)/MEHP, MEOHP, MEHHP | PVC in household products, floor tile, wall coverings), children's toys, medical devices | Prematurity, respiratory disease (children) |

Semen quality and related outcomes in males

Duty and colleagues have published four studies that examined urinary phthalate levels and semen characteristics, sperm DNA damage, and serum reproductive hormones (Duty, S.M. et al. 2004; Duty et al. 2005; Duty et al. 2003; Duty et al. 2003)

A number of studies have examined respiratory function (as defined by a range of symptoms and diseases) in children in relation to the use of PVC products in the home or phthalates in house dust (Bornehag, C.G. et al. 2004; Jaakkola, J.J.K. et al. 2000; Jaakkola, et al. 1999). These identified significant associations with PVC flooring and wall material as well as DEHP in house dust. One study of adults found associations with respiratory function, which was only seen in males (Hoppin, J.A. et al. 2004).

## CONCLUSIONS

In the few years (5-6) that sensitive biomarkers for assaying human phthalate exposure in humans have been available, a rapidly growing literature has demonstrated human toxicity of this ubiquitous exposure. Because epidemiology is limited in its ability to draw causal inferences, it must be informed by, and closely linked to animal studies in which mechanism and causality can be demonstrated. While several links between rodent and human studies have been made in this area, as discussed above, these are limited by several discrepancies between human and animal testing as shown in Table 3.

**Table 3: Discrepancies between human and animal testing.**

| Factor | Rodent studies | Human studies |
|---|---|---|
| Route of exposure | Oral | Oral, inhalation, dermal, parenteral |
| Dose of single phthalate | High or medium | Very low to low |
| Exposure | Single phthalate | Multiple phthalates |

Because of these important differences in route, dose level and multiplicity of agents, it is difficult to extrapolate across species. The magnitude of human health consequences cannot be accurately assessed studying chemicals by only one route of exposure or by one chemical at a time. For this reason, these new and more environmentally relevant scenarios should from the basis of future chemical risk assessment.

REFERENCES

Agency for Toxic Substances and Disease R. 1995. Toxicological profile for diethyl phthalate. Atlanta, GA:Agency for Toxic Substances and Disease Registry, Division of Toxicology. Available: http://www.atsdr.cdc.gov/toxprofiles/tp73.html [accessed 2005/02/23/.

---. 2001. Toxicological profile for di-n-butyl phthalate. Atlanta, GA:Agency for Toxic Substances and Disease Registry, Division of Toxicology. Available: http://www.atsdr.cdc.gov/toxprofiles/tp135.html [accessed 2005/02/23/.

---. 2002. Toxicological profile for di(2-ethylhexyl)phthalate (dehp) Atlanta, GA:Agency for Toxic Substances and Disease Registry, Division of Toxicology. Available: http://www.atsdr.cdc.gov/toxprofiles/tp9.html [accessed 2005/02/23/.

Bornehag, C.G., Sundell, J., Weschler, C.J. Al. E. 2004. "The association between asthma, and allergic symptoms in children and phthalates in house dust: A nested case-control study." *Environ Health Perspect* **112**:1393-1397.

Callegari, C., Everett, S., Ross, M., Brasel, J.A.. 1987. "Anogenital ratio: Measure of fetal virilization in premature and full-term newborn infants." *The Journal of Pediatrics* **111**(2):240-243.

Centers for Disease Control and P. 2003. Second national report on human exposure to environmental chemicals. Atlanta, GA:U.S. Department of Health and Human Services, Centers for Disease Control and Prevention, National Center for Environmental Health, Division of Laboratory Sciences.

---. 2005. Third national report on human exposure to environmental chemicals. Available: http://www.cdc.gov/exposurereport/3rd/ [accessed November 22, 2005.

Duty, S.M., Ackerman, R.M., Calafat, A.M., R.H. 2005. "Personal care product use predicts urinary concentrations of some phthalate monoesters." *Environ Health Perspect* **113**:1530-1535.

Duty, S.M., Calafat, A.M., Silva, M.J.. 2004. "The relationship between environmental exposure to phthalates and computer-aided sperm analysis motion parameters." *J Androl* **25**:293-302.

Duty, S.M., Calafat, A.M., Silva, M.J., Ryan, L., Hauser, R. 2005. "Phthalate exposure and reproductive hormones in adult men." *Human Reproduction* **20**(3):604-610.

Duty, S.M., Silva, M.J., Barr, D.B., Brock, J.W., Ryan, L., Chen, Z., et al. 2003. "Phthalate exposure and human semen parameters." *Epidemiology* **14**(3):269-277.

Duty, S.M., Singh, N.P., Silva, M.J., Barr, D.B., Brock, J.W., Ryan, L., et al. 2003. "The relationship between environmental exposures to phthalates and DNA damage in human sperm using the neutral comet assay." *Environmental Health Perspectives* **111**(9):1164-1169.

FDA. 2001. Safety assessment of di(2-ethylhexyl)phthalate (DEHP) released from PVC medical devices. Rockville, MD: U.S. Food and Drug Administration.

Gray, L.E. Jr., Wilson, V.S., Stoker, T., Lambright, C., Furr, J., Noriega, N., et al. 2006. "Adverse effects of environmental antiandrogens and androgens on reproductive development in mammals." *International Journal of Andrology* **29**(1):96-104.

Hoppin, J.A., Ulmer, R., London, S.J. 2004. "Phthalate exposure and pulmonary function." *Environ Health Perspect* **112**:571-574.

Jaakkola, J.J.K., Verkasalo, P.K., N.J. 2000. "Plastic wall materials in the home and respiratory health in young children." *Am J Public Health* **90**:797-799.

Jaakkola, J.J.K., Oie, L., Nafstad, P., et. al. 1999. "Interior surface materials in the home and the development of bronchial obstruction in young children in Oslo, Norway." *Am J Public Health* **89**:188-192.

Koch, H.M., Bolt, H.M., J.A. 2004. "Di(2-ethylhexyl)phthalate (dehp) metabolites in human urine and serum after a single oral dose of deuterium-labeled DEHP." *Arch Toxicol* **78**:123-130.

Main, K,M., Mortensen, G,K., Kaleva, M., Boisen, K., Damgaard, I.N., Chellakooty, M., et al. 2005. "Phthalate monoester exposure from human breast milk and alterations in endogenous reproductive hormones in 3-months old infants."

Moore R, Rudy T, Lin T, Ko K, Peterson R. 2001. Abnormalities of sexual development in male rats with in utero and lactational exposure to the antiandrogenic plasticizer di(2-ethylhexyl) phthalate. Environ Health Perspect 109(3):229-237.

Phillip, M., De Boer, C., Pilpel, D., Karplus, M., Sofer, S. 1996. "Clitoral and penile sizes of full term newborns in two different ethnic groups." *Journal of Pediatric Endocrinology and Metabolism* **9**(2):175-179.

Rhees, R.W., Kirk, B.A., Sephton, S., Lephart, E.D. 1997. "Effects of prenatal testosterone on sexual behavior, reproductive morphology and lh secretion in the female rat." *Developmental Neuroscience* **19**(5):430-437.

Rudel, R., Camann, D., Spengler, J.D., et al. 2003. "Household exposure to phthalates, pesticides, alkylphenols, pbdes, and other endocrine active compounds." *Toxicol Sci* **72**:184.

Salazar-Martinez, E., Romano-Riquer, P., Yanez-Marquez, E., Longnecker, M.P., Hernandez-Avila, M. 2004. "Anogenital distance in human male and female newborns: A descriptive, cross-sectional study." *Environ Health* **3**(1):8.

Silva, M.J., Slakman, A.R., Reidy, J.A., Preau, J.L., Jr., Herbert, A.R., Samandar, E., et al. 2004. "Analysis of human urine for fifteen phthalate metabolites using automated solid-phase extraction." *Journal of Chromatography B: Analytical Technologies in the Biomedical & Life Sciences* **805**(1):161-167.

Swan, S., Main, K., Liu, F., Stewart, S., Kruse, R., Calafat, A., et al. 2005. "Decrease in anogenital distance among male infants with prenatal phthalate exposure." *Environmental Health Perspectives* **113**:1056-1061.

Swan, S.H., Brazil, C., Drobnis, E.Z., Liu, F., Kruse, R., Hatch, M., et al. 2003. "Geographic differences in semen quality of fertile U.S. Males." *Environmental Health Perspectives* **111**(4):414-420.

Wormuth, M., Scheringer, M., Wollenweider, M., K.H. 2006. "What are the sources of exposure to eight frequently used phthalic acid esters in Europeans?" *Risk Analysis* **26**:803.

# 7. INFORMATION SECURITY

## FOCUS: RELEVANCE OF CYBER SECURITY

# THE GROWING RELEVANCE OF CYBER INSECURITY

HENNING WEGENER
Ambassador of Germany (ret.), Madrid, Spain

I am grateful that this plenary session on Information Security could be scheduled; for the last five years we had to forego such privilege, and our work on cyber security was confined to the more intimate venue of a small working group on the subject. Yet, the World Federation of Scientists had, early on, identified the threats emanating from cyberspace as a major indicator of the fragility of modern, integrated societies and as an issue of major relevance to the functioning and security of the world system. It had therefore, as early as 2000, included the challenges of information security among the Planetary Emergencies which call for urgent and coordinated international responses on the basis of interdisciplinary efforts. Later, in 2001, the Permanent Monitoring Panel on Information Security was established to analyze the threat panorama and make appropriate recommendations. Since then, the threat has grown by orders of magnitude. It is therefore fitting that the subject return to the Plenary level in order to allow us to take stock of the new relevance of cyber insecurity. We are proud that we have been able to include among the speakers two prominent experts who have come to Erice for the first time: Dr. Udo Helmbrecht, President of the German Federal Office for Information Security, and Professor Pradeep Khosla from Carnegie Mellon University.

The benefits of Information and Communication Technologies (ICTs) which increasingly pervade all aspects of human endeavour need hardly to be underlined. They have ushered in a new era of opportunity in terms of wealth creation, government efficiency, human development, and the emergence of a new type of knowledge society. Information has become the decisive raw material of all human endeavours. Telecommunications, the Internet and the capabilities of broad-band networks negate the relevance of frontiers and distances, and increasingly enable the vision of a global society with a new division of labour and shared benefits including for developing countries, and, within national societies, for more inclusiveness and integration.

The range of stakeholders in a functioning information society is huge. These benefits can, however, be undercut by digital disruption, by the negative use of the new technologies in the form of cyber attacks, viruses and other malware, sabotage of data and systems, etc. These perils are now increasingly recognized at the multinational level, including in several resolutions of the United Nations General Assembly (UNGA). The recent World Summit on the Information Society (WSIS) has strongly emphasized the value of confidence and security in the use of ICTs, and outlined measures to create a *global culture of cybersecurity.*[1] The alarming fact is that the negative techniques that operate against this culture are available to individuals and small groups with criminal intent as well as to State actors. This takes on special significance in conjunction with the emerging dangers of international terrorism, as it widens the panoply available to the perpetrators of terrorist acts. Cyber attacks display a growing sophistication commensurate with the evolution of ICTs, but the tools are also increasingly easy to acquire, thus lowering the skill level needed to launch attacks. There is an age-old and perpetual race between attack and defense—and information security provides no exception—where attackers are tendencially ahead. And, in the cyber world, the fragile balance between attack and defense is further tilted by the fact that the attacker can choose his target from anywhere, totally independent of time and place, but with potentially global effects. The dimensions of the damage and destabilization that can be caused are colossal. Awareness of these damage potentials is growing, but unevenly, and, even among scientists, not yet to a sufficient extent.

The damaging potential of cyber attacks affects the personal, private use of ICTs, shaking confidence in the reliability and confidentiality of private communications. More important than the consequences for human development, however, are the effects on the triple target spectrum of the economy, vital societal infrastructures, and national and international security.

It has been calculated that today 80% of all business assets are, in one form or the other, managed in digital form. The management and protection of IT-processed knowledge and the secure acquisition and conservation of information are thus key ingredients of successful business. Yet, in spite of a wide array of defense and response mechanisms, the number of harmful incidents has risen dramatically, even exponentially. As we will hear in gruesome detail, annual damage to the economy in the U.S. and other ITC-heavy industrial countries, to the extent that companies disclose it, is now in the range of billions of dollars annually. Safeguarding the legitimate economy from cyber damage will need increasing emphasis, even beyond the current growth in information security industries.

Cyber attacks against critical infrastructures that increasingly depend on ICTs also pose a serious and, in the last analysis, existential problem. They are typically in private hands and especially vulnerable because many of their distributed control systems (DCS) and supervisory control and data acquisition systems (SCADA) are connected to the Internet from where they can be disrupted. Given the growing interdependencies in interwoven modern societies, cyber attacks on these infrastructures can produce immediate grave repercussions throughout the national economic and political systems, but also momentous transfrontier effects. Cumulative attacks on various structures may, through instant chain reactions, cause the damage to grow exponentially. If one projects these vulnerabilities onto the current menace of international terrorism, ominous scenarios of the effects of cyberwar and cyberterrorism become menacingly plausible.

State and non-State actors are now in a position of directly or indirectly committing cyber attacks against national defense assets of another State, disabling its electrical and communication systems, interfering with the acquisition of intelligence, the functioning of weapon systems, or command and control procedures. Cyberwar, the use of (as they are sometimes labelled) "information weapons," is a very real technique of war and likely to be used more and more as time passes. Here again, an overlay of terrorist activities offers worrisome perspectives. They will be made explicit during this session.

Even more disconcerting is the fact that there is not only the potential—but the likelihood—of a combination of attacks that would *simultaneously* impair economic interests, critical infrastructures and military and defense capabilities. Information security and the concepts of national and international peace and stability are intrinsically linked.

There are new trends in the ICT field that render the problem of information security even more topical, and require heightened responses:

- The sheer increase in volume of ICT devices: worldwide, there are now more than 2 billion computers and tens of billions of other—equally vulnerable—processors and microprocessors in operation, the latter as embedded systems that invisibly govern vital controlling, monitoring and steering equipment .
- The exponential growth in connectivity, revolutionary computing advances like breakthroughs in miniaturization, speed and storage, the advent of intelligent systems and robotics—growing human-computer interaction. These emerging technological trends not only pervade our environment in an unprecedented way—linking people, objects and information in a novel manner—they also bring with them the next generation of digital disruption possibilities and, indeed, a sea change in how we must view and deal with information security.
- The rapid progress of a huge range of wireless techniques, including all-pervasive RFIDs, adds to the dimension of new vulnerabilities. This session will explore in some necessary detail the security challenges of the new digital networks.

The characteristics of the cyber emergency are peculiar and set it off from the other planetary emergencies which we study and combat in these Seminars. The cyber threat potential is asymmetric; it is inherently invisible, "virtual," and non-linear. Cyber attacks may have debilitating in-depth effects, disrupting the social fabric and essential assets, all with a minimum of input and investment. And information security is inherently a problem of universal and transnational character. The challenges it offers will thus not be resolved by the efforts of just one State or group of States, or even on a regional base. As the Internet knows no frontiers, and attacks may come from distant and undisclosed locations, or out of countries where the regulatory framework, including penal sanctions, is insufficient, a united effort by the international community, and harmonized or compatible measures by all States are required. Safeguarding information security is a universal challenge. There must be no loopholes in legal prescription and practices of law enforcement that could afford safe havens to cybercriminals.

To the members of the Permanent Monitoring Panel it has, therefore, always been evident that the thrust of their work would have to be in the political-institutional and legal domain. Technical issues, although their grasp is a prerequisite to useful and practical recommendations, are essentially and effectively taken in hand by industry and academia. In any event, in the course of the Panel's work it has become ever clearer that the challenge of information security requires an interdisciplinary input. Indeed, this Planetary Emergency is perhaps one of those that make the multiple contribution and dialogue of many disciplines—politicians, diplomats, military experts, lawyers, system analysts, economists, logistics experts, computer and information scientists, social scientists—particularly necessary. Science, advanced technology and political analysis and judgment need to interact to arrive at purposeful solutions.

In its comprehensive Recommendation packages in 2003 and 2005, the Permanent Monitoring Panel has, therefore, made the case for urgent international action and placed the emphasis on steps required to gradually establish *a universal order of cyberspace*. The United Nations system, because of its unique universal character, is assigned a key role in intergovernmental activities for the functioning and protection of cyberspace, but some of our specific recommendations are also directed to civil society and the private sector including industry, international law enforcement agencies, and, of course, national governments.

These recommendations have found considerable echo in the international community, especially at the two sessions of the World Summit on the Information Society where they figured as conference documents. The Panel is about to make an input to the follow-up bodies of the World Summit and maintains cooperative relationships with other multilateral bodies. I will comment on these aspects in more detail in my annual report as the Chairman of the Panel in a few days.

The presentations at this plenary meeting cannot possibly exhaust even the most important aspects of global cyber insecurity. Yet, they touch on four topical features of the new overall threat landscape: (1) the dangers emerging from the vast array of new digital networks—by no means a speculative glimpse into the future, but already a real-world analysis; (2) the evolving face of cyberwar cum cyberterrorism; (3) the mind-boggling economic damage potential of cybercrime in the global, and in individual national economies.

In closing, let me point to four *additional* problem areas in which the Panel is currently attempting to make its contribution.

The first one concerns what is now aptly referred to as cyber-repression: the denial of information access through restricting the Internet. Upholding its view that governmental Internet censorship via advanced content filtering and monitoring technologies, designed to suppress political information, and to inhibit the freedom of expression guaranteed by international law, constitutes a severe deprivation of the full benefits of the information society, the Panel is working on a comprehensive analysis and recommendations. Massive filtering practices by governments, but also assistance lent to them by corporate providers of the requisite technology, require public monitoring and analysis as well as a coordinated multilateral response. Cyber-repression has risen dramatically over the last few years, and the resulting information insecurity needs to be addressed forcefully.

The World Summit on the Information Society has given new focus and impetus to the digital needs of the developing world. The contribution of the Panel to the World Summit has been based on the conviction that, as yet, nascent information societies are specially vulnerable to cyber crime, cyber terrorism and even cyberwar, and are thus in special need of protection against cyber insecurity. They depend vitally on reliable and confidence-inspiring information structures that can foment development and investment. The recommendations of the Panel, arguing that capacity-building in these fragile societies and security-building must go hand in hand, have addressed these needs through concrete proposals. The Panel, in its advisory capacity to the new UN Global Alliance for ICT and Development, wishes to make further inputs stressing the information security requirements, thus making a contribution to the required *global culture of cyber security.*

Adding to the economic dimension of cyber attacks, there is another troubling consequence of the evolving structure of the global economy: its new divisions of labour via transcontinental outsourcing urgently require legal and administrative frameworks in which information remains secure and protected. This is an area in which the Panel will seek to elaborate recommendations aiming at bridging the existing Legal Divide.

Finally, the Panel is exploring in more depth the privacy-security dilemma that has emerged so strongly in the wake of the 9/11 events. Traditional solutions to this obvious dilemma in the realm of information technology have to be revisited, for instance by asking whether there are new technologies that offer *more* public security, while safeguarding individual rights and confidence in the privacy and reliability of ICTs for the user.

In addition to these four additional problem areas, the Panel will, of course, remain alert to any new development in this rapidly moving field, constantly reassessing the rising relevance of cyber security.

REFERENCES

1. Tunis Commitment, § 15, WSIS-05/TUNIS/DOC. 7; Tunis Agenda for the Information Society, § 39 et seq., WSIS-05/TUNIS/DOC/Rev. 1.

# PERFORMANCE LIMITS OF SENSOR NETWORKS FOR LARGE-SCALE DETECTION APPLICATIONS

YARON RACHLIN, ROHIT NEGI, AND PRADEEP KHOSLA
Carnegie Institute of Technology, Pittsburgh, USA

## INTRODUCTION

A sensor network is deployed in order to obtain information about the state of an environment. In many sensor network applications, such as pollution monitoring and border security, the phenomena under observation has a large scale that exceeds the range of any one sensor. As a result, collecting measurements from multiple sensors is essential to the sensing task. Obtaining information about an environment can be cast as either a 'detection' or an 'estimation' problem. In estimation problems for example, the problem of estimating a continuous field to within a desired accuracy, the state of the environment is continuous. In detection problems, such as binary hypothesis testing, the state of the environment is one of a finite set of hypotheses. In this paper we study the problem of 'large-scale detection' where the state of the environment belongs to an exponentially large, structured set of hypotheses. Large-scale detection problems characterize many applications where a sensor network is deployed in order to monitor a large-scale phenomena. In previous work,[1] we exploited the structure of large-scale detection problems to demonstrate a fundamental information-theoretic relationship between the number of sensor measurements and a sensor network's ability to detect the state of the environment to within a desired accuracy.

In a large-scale detection problem and for a fixed sensor configuration, each state of the environment produces a corresponding set of sensor outputs. This correspondence can be thought of as a code, where the sensors act as an encoder. To motivate this analogy we consider several large-scale detection applications.

## MOTIVATING APPLICATIONS

Large-scale detection problems characterize many important sensor network applications. We first consider an example from robotics. One of the main areas of research in robotics is mapping.[2] In mapping, a set of sensor measurements obtained from possibly multiple robots is used in order to construct a map of an unknown environment. One of the most popular techniques for combining such sensor measurements into maps is known as occupancy grids.[3] In occupancy grids, the world is modeled as a discrete field. A robot traversing an unknown environment takes a sequence of sensor measurements (e.g., using sonar range sensors). The state of the environment (e.g., the location of obstacles and free space) is encoded by these noisy sensor measurements. The number of possible states of the environment is exponential in size of the field. The occupancy grid algorithm detects the state of the environment using the noisy sensor measurements. How many sensor measurements must a robot collect in order to detect the state of the environment to within a desired accuracy? How does this number vary with sensor type (e.g., laser vs. sonar range sensors)? Do occupancy grids require significantly more measurements to achieve a given accuracy than is necessary for an optimal sensor fusion scheme? While robotic mapping has been implemented in practice, such questions remain unanswered.

Sensing complex chemical substances provides another example of a large-scale detection application. Interestingly, this multi-sensor application does not have a spatial component. Chemical sensor arrays, consisting of an array of semi-selective chemical sensors, can distinguish among a set of substances.[4] A complex chemical can be modeled as a mixture of multiple constituent chemicals at various discrete concentrations. The number of complex chemical, i.e., the number of states, is therefore exponential in the number of constituent chemicals. Each sensor reacts with a subset of these constituent chemicals. The output of a chemical sensor array encodes the complex substance being sensed. There are a wide variety of chemical sensors based on different technologies with various associated noise levels. A theoretical analysis is necessary to provide insight into the design of such chemical sensor arrays.

Target detection and classification, an important class of applications for sensor networks,[5] provides another example of a large-scale detection application, as illustrated in Figure 1. We consider the problem of detection and classification based on seismic sensors, as demonstrated in Li et al. 2002[5] and Tian and Qi, 2002.[6] We model the environment as a discrete field, where each entry represents the presence and class of a target at a corresponding location. As in the other examples, the number of possible target configurations in this field is exponential in the size of the field. Seismic sensors are scattered randomly on this field. A sensor is affected by targets in a localized region in the field, whose extent is defined by random variations in soil composition and the limits of the sensor's range. The intensity of vibration is dependent on the target's distance from the sensor, and therefore the sensor observes a function of target vibrations. The set of seismic sensor outputs encode the locations and type of targets in the field.

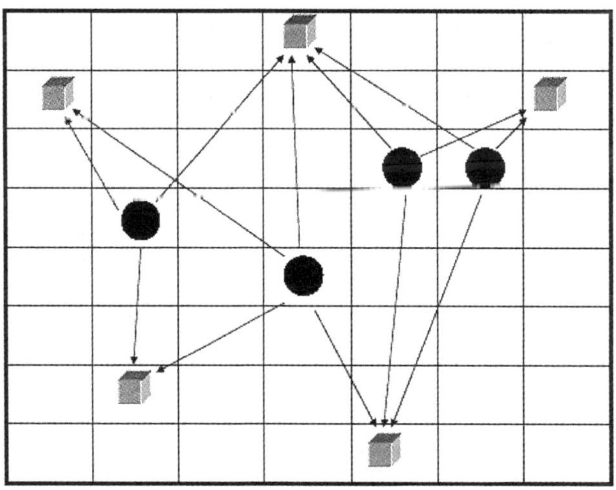

*Figure 1. Seismic sensor network with sensors (gray cubes) sensing vibrations from multiple vehicles (black circles).*

All of these examples share common elements. They are large-scale detection problems where the number of possible states of the environment is exponentially large. The sensors in each example produce an output that is a function of some subset of the environment. The sensor measurements must be considered jointly in order to detect the state of the environment. To understand the fundamental performance limits of sensor networks for such applications, we use the insight that such sensor networks can be modeled as channel encoders.

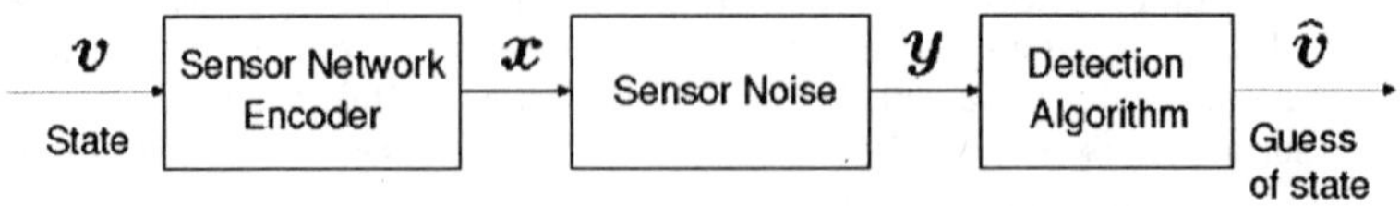


*Figure 2. Sensor network model.*

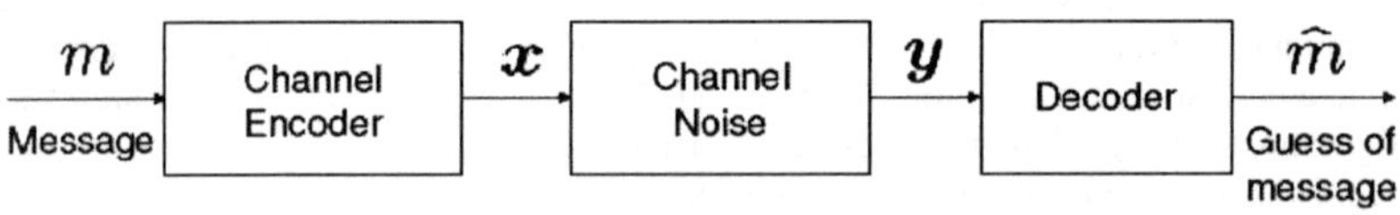


*Figure 3. Communication channel model.*

SENSOR NETWORK AS AN ENCODER

The examples discussed in Section II motivate the following sensor network model. In each of the motivating examples, the state of the environment being sensed can be modeled as a discrete vector $v$. A set sensors encode the state of the environment by producing functions of subsets of the environment $x$. The sensor outputs are corrupted by noise so that we observe $y$. These noisy sensor outputs are then used to produce a guess of the true state of the environment $\hat{v}$. This process comprises the sensor network model shown in Figure 2.

This sensor network model is similar to the classical model of a communications channel shown in Figure 3. The message $m$ being transmitted corresponds to the state of the environment $v$. The sensor network acts as a channel encoder, assigning the codeword $x$. The channel decoder estimates the message sent, and similarly the detection algorithm estimates the state of the environment. The fundamental limits of a communication channel is described by Shannon's celebrated channel capacity results.[7] The channel capacity characterizes the maximum data transmission rate at which communication with arbitrarily small error is feasible. In Rachlin, Negi and Khosla, 2004[1] we defined and analyze an analogous limit for a sensor network model. This limit is called the 'sensing capacity.' The sensing capacity bounds the smallest number of sensor measurements required to detect the state of the environment to within a desired accuracy. The sensing capacity differs significantly from the classical Shannon capacity due to differences between the two models.

The most significant difference between the sensor network model and the communication channel model arises due the fixed sensor configuration of a sensor network. In communications, a message can be mapped to an arbitrary codeword since a channel encoder can implement any mapping. In a sensor network, the state of the environment and its codeword representation are coupled by the sensor type and configuration. Using the analogy of a communications channel, a sensor network corresponds to a constrained channel encoder that is constrained in its codeword selection by the content of the message. As a result, widely different states of the environment correspond to highly dissimilar codewords, and similar states of the environment correspond to similar codewords. As a result, it is not possible to distinguish between two similar states of the environment to within an arbitrarily high accuracy.

## RELATED WORK

Work on the sensing capacity of sensor networks is most closely related to research about detection in sensor networks. Varshney, 1997[8] describes a large body of work in distributed detection which focuses on hypothesis testing problems where the number of hypotheses is small. Chamberland and Veeravalli, 2003[9] and 2004[10] extend this work to consider a decentralized binary detection problem with noisy communication links to obtain error exponents. D'Costa, Ramachandran, and Sayeed, 2004[11] analyzes the performance of various classification schemes for classifying a Gaussian source in a sensor network, which is an m-ary hypothesis testing problem where the number of hypotheses is small. Kotecha, Ramachandran, and Sayeed, 2005[12] analyzes the performance suboptimal classification schemes for classifying a fixed number of targets. While in this work the number of hypotheses is exponential in the number of targets, the fundamental limits of sensing for a large number of targets, and therefore an exponentially large number of hypotheses, are not considered. Chakrabarty et al. 2001[13] considers the problem of sensor placement for detecting a single or few targets in a grid. This problem is similar to a large scale detection problem. However, due to the restrictions on the number of targets, the number of hypotheses is comparatively small. A coding-based approach was used to propose specific sensor configurations, and to propose bounds on the minimum number of sensors required for discrimination using this structured approach. Sensors were noiseless, and of limited type, and no notion of sensing capacity was considered. In contrast to existing work on detection and classification in sensor networks, the sensing capacity provides fundamental performance limits for large-scale detection applications.

## SENSING CAPACITY OF SENSOR NETWORKS

How many sensor measurements are necessary to distinguish among the exponentially large number of state of the environment? To answer this question, we consider an example where the state of the environment is a discrete $k$-dimensional binary vector. Each entry of the vector corresponds to a spatial location which may or may not contain a target. There are $2k$ possible vectors. We define the distortion $D$ as the fraction of target positions which can be misclassified without considering the guess of the state of the environment to be in error. The rate $R$ of a sensor network is defined as the ratio of $k$ target positions being sensed to the number of sensor measurements $n$, $R = kn$. The sensing capacity of a sensor network, $C(D)$, is defined as the minimal rate $R$ such that below this rate there exists a sensor network that can distinguish among all target vectors to within a distortion $D$, for sufficiently large $k$ and $n$.

In Rachlin, Negi, and Khosla, 2004[1] we introduced a simple but useful sensor network model. We analyzed the sensing capacity of this model and demonstrate important differences between the channel capacity and the sensing capacity. A prominent difference is that the sensing capacity is not a mutual information. This is an important observation due to the use of mutual information as a sensor selection heuristic.[14] In Rachlin, Negi, and Khosla, 2005[15] we extended this model to account for contiguity in sensor observations and for arbitrary sensor types. Other extensions of this model account for nonbinary vectors (e.g., classification of spatially distributed targets), sensor heterogeneity, and target sparsity. In Rachlin, Negi, and Khosla, 2005[16] we examined the effect of structure in the environment on the sensing capacity. We bounded the sensing capacity for a sensor network sensing an environment modeled as a two-dimensional Markov random field. The Markov random field assumption allowed us to investigate the impact of spatial dependencies such as target clustering (e.g., groups of people in a surveillance application) on the sensing capacity. In Rachlin, Negi, and Khosla, 2006[17] we extended sensing capacity results to account for large-scale detection problems where the environment is evolving in time. Examples of such applications include pollution, traffic, agricultural monitoring and surveillance.

## DISCUSSION

The sensing capacity results discussed in this paper provide performance limits for sensor networks in large scale detection applications. Given a large-scale detection problem, the sensing capacity bounds the number of sensor measurements required to detect the state of the environment to within a desired accuracy. As importantly, these results provide a strong connection with results in communications. As a first step towards demonstrating the benefit of this connection, we applied the idea of sequential decoding from communications to sensor networks in Rachlin, Negi, and Khosla, 2006.[18,19] Sequential decoding is used to efficiently decode convolutional codes. This heuristic algorithm works at communication rates sufficiently below channel capacity, exploiting an interesting connection between decoding complexity and the channel capacity. Applying this insight and adapting the sequential decoding algorithm, we demonstrate that the same idea can be applied to sensor networks. When a sufficiently large number of sensor measurements is available, sequential decoding provides a computationally efficient alternative to more complex algorithms such as belief propagation. Our empirical results demonstrate an interesting relationship between the number of sensor measurements and the computational complexity of detection.

A large number of questions remain open in the theory of sensing capacity. Further, the potential for exploiting insights from communications and coding for large-scale detection problems remains largely unexplored. The development of the theory and practice for sensor networks in large-scale detection applications is a promising and important area of future research.

REFERENCES

1. Y. Rachlin, R. Negi, and P. Khosla, "Sensing capacity for target detection," in *Proc. IEEE Inform. Theory Wksp.*, Oct. 24-29, 2004.
2. S. Thrun, "Robotic mapping: A survey," in *Exploring Artificial Intelligence in the New Millenium*, G. Lakemeyer and B. Nebel, Eds. Morgan Kaufmann, 2002.
3. A. Elfes, "Occupancy grids: a probabilistic framework for mobile robot perception and navigation," Ph.D. dissertation, Electrical and Computer Eng. Dept., Carnegie Mellon University, 1989.
4. M. Burl, B. Sisk, T. Vaid, and N. Lewis, "Classification performance of carbon black-polymer composite vapor detector arrays as a function of array size and detector composition," *Sensors and Actuators B*, vol. 87, pp. 130–149, 2002.
5. D. Li, K. Wong, Y. Hu, and A. Sayeed, "Detection, classification and tracking of targets in distributed sensor networks," *IEEE Signal Processing Magazine*, pp. 17–29, March 2002.
6. Y. Tian and H. Qi, "Target detection and classification using seismic signal processing in unattended ground sensor systems," in *International Conference on Acoustics Speech and Signal Processing (ICASSP)*, vol. 4, May 2002.
7. C. Shannon, "A mathematical theory of communication," *Bell System Technical Journal*, vol. 27, pp. 379–423 and 623–656, July and October 1948.
8. P. Varshney, *Distributed Detection and Data Fusion*. Springer-Verlag, 1997.
9. J. Chamberland and V. Veeravalli, "Decentralized detection in sensor networks," *IEEE Transactions on Signal Processing*, vol. 51, no. 2, pp. 407–416, 2003.
10. ——, "Asymptotic results for decentralized detection in power constrained wireless sensor networks," *IEEE JSAC Special Issue on Wireless Sensor Networks*, vol. 22, no. 6, pp. 1007–1015, 2004.
11. A. D'Costa, V. Ramachandran, and A. Sayeed, "Distributed classification of gaussian space-time sources in wireless sensor networks," *IEEE J. Selected Areas in Communications (special issue on Fundamental Performance Limits of Wireless Sensor Networks)*, pp. 1026–1036, Aug. 2004.
12. J. Kotecha, V. Ramachandran, and A. Sayeed, "Distributed multi-target classification in wireless sensor networks," *IEEE JSAC Special Issue on Self Organizing Distributed Collaborative Sensor Networks*, 2005.
13. K. Chakrabarty, S. S. Iyengar, H. Qi, and E. Cho, "Coding theory framework for target location in distributed sensor networks," in *Proc. Int. Conf. on Inform. Technology: Coding and Computing*, April 2001.
14. J. Manyika and H. Durrant-Whyte, *Data Fusion and Sensor Management: A Decentralized Information-Theoretic Approach*. Prentice Hall, 1994.
15. Y. Rachlin, R. Negi, and P. Khosla, "Sensing capacity for discrete sensor network applications," in *Proc. Fourth Int. Symp. on Information Processing in Sensor Networks*, April 25-27 2005.
16. ——, "Sensing capacity for markov random fields," in *Proc. Int. Symp. on Information Theory*, 2005.
17. ——, "Temporal sensing capacity," 2006, to appear in Proceedings of the Fourty-Fourth Annual Allerton Conference on Communication, Control, and Computing.

18. ——, "On the interdependence of sensing and estimation complexity in sensor networks," in *Proc. Fifth Int. Conf. on Information Processing in Sensor Networks*, April 19-21, 2006.
19. ——, "Sensor networks: Estimation complexity vs. the number of measurements," 2006, to appear in *Proceedings of the Military Communications Conference*.

# ECONOMIC DIMENSION OF CYBER SECURITY

UDO HELMBRECHT[1]
Federal Office for Security of Information Technologies
Bonn, Germany

## ABSTRACT

The Information Communication Technology (ICT) is facing an increasing number of threats caused more and more by organised criminals. These threats, emanating from cyberspace, are a major indicator of the fragility of modern, integrated societies and are an issue of major relevance to the functioning and security of the world system. Amongst the global uses and misuses of *the web*, information security is one of the Planetary Emergencies which calls for urgent and coordinated international responses on the basis of interdisciplinary efforts. International strategies have to be developed to secure a certain standard of IT security for enterprises as well as for the consumers.

## INTRODUCTION

Information Communication Technology (ICT) has become the outstanding social factor of our time. Its world-wide application is on the rise, and so is our dependence on IT. Computers, mobile telephones and the Internet have evolved into the basis of a mobile, knowledge- and networked-based information society. Most of us hardly ever think about how we take the IT for granted in our day-to-day life. There is the *obvious IT*—such as the Internet and all the opportunities arising with the web: e-mail, electronic data transfer, or even Voice-over IP, just to name a few. Additionally a more *hidden IT* exists, which is also very important. Without an extensive use of various and wide-ranging IT components, we could not guarantee our supply of energy, and even of drinking water. Financial transactions would only be possible to a limited degree or not at all, and government proceedings would be severely hampered.

There is no doubt that IT has given our society many advantages and that IT has made our lives easier. But we cannot ignore the risks: unprecedented threats through malicious codes such as viruses, worms or Trojan horses. To the extent that society is increasingly dependent on information technology, an effective protection against these threats is becoming more and more important.

Today, attackers are developing highly sophisticated destructive programs, and they are doing it with an ever increasing rapidity. These threats affect all those who use networked IT-systems on a daily basis. National boundaries are no longer relevant to these types of threats. Increasing danger and new types of IT threats force us to develop new ways of thinking and acting. Today, it is impossible to confine the protection of information technology and the IT infrastructure by domestic policies. This has to be addressed on an international level. For instance, the authors of computer viruses are operating globally. All they need is access to the Internet. It was easy to identify a person responsible for polluting the Rhine; whereas the author of a virus is lost in the vastness of the Internet. The chances of getting hold of him or her are low to non-existent. A 19-

years old virus writer, who was sentenced to a prison term about two months ago, is unfortunately still the exception that proves the rule.

And it is becoming increasingly clear that the term *security* has taken on an entirely new meaning: Historic national boundaries provide less protection today than ever before. Such terms as internal and external security are increasingly difficult to define or might indeed merge in some cases.

THREATS

Rather than the old-style crime of merely going on-line, criminals are taking advantage of the increasing dependence of our day-to-day life on IT by developing creative strategies to exploit the vulnerabilities of IT-systems.

The prevalence of standard software results in *monocultures* of operating systems. Since a single software product dominates the market, the vulnerabilities inherent in this product are particularly widespread and, if exploited, lead to substantial damages. The main source for distributing computer viruses are, therefore, innocent users of personal computers and company computers, who often are not aware of the risks inside *the web*.

For BSI's[2] report on *The IT Security Situation in Germany in 2005*,[3] published in July 2005, we registered more than 7,360 new variants of viruses and worms in the second half of the year 2004. This constitutes a 64-percent-increase compared to the first six months. It is impossible to completely avoid security gaps in complex software, but security is also a matter of quality. Today we realise a yearly doubling of the number of vulnerabilities in IT-products. For example, in 2006, Microsoft patched more critical vulnerabilities by August than 2004 and 2005 combined. Some of these security gaps are used for industrial espionage. We still register a tendency towards inconspicuous spying programs and Trojan horses that are programmed for single-targeting purposeful espionage.

Malicious code is less and less programmed to directly cause irreparable damage. Rather, attackers try to bring infected computers under their control so that they can continue to misuse them. For this purpose, *backdoor-programs* are installed by way of Trojan horses so that the attacker may control such computers from the Internet by remote control. These *bot-nets* provide an effective and increasingly used infrastructure to distribute spying programs in a wider range, e.g., to commit identity theft in online-banking. Almost every 10$^{th}$ e-mail is infected by viruses, approximately 60 to 90 percent of the e-mails world-wide are *spam*.

Another trend can be observed: up until now, attacks on IT systems were presumably motivated by *sporty ambitions* and conducted by *ethical hackers*, but this aspect is becoming increasingly less significant. By contrast, Internet criminality is being conducted more and more in a professional and commercial manner. Instead of an isolated or spotty-faced computer hacker, targeted attacks are increasingly carried out by organised criminals using *tailor-made malware*, such as the Trojan horse *Pinka*[4] discovered in 2005, against which common anti-virus software may prove ineffective. Financial interests are the decisive driving power. Already in 2004, sixteen percent of hacking activities were aimed at E-commerce companies. This represents a 400-percent-increase compared to the previous year. It is feared that this trend will continue in the future.

The Internet and the increasing use of standard software in critical areas thus open up new dimensions for industrial crimes. Primary targets are technology and know-how theft as well as getting the competitive advantage, e.g., through spying-out tenders, contracts, or price lists.

At the same time, the significance of new transmission technologies such as Bluetooth, WLAN or UMTS is on the rise. Thanks to those technologies, communication will become more and more mobile. Yet, the future of mobile applications decisively depends on whether security challenges can be overcome. Without security mechanisms, attackers can easily follow, tamper with, or otherwise manipulate data transfers.

Companies with large valuable development departments, such as pharmaceutical enterprises, companies in the automotive industry as well as the software industry, are especially vulnerable to such threatening scenarios. Protecting data confidentiality is not given the proper attention in many areas, the reason being that on the management level, a consciousness for IT-based industrial espionage has not yet developed adequately.

Targeted DDoS (*Distributed Denial of Service*) attacks against companies also pose a major security problem. In this attack method, the attacker floods the server with useless data packages, thus overloading the system in order to provoke business interruptions in the targeted company. By using Trojans, hackers often misuse several thousands of computers and then rent out these so-called Bot Networks for use as platforms for DDoS attacks. Such attacks, which may be launched by competitors, dissatisfied personnel, or otherwise motivated groups of people, obstruct the smooth operation of web sites massively, which can result in considerable economic consequences, especially for E-commerce companies.

Not only businesses, but also governments, are the target of cyber attacks. Last year, according to the Guardian, the English parliament nearly fell victim to a sophisticated hacking fraud. Experts expressed the view that such attacks had even had the support of foreign countries' authorities.

## ECONOMIC IMPACT

Surveys on the economic impacts of IT-based criminality suggest an overall cost of world wide USD 250 billion per annum.[5] Estimates induce costs for damages, down time and repairs caused by viruses Love Bug, Code Red and other viruses of USD 54 Billion.[6] A survey led by Ernest & Young among 500 U.S. companies in 2002 speaks of financial losses of USD 455.8 million.[7] U.S. Justice Department's Operation Web Snare has identified 150,000 victims of cybercrime with losses in excess of USD 215 million.[8] The U.S. Federal Trade Commission has published a report according to which 9.9 million Americans were victims of identity theft and calculated an average loss of about USD 1,200.[9]

Though no 100 percent reliable data exist, various surveys show that a large number of business and households have fallen victim to diverse forms of cybercrime. According to a recent study in the United States,[10] one in two of the respondents had experienced high levels of spam. Moreover, one in four respondents had had a major problem with viruses (cost per incident: USD 109; total damage: USD 5.2 billion). As to spyware, one in eight respondents have reported a major problem with a cost per incident of USD 100 and a total damage of USD 2.6 billion. Finally one in 115 respondents had fallen victim to password-fishing, those activities summing up to USD 630 million by a cost per incident of USD 850.

According a 2006 survey,[11] an 8th survey since 1991, and based on telephone interviews with 1,000 businesses of all sizes in the United Kingdom, the overall cost of IT security cost may be broken down to the following positions:

- Business disruptions
- Time spent responding to incident
- Direct cash spent responding to incident
- Direct financial losses
- Damage to reputation.

Moreover, large enterprises in the United Kingdom learnt to protect themselves against IT-based security incidents and to reduce the damage. Whilst their overall costs are partly falling, for small and medium business the costs for security incidents have risen 50 percent since 2004.

For the U.S., the Computer Crime and Security Survey[12] of the Computer Security Institute and the Federal Bureau of Investigation's Computer Intrusion Squad identified the main reasons for losses caused by cybercrime by interviewing 5,000 computer practitioners in the United States. It is remarkable that only 25 percent of the computer incidents are reported to law enforcement.

639 respondents in 2005 estimated total losses of USD 130,104,542 and an average loss per respondent of USD 203,606. In 2004, the average loss per respondent was USD 526,010. In the fourth consecutive year of the research, these loss estimates have dropped, but this year's decline marks the smallest percentage drop of the four years. 313 respondents were willing and able to estimate losses with a total amount of USD 52.494.290 and USD167.713 average loss per respondent.

## THE GERMAN APPROACH

For 15 years, BSI[13] has strongly committed itself to secure information technology in Germany. During this time, the office—which is a division of the Federal Ministry of the Interior—has been continuously evolving.

In 2006, BSI's overall budget reached 62 million Euros; 20% is invested into research and development. BSI employs nearly a total of 500 members of staff. Apart from IT experts and natural scientists—who still form the largest group—the work of jurisprudence as well as administrative, economic and social scientists is also indispensable for BSI's wide range of tasks, since an issue as multifaceted as IT security needs to be viewed from many different perspectives.

Our technical competence is steadily increased through continuous education. Based on this broad expertise, we offer our customers services in the four central areas of information, consultancy, development and certification:

*Information*:
We provide information about all important IT security issues.

*Consultancy*:
We give advice in questions of IT security and offer support for appropriate action.
*Development*:
We conceive and develop IT security applications and products.
*Certification*:
We test, value and certify IT systems with regard to their security qualities.
*Approving* of IT systems for the processing of classified information is also one of our tasks.

The BSI examines security risks in the application of information technology and develops corresponding security measures in individual cases. It informs those affected, about risks and dangers related to the use of information technology and provides assistance in finding solutions to concrete problems. In order to minimise or even avoid the risks mentioned, the BSI addresses a large number of target groups, by providing advice for manufacturers, distributors, and users of information technology.

While the Computer Emergency Response Team of the Federal Authority (CERT-Bund) provides comprehensive information about new vulnerabilities and threats through its warning and information service, private IT users are being informed by means of the following portal: www.bsi-fuer-buerger.de. With its competencies, among other things, in the area of certification, basic IT protection, and decryption technologies, the BSI creates the foundation necessary to effectively meet future IT-security challenges in Germany head-on.

Certification of IT products based on internationally acknowledged IT security criteria is of growing importance to BSI. The purpose of certification—to make IT products and systems transparent and comparable as to their security qualities—in the past year caused a strongly rising demand particularly for internationally agreed certificates on the basis of the Common Criteria (ISO/IEC 15408:1999).

German National Plan for Information Infrastructure Protection

In 2004 the Federal Government decided on an IT security strategy for Germany: the National Plan for Information Infrastructure Protection (NPSI). This was drawn up under the auspices of the Federal Ministry of the Interior, in close collaboration with BSI.

The National Plan is aimed at three strategic goals:

*Prevention*:
We have to work on prevention. We need to sensitize the responsible people to the existing threats and problems. We also need adequate technologies, such as early warning systems.
*Response/Readiness*:
We have to improve our response to incidents. Responding effectively to IT security incidents needs an accurate and up-to-date status report as well as elaborate and well practised crisis response concepts and emergency plans.

*Sustainability*:
The way we protect our information infrastructures must be sustainable. This means promoting training in IT security as well as in research and development.

CHALLENGE FOR THE INTERNATIONAL COMMUNITY

The cyberworld has lost its virginity. We are today faced with cybercrime and cyberterrorism. Internet technology is cheap and easily accessible by anyone who has an internet access. National borders are no barrier. Thus we need:

- International standards (i.e., common criteria)
- Advanced early warning systems
- Trusted computer platforms
- Use of biometrics (i.e., in passports)
- Research and development in secure internet technologies.

I appreciate the participation at this World Federation of Scientists International Seminars and count myself lucky to contribute to the 36th Seminar on Planetary Emergencies among so many high-ranking international scientists.

REFERENCES

1. Dr. Udo Helmbrecht, President of the Federal Office for Information Security (BSI), Bonn, Germany and associate member of the World Federation of Scientists' Permanent Monitoring Panel on Information Security. (www.federationofscientists.org; www.itis-ev.de/infosecur/).
2. BSI = Federal Office for Information Security www.bsi.bund.de
3. www.bsi.bund.de/english/publications/securitysituation/lagebericht2005_englisch.pdf
4. www.washingtonpost.com Glenn Frankel; The Washington Post; May 31, 2005; E.01; "18 Arrested In Israeli Probe Of Computer Espionage".
5. www.mi2g.com
6. Geralds, 2003. in www.vnunet.com
7. Survey by Ernst & Young
8. U.S. Justice Department
9. Federal Trade Commission
10. www.consumerreports.org
11. PWC UK, 2006, www.security-survey.gov.uk
12. CSI/FBI 2006
13. BSI = Federal Office for Information Security www.bsi.bund.de

# THE EVOLVING FACE OF CYBER-CONFLICT AND INFORMATION WARFARE

WILLIAM A. BARLETTA
Department of Physics, Massachusetts Institute of Technology
Cambridge, USA

ABSTRACT

This paper analyzes the potential for organized malicious behavior on the part of nation-states and non-governmental organizations and networks against the social, economic, political, or military assets and interests of local state and national governments involving computerized information technologies. The object of the analysis does not differentiate between cyber warfare and cyber terrorism as in both cases a principle measure of merit for the attacker is the deleterious effect on the *objective of the attack* rather than the proximate gain by the miscreant—a motivation that is the converse of the aim of the cyber-criminal who seeks his own gain or gratification regardless of the effects on the party attacked. Nonetheless, the technical considerations, instruments and methods are the same as those found in "peer-on-peer" criminal activity. The paper discusses some necessary aspects of an international social and legal framework and offers recommendations for the United Nations to follow-up the processes begun in the World Summit on the Information Society (WSIS).

## ASYMMETRIC VULNERABILITY OF INFORMATION SOCIETIES

For millennia, dominant nation states have had the ability to inflict widespread catastrophic physical damage on the well being of other nations. With the advent of nuclear technologies, the possibility of devastating attack is becoming open to relatively weak pariah states and perhaps even well-organized and financed non governmental organizations. These "facts of life" drive the concerns about nuclear non-proliferation and terrorism using weapons of mass destruction. Likewise the widespread advance of computer (cybernetic) technologies coupled with high bandwidth, digital links has dramatically transformed[1] an old tool of conflict, information warfare, with its aim of disrupting the social fabric of the target country.

### The information age offers new promise and new perils.

Progressing into the information age, industrialized nations are exploiting, at an ever-increasing level, pervasive networks of major economic, physical, and social assets[2] connected via Information and Communication Technologies (ICT) to advance their national prosperity, influence, and power. Likewise developing nations see information technology as an economic fast track. Smart devices (containing both sensors and microprocessors) abound. As the price associated with radio frequency tags (RFID) drops, even packages of soap "communicate" automatically with the cashier or inventory clerk stocking shelves. Already major hospital systems are combining RFID patient tags with real-time patient data entry by physicians to open the possibility of real-time epidemiology that can identify incipient epidemics or bio-terrorist attack. Extensive communications networks now permit the intensive application of information resources to facilitate commerce, provide services, monitor the environment and address complex societal problems.

*Connectivity is non-linear. Moreover, the nature of information is to increase the degree of non-linearity in the fabric of technological societies because information invites and facilitates new causal relationships throughout the society. Such relationships are pregnant with economic utility and potential for undermining governmental repressions.*

Those whose interest is command and control enjoy greatly facilitated top-down communication, but more importantly—and especially with respect to expanding human rights and economic well-being[3]—the streams of bottom-up and horizontal information flows have expanded to great rivers. The developmental pattern of information societies is to augment both number and nature of information nodes (where information is generated and consumed) and the number and strength (bandwidth) of links.

A further complexity of the information society as network derives from the fact an increasing percentage of both nodes and links carry autonomic sensors of status. If appropriately configured and equipped,[4] the network as a whole can both compute and test its status plus sense the level and sources of "uncertainty associated with the unpredictable presence of obstacles to sensing that appear in the environment"[5] to such an extent that it becomes able to heal or circumvent flaws in its connective structure. We are gradually witnessing the information network accrete—to use a loaded phrase—"self-awareness"[6] with the ability to reduce its level of uncertainty, manage,[7] modify and repair itself. Just as the propagation and consequences of widespread malicious code (malware)[8] attacks in cyberspace are frequently analyzed in biological and epidemiological terms, so can the means of protecting information systems and networks be designed as analogs to biological immune system responses.[9]

*The vexing paradox of highly non-linear connectivity is that it simultaneously increases both the resilience of the information network and risks and consequences of debilitating attacks on the nodes[10] and backbone links, and the difficulties of anticipating the consequences[11] of network failures. Consequently as the complexity of information networks evolves (in an unplanned manner), the potential of cyber information warfare is evolving toward putting ever-greater societal value (or utility[12]) at risk.*

Without question, the very connectivity of smart technologies—and later the complexity of artificial network intelligence—that imbues it with such potency to benefit society also opens highly networked societies to new forms of asymmetric attack[13] on information systems, in which the probability[14] of at least partial success is significant. The aims of asymmetric attacks may range from a comprehensive level of disruption so great as to be properly called warfare[15] to the level of terrorizing civilian populations to, at the low end, cyber criminality and hooliganism. Indeed, a Rand Corp. study has labeled as netwar[16] an intermediate level of networked attacks on a society via its information networks. With regard to the level of the threat and what can and should be done about it, expert opinions differ.[17]

As military and intelligence agencies of the United States and other nation states already "reconnoiter and probe to identify exploitable digital network weaknesses among potential adversaries,"[18,19,20,21] the decision-makers in these countries act as if the age of cyber warfare is now.[22] In fact, it is just these countries that have the capability and capacity[23] to launch or sponsor cyber attacks (especially as covert operations) upon countries less able to respond in kind.

In practice, the damage potential of a given type of attack can vary greatly depending on the degree of preparedness of the society and built-in security of the system under attack. From the point of view of the political or military decision-maker, the "important issue in countering any form of cyber attack is to quickly discern the type of attack and adversary and respond appropriately. Currently, tracking down computer intrusions is a law enforcement function. ...The traditional war fighting military is prohibited from executing this mission domestically… [therefore] domestic law enforcement has a critical role in national security and national defense."[24] It follows that nation-states in both their military and law enforcement agencies require powerful digital forensic tools, an appropriate legal structure to use them, credible approaches to preserving the integrity of evidence, and penalties for transgressors that have real deterrent value. Given the transnational character of the Internet, a high degree of compatibility among nations in the relevant legal frameworks is highly desirable.

At the level of criminality or "hacktivism"[25] the concept of deterrence through civil and criminal penalties may be operable if a suitable network of international homogeneity in criminal codes can be established. Unfortunately, at the level of cyber-attacks by nation-states, the concepts of deterrence developed during the Cold War may have little value, as a counterattack-in-kind may damage the international social and physical connectivity at a level that is unacceptable to third parties and the counter-attacker alike. In cyberspace collateral damage can be worldwide. In the intermediate case of cyber-terrorism, the recent behavior of the United States with respect to "illegal combatants" in its "war on terrorism" suggests that the model of deterrence at the level of civil and criminal penalties fails here also. While the difficulties of deterrence may encourage the pursuit of perfect technological defense against cyber-attack, the history of every other kind of weaponry cautions that what is at heart a socio-political problem must ultimately be dealt with at a socio-political level.

Consequently, in its report to the World Summit on the Information Society, the World Federation of Scientists Permanent Monitoring Panel (WFS PMP) on Information Security has recommended[26] that, "[The] WSIS should incorporate in its work programme an in-depth discussion of the potential adverse impact of cyberwar activities, in order to heighten the understanding and consciousness of ICT users in the public and private sectors. Given the potential of cyber attacks to constitute a breach of international peace and security, WSIS should support the urgent initiation of work at the UN to study and clarify the scenarios, criteria, and international legal implications and sanctions that may apply, and, in particular, to examine how traditional principles of international law relating to armed conflict are applicable to conflicts in the information age."

In substance this exhortation remains valid, but it must now be redirected to the Secretary General of the United Nations in the framework of Resolution 60/45,[27] requesting that urgent action be taken. The urgency of action is underscored by the recent history of the Israeli-Palestinian cyber-conflict, [28] which demonstrated the potential for rapid horizontal and vertical escalation by non-governmental organizations or networks (perhaps with state encouragement or support) and the proliferation of potent hacking tools.

## OBJECTIVES AND LOCI OF CYBER ATTACKS

In understanding the nature of cyber conflict, it is useful to draw parallels and contrasts with physical (or kinetic warfare). In their comparative analysis[29] Parks and Duggan begin with the principle that "Cyber-warfare must have kinetic world effects." Those effects derive from the locus or objective of the attack and serve to provide a starting point of the assessment of damage[30] from a cyber attack. The potential objective or locus of a cyber information attack may be differentiated according to the characteristics of the information or the component of the information system being compromised. Information as a communication from one party to another is described by internal characteristics—technical form (medium) and contextual form—as well as by externals: source, ownership, credibility or authority.

A. *Technical form or medium* can be destroyed, corrupted, compromised in the cyber attack. Commonly used forms are magnetic and optical records; emerging forms that promise to store higher information density are spintronic and biophysical. The attack may focus on 1) the stored information itself or 2) upon the communications link or traffic between the data storage system and authorized users.

B. *Contextual form* is the paradigm, in which information is accumulated, structured, accessed, transferred, and used. Context gives meaning to data; it is the Rosetta stone of the content. Context confers meaning and makes possible understanding.[31] Attack on (elimination or compromise of) contextual form may lower the credibility of information, may make information non-actionable (either by revealing or hiding source of authority) or otherwise reduce its utility.

C. *Data (raw) content* describes the digital "facts" information, enormous strings of ones and zeroes (bit patterns) that encode the information. Attacks on content may alter or destroy the bit patterns. In those attacks in which bit patterns are carefully altered, *stealth and surprise* have particular value for the attacker as the authorized user of the information may continue to use the data as if it were valid, even though he may possess or have access to uncorrupted data. For the owners and uses of databases adequate *testing & evaluation of data integrity*[32] becomes an essential fiduciary activity. Alternatively, the alterations by the attacker may be intentionally clear as in the case of the defacements of high-profile governmental websites.

D. *Nature of the information source* is subject to attack in multiple manners. Attacks on the information source have at least three potential loci: 1) the information generator, 2) computing infrastructure, 3) ownership of the information, 4) ownership of information media, and 5) access to and trustworthiness of the information.

*Generator* – Scientific, commercial, and security organizations worldwide generate huge databases of technical information. Whether the generation is episodic (though at regular intervals) or continual, the existential fact of the data carries information, namely, the apparent functioning or well being of the generator.

The availability of current information to the user depends upon ability of the generator to continue to provide and broadcast information.[33] Therefore effective attacks on the generator may consist of reducing the reliability of the movement and archiving of data from its source or of creating load conditions that slow, prevent, compromise, or corrupt transmission.[34,35] The impracticality or impossibility of reproducing huge technical databases such as collected climate global sensor data, has led to the development of redundant mass storage and multi-casting technologies.

*Computing infrastructure* – An critical operational and strategic asset of the information society is its collection of the computer clusters and mainframe supercomputers that are essential to its scientific, financial, military and other governmental enterprises. Attacks on the computing infrastructure both dramatically reduce capacity and eliminate computing capability. The vulnerability of this critical information resource is not hypothetical. In 2005 "...half a dozen U.S. supercomputer centers were knocked off-line—some for weeks. And supercomputers are typically more carefully tended than Grid based clusters."[36,37]

Connecting the ensemble of computing capability is a rapidly expanding net of ultra-high speed (10–40 Gbits per second) transmission links. These links themselves constitute a potential target[38] for attack.

A vital tool in the protection of information and computing systems is robust *active intrusion detection* (IDS) that aims at immediate detection of unauthorized intrusion and activity in ICTs[39] and thereby extends the "intelligence" of the defenses by piecing together the traffic (packets) seen on the system's network devices, comparing the characteristics network events with the signatures of known types of attack (signature detection), and finding anomalies in activity patterns.

*Ownership* – The [putative] owners of information often claim legal protection of rights over dissemination and use of information. The owner may set the criteria or even the control the access to information. Such criteria may include rights to further dissemination by the authorized user (or user organization). Such control is the practice with respect to state security information, proprietary information, and personal confidential information. Oblique [and legalistic] attacks on ownership rights can lower the utility of information even to the point of making information non-actionable.

Ownership of information may also figuratively refer to the person with the greatest interest[40] in the maintaining the utility and integrity of the information. That person will ideally take the responsibility to ensure the security of the information

*Use rights* – The owner of the information may set the criteria for the use of information or may even the control the access to information. Such control is normal when the information deemed legally protected intellectual property.

*Credibility* – The user of the data should (and may be legally required to) assess (and document) his level of confidence of the data generator, source (provider), and the actual uncertainties in the data content (ranging from measurements, transactional records, statistics, etc. to news reports, war zone photos). Attacks on the credibility of information aim at reducing the utility of data, and undermining the confidence[41] of stakeholders in the competence of the parties (and institutions) using that data. News media can be both unwitting accomplices in and targets[42] of such attacks.

*Ownership of information media* – "The owners of information media have control over what information appears on their media. They can ensure that particular information is present on the information front by publishing it on their media. Conversely, while they cannot prevent other media from carrying particular information, they can at least keep it off their own. If they are the sole source of the information, their ability to prevent its disclosure will be greater than if the information is shared by several parties."[43] The techniques of attacks on ownership of media may include foreign takeover of critical media, legal challenges to exercise of ownership rights, legislative attacks on ownership rights—actions which may be legitimate and justifiable in some contexts—in addition to more obviously hostile action such as insider sabotage of media corporations.

A particularly troubling form of attack on effective ownership is *social engineering.*[44] This type of attack is most commonly the province of both individuals and transnational networks[45] of cyber criminals engaged in *phishing*[46] schemes and identity theft[47] against individuals and *pretexting*[48] against telecom carriers. An attacker could use social engineering on a massive level aiming to destroy the fabric of social trust.

What makes social engineering attacks so dangerous is that they enable[49] the attacker (perpetrator) to launch more damaging attacks or commit other more damaging crimes. Within cyberspace, identity theft is the essential first step for an attacker to gain apparently authorized access to an information system. From that point on system firewalls are essentially useless, as the attacker appears to be a legitimate user. The attacker is then able to plant Trojan horses, password sniffers and eventually root kits to seize control of the system. With respect to enabling physical threats, pretexting and identity theft can be used to provide transient background identities and funds for terrorist cells, or they may aid in money laundering operations of such cells.

*Access* – Information that cannot be accessed by the user has minimal utility beyond that given by the fact of its existence. Attacks on user access may be direct via corruption (no access is possible), indirect via incapacitation of access controls such as in denial of service (DoS) attacks,[50] or elimination of authorized status (access is opened to anyone). Attacks on access via denials of service, e-mail bombs,[51] and web sit-ins[52] have been the most commonly cited examples of hacktivism or low-grade cyber terrorism.

Because of the ease of launching DoS attacks, they provide the most commonly cited examples of information conflict in cyberspace. For example a DoS attack by Israeli teenagers on Hamas and Hezbollah sites in September 2000 triggered a wave of attacks and counter-attacks between Israeli and Palestinian hackers. "By the end of January 2001, the conflict had struck more than 160 Israeli and 35 Palestinian sites, including at least one U.S. site."[53]

The flip side of denial of access to authorized users is the access to information by unauthorized users, the cyber-analogue of physical trespass. Legal systems throughout the world differ as to whether such trespass is unlawful.[54] In some nations, the cyber-trespass is illegal only if the data accessed is legally protected as being confidential information. Some legal scholars argue that unauthorized access constitutes data theft, while others argue that the data theft occurs only when authorized users are denied access to the data. Translated to the context of information warfare, unauthorized access is a form of espionage; in that case, the act is already a crime.

*Trust* – Information that is not trusted by the authorized user (however, unjustifiably) is of diminished utility and may be rendered legally non-actionable. For example, the widespread defacement and alteration of e-commerce and transactional governmental sites leading to an erosion of consumer trust[55] in internet commerce could lead substantial economic loss. Likewise, lack of trust in the security and confidentiality of communications undermines the broader public confidence in the information infrastructure, in which case, through abuses of privacy rights and expectations, governments may be their own worst enemies.

E. *Information system users* – Users of information are nodes in the network; information must flow in two directions for authorization of use to be authenticated.[56,57]

Toll records can reveal the fact of use, allowing an attacker the possibility of mining other data available on the Internet to build a personal profile of users. "Once personal information reaches the liquid realm of cyberspace, the opportunities to exploit it are endless... New experimental wireless tracking technology could one day meticulously monitor everything from the clothing on your back to the currency in your pocket. ... With so much personal information readily available in the public domain, information and communications technologies have managed to blur the once obvious line between public and private realms."[58] It is easy to imagine malicious exploitation of personal information to intimidate citizens to forego the benefits of ICTs and thereby weaken the society economically, politically, or even militarily.

F. *Information systems* – Data acquisition systems, storage systems, means of transmission and latent bandwidth, and computing or processing capacity are all subject to direct physical attack using kinetic weapons or electronic means such as jamming and burn-out with high power microwaves or electrical surges. With respect to electronic attack, information and communications technologies are harder than are generally assumed (if generally recommended precautions are taken) as they are designed to withstand nearby lightning strokes that generate large electromagnetic pulses and surges. Information systems are also vulnerable to cyber-attack. As great economic and operational utility derives from linking information and physical systems via ICTs, linked systems may be especially attractive targets for the cyber aggressor.

*Digital control systems* – The most common example linked information (both digital and analog) and high-value physical systems are open-loop supervisory control and data acquisition (SCADA) systems such are common in the utilities industries. SCADA networks may have tens of thousands of nodes and may physically span hundreds of kilometers. Since the hacker penetration[59] of a test network of a California electric power transmission company in 2001, the cyber-security challenges[60] for SCADA operators have received considerable attention. At the local or plant level, Distributed Control Systems perform analogous functions via closed loop control algorithms.

With respect to some of the loci of attack discussed above hiding the very fact of the attack raises its value to the attacker. In such cases it is difficult—or at least counter-intuitive—to speak of acts of terrorism, as no terror is created. When such surreptitious attack on the fabric of societal trust is engaged in by a non-governmental organization or network, one might justifiably consider the attacks to be acts of cyber guerrilla warfare. As noted previously, covert attacks may be seductively attractive to nation states seeking "regime change" in disfavored or pariah developing nations. The practicality of both surreptitious and surprise cyber attacks is greatly enhanced by anonymity on the Internet. In turn, the possibilities for anonymity grow with the connectivity of the network unless specific counter-measures are taken.

SOCIAL AND LEGAL FRAMEWORKS

Whether in the context of cyber crime or in the realm of international cyber conflict, the aggrieved state entity must be able to determine to appropriate evidentiary standards[61] to be actionable[62] a) what is damaged or lost, b) who launched the attack, c) from where and when was the attack launched, d) how was the attack accomplished.

*What is damaged or lost* and *how was the attack accomplished* – Determining the nature and extent of damage from an attack on information systems and computing infrastructure, while a primarily a technical issue[63] at the enterprise level, grows to include the legal issues of lines of investigatory authority[64] and determination of armed aggression[65] an "at the state/societal level.

*Who launched the attack* and *from where was the attack launched* – For a nation's response to attack to go beyond strengthening the defenses of its information infrastructure, it must discover the identity and location[66] of the attacker. To the aggrieved party, anonymity[67] in cyberspace is confounding in that it undermines both deterrence and redress of wrongs. "The analogy to hiding in cyber-warfare is the physical world use of camouflage.... the cyber-warfare protagonist must try to hide the evidence within the existing data streams."[68]

*Digital tracking* – Hackers have proven themselves adept at hiding their tracks aided by the difficulties of tracking information packets through transmission networks and aided by the technical security limitations of the present Internet Protocol version 4 (IPv4). As IPv4 limits length of an IP address to 32 bits, the dramatic increase in the number of users of the Internet has required the sharing[69] of Internet addresses. Under the next generation[70] Internet Protocol, IPv6, with 128-bit addresses, every network device can be assigned a unique, static IP address. This difference will make tracking and tracing of communications[71] far easier, assuming the storage[72] of packet contents (or a part thereof) for some limited but sufficient time.

As with other information technologies, IPv6 is dual use. With respect to potential abuses of human rights, the structured hierarchy of static IP addresses that increases network efficiency under IPv6 will make content filtering more efficient, more precisely targeted, and more difficult to circumvent. Secret monitoring by governments of the activities of their citizens becomes easier and less costly. If one presupposes that all cell phones will also have unique IPs, the stored logs of the cell pings could be mined, merged with surveillance of internet and phone usage to constitute nearly ubiquitous monitoring of millions of "suspect" individuals. Controlled under circumstances and transparent standards that are fully and clearly spelled out in legislation available to the public, surveillance in cyberspace is a legitimate tool of law enforcement and national security; arbitrarily applied, it has enormous potential for abusing human rights.

Those who seek to undermine the legitimate protections that can be afforded by tracking and surveillance procedures under proper, transparent judicial review may find it expedient to be proponents of extreme libertarian positions and use the freedom of the internet to incite the public to block the institution of any tools that could be used to undermine the more nefarious activities of these parties. Their legitimacy is made more credible by the fact that even democratic governments have demonstrated their willingness to use information technology to intrude upon the legitimate privacy expectations of their citizens.

Legislative expressions concerning unlimited rights to anonymity in cyberspace are either Pollyanna sentiments[73] from a "more innocent" time before the avalanche of cyber crime costing hundreds of billions of Euros or before cyber warfare was potentially powerful tool of international conflict, or they are more sophisticated, expressions of the tensions between compelling state interests and conflicting personal rights to anonymous communication with (unstated) delimitations.[74]

Assuming the later interpretation,[75] widespread, transparent international agreement on these delimitations and on what constitutes compelling state interests is in order, especially as it applies to the actions of nation states and state-sponsored non-governmental organizations or networks. Accordingly, in crafting the international framework for the rule of law in cyberspace, legislators must be aware of the opportunities left open to miscreants and aggressors in cyberspace when they balance issues of anonymity, privacy, and free expression against ease of applying strong digital forensics[76] and screening of the content of internet traffic. These prudential matters require informed examination, debate and striving for consensus in the broadest international forum—a practice that the World Federation of Scientists strongly encourages.

*Cyber warfare arms race* – Both nation states and their surrogates among non-governmental organizations or networks continue to increase their preparedness for information warfare. On the technological side acquiring such preparedness is inevitable as the owners of information assets have the right and typically the legal obligation to exercise due care to protect those assets. Unfortunately the tools of cyber security specialists are also the tools of the cyber aggressor. What is not dual use are information warfare doctrines and their associated implementation plans. Such plans by nation states are not open to public scrutiny and therefore are not subject to effective international control.

Military doctrines concerning information warfare can evolve and mature rapidly, especially in sets of countries (such as India and Pakistan) with a long and continuing history of conflict and with cadres of non-governmental surrogates[77] to implement concepts for structured conflict. Such situations exemplify the difficulties of distinguishing combatants from non-combatants[78] in cyberspace. Moreover, as the Israel-Palestine skirmishes show, little attention seems to be paid to the *jus ad bellum*[79] principle of military necessity, that "noncombatants and civilian objects making no direct contribution to the war effort, and whose destruction would provide no significant military advantage to the attacker, are immune from deliberate attack."[80]

The use of jurisdiction hopping by cyber-criminals and "hackivists" illustrates that the *jus ad bellum* principle of neutrality is emptied of most meaning by the intrinsically transnational character of the Internet. In light of the extensive limitations of applying the usual principles of armed conflict in cyberspace, one sees that information warfare has the potential to become a weapon of mass social and economic disruption. Especially in nascent information technologies, the level of disruption could be sufficient to markedly slow the economic and social welfare of the nation.

The grave potential of international cyber conflict calls for immediate attention. The dual use nature of the technology precludes the kind of international control regime used to control nuclear technology. What one can hope for is the creation of transnational legal framework that lays down the rules and penalties for cyber conflict in a set of structured, internationally negotiated binding agreements. Such rules must specify the obligations of the signatory nations with respect to controlling non-governmental organizations or networks that physically operate within their borders.

RECOMMENDATIONS

In the framework of Resolution 60/45, Secretary General of the United Nations should promote urgent action by the relevant UN entities to accomplish the following objectives:

1. Heighten awareness of end users, especially in countries with nascent information infrastructures, as they acquire or upgrade ICT capabilities, of major risks, and of the importance of security policies, practices, and legal responsibilities.
2. Articulate uniform, transnational legal guidelines for enterprise managers that can be embodied in the laws of nascent information societies so as to reduce opportunities for jurisdictional arbitrage.
3. Employ network technologies and security tools including strong forensic capabilities at the early installation phases of networking hardware in nascent information societies. In parallel develop a strong legal framework protecting citizens against repressive levels of digital surveillance, search and seizure.
4. Encourage and promote additional scholarly and legal study on applying the *jus ad bellum* to the information warfare domain.
5. Encourage vigorous international dialog and negotiation through the United Nations aimed at developing a transnational legal framework that lays down the rules and penalties for cyber conflict in an internationally negotiated binding agreement. Specify the obligations of the signatory nations with respect to controlling non-governmental organizations and networks that physically operate within their borders.

ACKNOWLEDGEMENTS

I thank Prof. Antonino Zichichi and the staff of the Ettore Majorana Centre for Scientific Culture for their hospitality during the initial period of preparation of this manuscript. I am grateful to Amb. Ahmed Kamal of Pakistan for his suggestions for improving the clarifying and focusing the text. I also thank Amb. Henning Wegener of Germany and Ms. Jody Westby, Esq., for their encouragement and helpful discussions during the preparation of this work.

REFERENCES

1. "…these same technologies have been adopted and adapted by militaries and quasi-military movements, thus contributing to what some might call a 'revolution in military affairs.' Thus, ICT [Information and Communication Technologies] are helping to change the way warfare is planned, organized, and conducted. This 'Revolution' encompasses developments in the ability to conduct

Intelligence, Surveillance, and Reconnaissance; to command and control forces and their operations; to optimize logistical movements; to enable precision navigation and the employment of "smart" and "brilliant" weapons.... Very significantly, it also allows for the use of the 'network' as a medium from which, through which, and in which to conduct military operations." Gen. J. Casciano, "Threat Considerations and the Law of Armed Conflict," Aug. 2005, available at http://www.itis-ev.de/infosecur/

2. "We, as a country, have put all of our eggs in one basket. The reason that we're successfully dominating the world economically and militarily is because of systems that we have designed, and rely upon, which are cyber-based. It's our Achilles heel." Richard Clarke, Interview for *PBS Frontline: Cyber War!*, March 18, 2003, http://www.pbs.org/wgbh/pages/frontline/shows/cyberwar/interviews/clarke.html
3. The more people who are connected to the information network, the greater its economic utility. "Metcalfe's law states that the value of a telecommunications network is proportional to the square of the number of users of the system ...First formulated by Robert Metcalfe in regard to Ethernet, Metcalfe's law explains many of the network effects of communication technologies and networks such as the Internet and World Wide Web.... Metcalfe's Law can be applied to more than just telecommunications devices. Metcalfe's Law can be applied to almost any computer systems that exchange data." http://en.wikipedia.org/wiki/Metcalfe's_law
4. "The simplest form of knowledge management in a computer system occurs when it is maintained. ... When a computer system monitors its own performance and is able to learn from it, it can guarantee longer periods of response without the need for maintenance. This self-monitoring also gives the system the ability to recognize when it fails and cannot learn, flagging its need for maintenance." R. Weber and D. Wu, "Knowledge Management for Computational Intelligence Systems," (2004)
5. R. Pon, M. Batalın, M. Rahimi, Y. Yu, D. Estrin, G. J. Pottie, M. Srivastava, G. Sukhatme, W. J. Kaiser, "Self-Aware Distributed Embedded Systems," Proceedings of the 10th IEEE International Workshop on Future Trends of Distributed Computing Systems (FTDCS'04), http://www.ee.ucla.edu/faculty/papers/kaiser_ftdcs2004.pdf
6. The evolution foreseen here is beyond what is embodied in the motto of Sun Microsystems, "The Network is the Computer," and more akin to the concept of Goetzel, "The Network is the Computer is the Mind." http://www.goertzel.org/books/wild/chapNC.htm For an discussion of network "self-awareness, see M. Mowbray and A. Bronstein, "What kind of self-aware systems does the Grid need?," online at www.hpl.hp.com/techreports/2002/HPL-2002-266R1.pdf
7. Such systems are termed autonomic computing systems.
8. Malware includes such malicious software as viruses, worms, Trojan horses, and spyware.
9. See for example, Kim, J., & Bentley, P. (1999). "The Human Immune System and Network Intrusion Detection", 1999. Online, via

http://citeseer.nj.nec.com/kim99human.html

10. Approaches to protecting complex networks against fault propagation have been studied in great detail in the context of the electrical power grid by the Consortium for Electric Reliability Technology Solutions, and funded by the Office Energy Efficiency and Renewable Energy, of the U.S. Department of Energy.

11. J.O. Kephart and D. M. Chess, "The Vision of Autonomic Computing," Computer, January, 2003, http://www-3.ibm.com/autonomic/pdfs/AC_Vision_Computer_Jan_2003.pdf

12. For example, a huge societal utility of distributed sensor networks linked with extensive databases is in the monitoring of public health and the detection of epidemics at an early stage. The more a society comes to depend of such functionality, the more widespread the damage from a cyber-attack on it can be.

13. [So] information warfare looks rather like air warfare looked in the 1920s and 1930s. Attack is simply easier than defense … Another possible relevant analogy is the use of piracy on the high seas as an instrument of state policy by many European powers in the sixteenth and seventeenth centuries. Until the great powers agreed to deny pirates safe haven, piracy was just too easy. The technical bias in favour of attack is made even worse by asymmetric information." R Anderson, "Why Information Security is Hard: An Economic Perspective," University of Cambridge Computer Laboratory report, (2001), www.acsac.org/2001/papers/110.pdf

14 Id. Sec. 4 and also RM Brady, RJ Anderson, RC Ball, *`Murphy's law, the fitness of evolving species, and the limits of software reliability'*, Cambridge University Computer Laboratory Technical Report no. 476 (1999); at http://www.cl.cam.ac.uk/~rja14

15. "the status of information operations under Article 51 of the UN Charter, i.e., the definition of what constitutes a "force" or "armed attack" is as yet undetermined, and that the justification of the use of legitimate self-defense is, as a consequence, equally unclear… new, extended criteria for the definition of weapons and armed aggression should be sought. Cyber attacks on other states could then be considered acts of armed aggression under the UN Charter, and, applying the principles of proportionality and necessity, thresholds for responsive actions in self-defense could be defined, taking into account the direct as well as the indirect damage cyber attacks can cause." *Information Security in the Context of the Digital Divide*, Information Security Permanent Monitoring Panel (ISPMP) of the World Federation of Scientists, Document WSIS-05/TUNIS/CONTR/01-E, Sept. 2005. P.35 http://www.itu.int/wsis/documents/listing-all-en-s|2.asp

16. "Netwar is the lower-intensity, societal-level counterpart to our earlier, mostly military concept of cyberwar. Netwar has a dual nature, like the two-faced Roman god Janus, in that it is composed of conflicts waged, on the one hand, by terrorists, criminals, and ethnonationalist extremists; and by civil-society activists on the other. What distinguishes netwar as a form of conflict is the networked organizational structure of its practitioners—with many groups actually being leaderless—and the suppleness in their ability to come together quickly in swarming attacks. The concepts of cyberwar and netwar encompass a new

spectrum of conflict that is emerging in the wake of the information revolution." Summary in *Networks and Netwars: The Future of Terror, Crime, and Militancy*, J. Arquilla and D. Ronfeldt, ed., National Defense Research Institute, RAND, 2001 In an interesting twist, the Hizballah tactics in the 2006 war with Israel have applied the paradigm of netwar back to the realm of kinetic warfare.

17. "Some experts maintain that cyber attacks with potential strategic national security effects, often referred to as an 'electronic Pearl Harbor,' are impossible. Others proclaim they are inevitable. Contemporary predictions on these matters run from the benign to the apocalyptic." C. Billo and W. Chang, "Cyber Warfare An Analysis of the Means and Motivations of Selected Nation States," Institute of Security Technology Studies at Dartmouth College, Dec. 2004, http://www.ists.dartmouth.edu/directors-office/cyber-warfare.php. As physical security measures make physical attacks by terrorists more difficult, the attractiveness of cyber attacks high value information networks will increase especially if those networks cause widespread societal discomfort or disruption. To those who relegate the cyber terrorism to the future, one replies that the future is near at hand.
18. Id., p. 3
19. "U.S. national security experts have included Iran on a published list of countries said to be training elements of the population in cyber warfare." Id., pp. 59 –74
20. "Well-documented hacker activity in Pakistan and possible ties between the hacker community and Pakistani intelligence services indicate that Pakistan appears to possess a cyber attack capability," and "Pakistan poses a threat to cyberspace with its growing army of young talented hackers. Regardless of state backing, these hackers have shown a penchant to involve themselves in real world situations such as the Kashmir conflict and countering anti-Islamic sentiments in the West following 9/11." Id., pp. 97 106
21. "Russia's armed forces, collaborating with experts in the IT sector and academic community, have developed a robust cyber warfare doctrine. The authors of Russia's cyber warfare doctrine have disclosed discussions and debates concerning Moscow's official policy. 'Information weaponry,' i.e., weapons based on programming code, receives paramount attention in official cyber warfare doctrine." Id., pp.107 – 118. Official U.S. doctrine is set forth in Joint Publication 3-13, Feb. 2006 which is available at www.fas.org/irp/doddir/dod/jp3_13.pdf A broad comparative perspective on information operations doctrines is given by T. Thomas, "Comparing U.S., Russian, and Chinese Information Operations Concepts" May, 2004 available at http://www.dodccrp.org/events/2004/CCRTS_San_Diego/CD/track03.htm
22. In a 1999 white paper for the U.S. Navy Center for Terrorism and Irregular Warfare, the authors argue that large scale damage from cyber attack upon "at large-scale command and control, industrial or infrastructure networks" requires a level of capability that they define as "Complex-Coordinated: The capability for a coordinated attacks capable of causing mass-disruption against integrated, heterogeneous defenses (including cryptography). Ability to create sophisticated hacking tools. Highly capable target analysis, command and control, and organization learning capability." W. Nelson, R. Choi, M. Iacobucci, M. Mitchell,

G Gagnon, "Cyberterrorism: Prospects and Implications," p. 87, Oct., 1999, www.nps.navy.mil/ctiw/files/Cyberterror%20Prospects%20and%20Implications.pdf

The authors further argue that such offensive capabilities are likely beyond the reach of terrorist groups. However, such capabilities are typical of the United States, Israel, most EU nations, Japan, South Korea, and perhaps India. Information technology is intrinsically dual use; the knowledge, skills, and equipment necessary for increased utility and security likewise increase the possessor's offensive capability. It is, therefore, difficult to imagine that such offensive capabilities will not typify dozens of emerging technological nations in the Middle East. Asia, and South America within the next five to ten years. See Billo and Chang, Op. cit.

23. Given the penchant of the present U.S. administration for "regime change" they may also have motivation and the will.
24. B. N. Adkins, "The Spectrum Of Cyber Conflict: From Hacking to Information Warfare: What Is Law Enforcement's Role?" Air Command and Staff College Air University document, AU/ACSC/003/2001-04, http://stinet.dtic.mil/oai/oai?&verb=getRecord&metadataPrefix=html&identifier=ADA406949
25. Hacktivism refers to writing or using computer code (hacking) to attack the target's ICT network with the purpose of promoting a political ideology or social goal. Hacktivists frequently defend their actions as acts of protest and civil disobedience. For an example see http://thehacktivist.com/hacktivism.php.
26. ISPMP, Op. Cit.
27. "Developments in the field of information and telecommunications in the context of international security, " Resolution adopted by the General Assembly [*on the report of the First Committee (A/60/452)*], A/RES/60/45, January, 2006, http://www.un.org/Depts/dhl/resguide/r60.htm
28. P. D. Allen and C, C, Demchak, "The Palestinian-Israeli Cyberwar," March - April 2003 l Military Review, p. 55 http://usacac.army.mil/cac/milreview/download/English/MarApr03/allen.pdf
29. Raymond C. Parks and David P. Duggan, "Principles of Cyber-warfare," IEEE report ISBN 0-7803-9814-9, www.shaneland.co.uk/ewar/docs/dissertationsources/educationalsource2.pdf
30. International standards on damage assessment methodologies would aid in developing an international consensus of the meaning of proportionality in cyber warfare exchanges.
31. For example, a recent U.S. Academy of Sciences study found that "Although there are many private and public databases that contain information potentially relevant to counter terrorism programs, they *lack the necessary context definitions (i.e., metadata)* and access tools to enable interoperation with other databases and the extraction of meaningful and timely information" National Research Council (2002), "Making the Nation Safer: The Role of Science and Technology in Countering Terrorism." Washington, D.C., National Academies Press. http://www.nap.edu/html/stct/index.html

32. As the sources of an institution's data become more complex and as data rates increase, the size of data files and information system complexity also grow. Automated testing checking routines become a necessity to ensure the validity and integrity of the data. With respect to data errors, checksums and automatic evaluation of data correlation can reveal faulty data. To examine data against clandestine alteration or corruption or to compare the purportedly identical copies of data, the checksums are comparisons of hash functions of the data files in question.
*One Way Hash Functions* are mathematical algorithms that transform an arbitrarily long input message into one of fixed length. To be useful and secure a hash must be computationally efficient, collision-free, and provably impossible to compute the inverse of the hash function. A hash function is collision-free if it is computationally infeasible to find two input messages that produce the same output. Modern hash algorithms produce hash values of at least 128 bits.
*Hashing* is also a pervasive tool in information forensics. "Examiners use hash values throughout the forensics process, from acquiring the data, through analysis, and even into legal proceedings. Hash algorithms are used to confirm that when a copy of data is made, the original is unaltered and the copy is identical, bit-for-bit. That is, hashing is employed to confirm that data analysis does not alter the evidence itself. Examiners also use hash values to weed out files that are of no interest in the investigation, such as operating system files, and to identify files of particular interest." R. P. Salgado, Fourth Amendment Search and the Power of the Hash, 119 Harv. L. Rev. F. 38 (2006), http://www.harvardlawreview.org/forum/issues/119/dec05/salgado.shtml
33. J. R. Westby and W. A. Barletta, "Consequence Management of Acts of Disruption," August, 2003, p. 11, http://www.itis-ev.de/infosecur
34. Widely applied, strong encryption can protect the confidentiality and integrity of disclosure-sensitive information during transmission to authorized recipients. Obviously, encryption must be applied in a way that does not compromise the strength of the procedure, and an institution must maintain the security of and access to its cryptographic keys. A discussion of the policies of governments around the world regarding encryption policies and responsibilities may be found in "Cryptography and Liberty 2000," Electronic Privacy Information Center, http://www2.epic.org/reports/crypto2000/overview.html#Heading7.
35. This category includes the kinds of attacks traditionally referred to as electronic warfare such as electronic suppression (jamming) of receivers, transmitters and other systems. "The application of international law to these traditional kinds of operations is reasonably well settled." "An Assessment of International Legal Issues in Information Operations" Department of Defense. Office of General Counsel, May, 1999, page 5, available at www.au.af.mil/au/awc/awcgate/dod-io-legal/dod-io-legal.pdf [hereinafter DOD General Counsel]
36. W. E. Johnston, E. Schultz, M. Livny, B. Miller, S. Canon, M. Helm, D. Olson, I. Sakrejdab, and B. Tierney, "An Integrated CyberSecurity Approach for HEP Grids and Clusters," http://dsd.lbl.gov/HEP-CyberSecurity/HEP-Cybersecurity.WP.pdf Hereinafter, Johnston et al.

37. "America's precious and powerful supercomputers are bound together by the "Grid/TeraGrid" which has now been proven to be extraordinarily vulnerable to intrusion. The recent hack of the Grid was most likely accomplished by a small group of young U.S. hackers." L. Z. Koch, "A quiet time bomb: The vulnerability of U.S. supercomputers" Raw Story, (2004) http://www.rawstory.com/exclusives/koch/vulnerable_computer_Grid.htm
38. "The new high speed networks make it impossible with today's technology to have a standard firewall and intrusion detection system implementation, as they cannot handle the volume of data being transported. New approaches to securing these networks are necessary. The technology is moving faster than security ever will." D. Agarwal, "A Distributed Science Cybersecurity Program," 15 August 2005, http://dsd.lbl.gov/~deba/publications/DOE-NSF-Cybersecurity-white-paper-draft-v3.1.pdf
39. Intrusions are defined as "attempts to compromise the confidentiality, integrity, availability, or to bypass the security mechanisms of a computer or network." Rebecca Bace and Peter Mell, "NIST Special Publication on Intrusion Detection Systems," Feb 12, 2001, http://www.dougmoran.com/tatzlwyrm/CACHE/nist_special_pub_on_ids.pdf
40. http://www.princeton.edu/~protect/BasicConceptsAndTips/SecurityIsRisk Management/WhyInformationOwners.shtm
41. Attack on the credibility of institutions and social fabric is documented aspect of Russian and Chinese information warfare operations. Thomas, Op. cit.
42. For example, "Israel hacks into Hezbollah TV," http://www.news.com.au/couriermail/story/0,,19992671-1702,00.html August 2, 2006 and "CNN's Anderson Cooper Exposes Hezbollah's Media Manipulations," August 7, 2006, http://www.infowar-monitor.net/
43. D. E. Denning, "Power Over the Information Front," Yale conference on the Global Flow of Information, March 10, 2005, http://islandia.law.yale.edu/isp/GlobalFlow/paper/Denning.pdf
44. "Social engineering is the practice of obtaining confidential information by manipulation of legitimate users. A social engineer will commonly use the telephone or Internet to trick people into revealing sensitive information or getting them to do something that is against typical policies… social engineers exploit the natural tendency of a person to trust his or her word, rather than exploiting computer security holes. It is generally agreed upon that "users are the weak link" in security and this principle is what makes social engineering possible." http://en.wikipedia.org/wiki/Social_engineering_%28computer_security%29
45. "Even when they are targeted by law enforcement, many criminal networks are inherently dispersed, with the result that they do not provide … loci for law enforcement attacks… networks, especially when they are transnational in character, can exploit differences in national laws and regulations by engaging in what might be termed jurisdictional arbitrage." P. Williams, "Transnational Criminal Networks," in *Networks and Netwars*, Op. cit. http://www.rand.org/pubs/monograph_reports/MR1382/index.html. What Williams writes about criminal networks is *a fortiori* true of international terrorist networks.

46. For an example, see "Online scams create 'Yahoo! Millionaires'" http://money.cnn.com/magazines/fortune/fortune_archive/2006/05/29/8378124/index.htm?cnn=yes
47. "Simply stated, identity theft occurs when a thief obtains confidential information about another individual, and uses it to defraud others. ... Identity theft is uniquely dangerous because it is an enabling crime—one that permits criminals to commit other crimes." Valetk, Op. Cit., ¶13
48. "Pretexting is to pretend that you are someone who you are not, telling an untruth, or creating deception... pretexting involves tricking the telecom carrier into giving up personal information, in most cases, with the scammer pretending to be the customer." http://en.wikipedia.org/wiki/Pretexting In the U.S., pretexting is illegal under the provisions of the Gramm-Leach-Bliley Financial Services Modernization Act, Pub. L. No. 106-102, 113 Stat. 1338. However, attempts to extend legal protections in the U.S. this spring, though supported by overwhelming bi-partisan majorities, have been thwarted by what must be overwhelming pressure from unknown sources, possibly including the White House. Statements by the Chair and ranking member, Proceedings of the House Sub-committee on Oversight and Investigations, "Internet Data Brokers and Pretexting: Who has Access to Your Private Records?" 21 June 2006. http://energycommerce.house.gov/108/Hearings/06212006hearing1916/hearing.htm
49. For example, Johnston et al. argue that identify theft was the initial, enabling step in the attack launched against the ESNet of U.S. supercomputers.
50. In a denial of service attack the attacker prevents access to or use of use of a cyber resource. An attack initiated by a single low-bandwidth source can be amplified into a distributed attack with several thousand innocent "host" computers compromised in under an hour. Distributed or amplified attacks can "flood" a network with large volumes of data or deliberately and completely consume limited resources, such as network connections so that no legitimate service requests can be processed.
51. E-mail bombing is the practice of sending huge volume of emails to the target's mailbox to disrupt the legitimate business or governmental interests of the target See T. Bass, A Freyre, D. Gruber, E-Mail Bombs and Countermeasures: Cyber Attacks on Availability and Brand Integrity, IEEE Network, Vol. 12, No. 2, pp. 10-17, March/April 1998. http://www.silkroad.com/papers/html/bomb/
52. "The theory behind the cyber-sit-in is that thousands of ordinary Internet users can 'picket' as site. They do this by making repeated connections to the server – equivalent to continually pressing the 'reload' button on your web browser – in order to tie up the server with more requests than it can handle." http://www.thing.net/~rdom/ecd/bushwar/help.html
53. P. D. Allen and C. C. Demchak, Op. cit., p. 52
54. Relevant Council of Europe Conventions can be found at http://conventions.coe.int/treaty/EN/cadreprincipal.htm
55. Id., p. 54
56. The most ubiquitous (and weakest) form of authentication is the password. As computer software can easily try many millions of passwords, a determined

attacker, a primary weakness of passwords is that they may be guessed. The use of one-time, transient password ameliorates this problem. A further increase in security obtains from adopting two-factor authorization through the addition of physical access tokens, than may also include a hash of a biometric characteristic of the user. For a comparison of the technical challenges of voice, fingerprint, face, iris and retinal recognition see "More Than a Pretty Face, Biometrics and SmartCard Tokens", Gregory Williams, The SANS Institute, Dec. 24, 2001, http://www.sans.org/rr/authentic/pretty_face.php

57. A lengthy legal analysis of authorization and access misuse is given in "Cybercrime's Scope: Interpreting 'Access' and 'Authorization Misuse Statutes", Orin Kerr, 78 New University Law Review, Nov., 2003
58. H. Valetk, "Mastering the Dark Arts of Cyberspace: A Quest for Sound Internet Safety Policies," 2004 STAN. TECH. L. REV. 2, ¶92, http://stlr.stanford.edu/STLR/Articles/04_STLR_2
59. Robyn Welsman, "California Power Grid Hack Underscores Threat to U.S.," www.newsfactor.com/perl/story/11220.html, 13 June 2001.
60. For an example of such analysis see, Understanding SCADA System Security Vulnerabilities, RIPTech report, Jan., 2001, www.iwar.org.uk/cip/resources/utilities/SCADAWhitepaperfinal1.pdf
61. Actionable information (evidence) must be sufficiently relevant, reliable, complete, accurate, and verifiable whether in a judicial or political sense. While there are now many national and international organizations devoted to developing standard procedures for the collection, retention, testing, and display of digital evidence, the legal framework of digital evidence is still evolving. In the U.S., "there is debate about whether digital evidence falls under the Daubert guidelines as scientific evidence or the Federal Rules of Evidence as nonscientific technical testimony." B. Carrier, "Open Source Digital Forensics Tools: The Legal Argument," Sept. 2003, www.digital-evidence.org/papers/opensrc_legal.pdf
62. "Actionable" is used here to refer to any substantive action that the aggrieved party may take. In the context of information warfare such actions span prophylactic-defense measures against future attack to legal prosecution to diplomatic or even military measures.
63. Even in this case an organization's investigations may be guided by overarching statute as in the U.S. where internal corporate investigations must fulfill the requirements of internal controls under Sarbanes-Oxley legislation. V. Limongelli, M. Abascal, "Internal Computer Investigations as a Critical Control Activity Under Sarbanes-Oxley," December, 2003, www.guidancesoftware.com/corporate/downloads/whitepapers/SarboxOnlineSeminarTranscript.pdf
64. For example, are military or national civilian intelligence or local police authorities the cognizant investigatory entity?
65. "'Is a computer network attack an 'armed attack' that justifies the use of force in self-defense?'... It might be hard to sell the notion that an unauthorized intrusion into an unclassified information system, without more, constitutes an armed attack. On the other hand, if a coordinated computer network attack ... causes

widespread civilian deaths and property damage, it may well be that no one would challenge the victim nation if it concluded that it was a victim of an armed attack, or of an act equivalent to an armed attack." DOD General Counsel, p. 18

66. "Attacks that cannot be shown to be state-sponsored generally do not justify acts of self-defense in another nation's territory... Only if the requested nation is unwilling or unable to prevent recurrence does the doctrine of self-defense permit the injured nation to act in self-defense inside the territory of another nation." Id. P. 22.

67. Council of Europe Convention on Cybercrime recognizes a legitimate right to internet anonymity. "In order to...enhance the free expression of information and ideas, member states should respect the will of users not to disclose their identity." Declaration on freedom of communication on the Internet (Strasbourg, 28.05.2003), (Adopted by the Committee of Ministers at the 840th meeting of the Ministers' Deputies) http://www.coe.int/T/E/Communication_and_Research/Press/News/2003/20030528_declaration.asp

68. Parks and Duggan, op. cit., p. 123.

69. "For example, a shortage of IP addresses has led to the increased use of dynamic IP addresses instead fixed IP addresses, and the use of network address translation (NAT) allowing multiple machines to share a single globally routable IP address", "Tracking and Tracing Cyber-Attacks: Technical Challenges and Global Policy Issues", Howard F. Lipson, p. 55

70. Unfortunately the adoption of IPv6 is significantly impeded by owners of information transmission networks that have large sunk investments in routers incompatible with IPv6 and that therefore have near-term economic interests often contrary to the long-tern benefit of their own enterprises. In the absence of legal and policy checks on perverse incentives, legal externalities can actually encourage, if not amplify negative effects of information attacks. Where such perverse financial incentives exist, fiduciary responsibilities of managers legally require them to exploit such incentives for proximate gain of their enterprise rather than to search for approaches more consistent with broader community interests.

71. Lipson, p. 61.

72. Such storage will require very large data storage resources, even if only a small fraction of the packet content is retained. Large-scale storage has significant privacy implications, and is clouded with jurisdictional, legal, and law enforcement considerations.

73. Ironically such expressions within nations with free and open social institutions are likely to find full support and even active promotion from those internal and external enemies seeking to launch cyber attacks against those very societies.

74. In terms of criminal law, it would be corrosive of public confidence in government if an unlimited right to anonymity for the criminal were to trump the right of the victim to seek the identity of the perpetrator. Translating this tension of conflicting rights the realm of cyber aggression by nation-states or state-sponsored terrorists, one would expect that insisting on the right of the aggressor

to remain anonymous would be highly deleterious to the rule of law among nations.

75. In fact, based on European case law the Council of Europe has developed extensive standards governing delimiting anonymity through interception of communications.

76. For extensive analysis of digital, search and seizure in the context of rights under the Fourth Amendment of the U.S. Constitution, see O. Kerr, "Searches and Seizures in a Digital World," Harv. L. Rev. 531 (2005) www.harvardlawreview.org/issues/119/Dec05/Kerr05.shtml and the discussions concerning this paper, R. P. Salgado, Op. Cit.. and P. Ohm, "The Fourth Amendment Right to Delete," 119 Harv. L. Rev. F. 10 (2005) www.harvardlawreview.org/forum/issues/119/dec05/ohm.shtml Also, O. Kerr, "Search Warrants in an Era of Digital Evidence," Mississippi Law Journal, 2005, Available at http://papers.ssrn.com/sol3/papers.cfm?abstract_id=665662

77. "Applying this principle to information warfare is problematic. One might assume that traditional nation-states would take the approach of organizing, planning, and operating their information warfare capabilities under traditional command authority and military rules of engagement. However, it is unclear whether this approach is generally accepted among those who possess such weapons capabilities, however desirable such an approach might be." Casciano, Op. cit.

78. A further difficulty is the common hacker practice of seizing clandestine control of thousands of computers that are used as unwitting accomplices (zombies) in a cyber attack.

79. For a synopsis of these principles see DOD General Counsel, pp. 5 – 7.

80. Id., p. 6

# COUNTERING TERRORISM WITH CYBER SECURITY

JODY R. WESTBY, ESQ.[1]
Global Cyber Risk LLC, Washington, USA

## THE PROBLEM

Terrorism is flourishing through terrorists' use of information and communication technologies (ICTs) in a globally-connected world with over one billion online users[2] and 233 countries connected to the Internet.[3] In part, this is due to (a) difficulties in tracking and tracing cyber communications, (b) the lack of globally-accepted processes and procedures for the investigation of cybercrimes, and (c) inadequate or ineffective information sharing systems between the public and private sectors. Although there are technological reasons why tracking and tracing are difficult and information security of shared information can be tricky, the problem largely rests with the plodding nature of governments around the globe in addressing these critical issues through legal frameworks and policy directives that would advance information security and improve the detection and prosecution of cyber criminal activities.

Following a global crackdown on terrorism following 9/11, which included a tightening of borders, increased surveillance, more restrictive visa requirements, military combat, and enhanced intelligence operations, terrorists quickly learned to leverage ICTs to keep their operations alive and advance their agendas. In short order, they developed a flat, global operation, with cells spread over more than 60 countries,[4] each pursuing their own local agenda while cooperating on a global scale.[5] Indeed, Jarret Brachman, Director of Research for the Combating Terrorism Center at the U.S. Military Academy, recently acknowledged that Al-Qaeda can no longer be considered just an "organization."

Despite the considerable resources that the United States has dedicated to combating *jihadi* terrorism since the attacks of September 11, 2001, its primary terrorist enemy, al-Qaeda, has mutated and grown more dangerous. Al-Qaeda today is no longer best conceived of as an organization, a network, or even a network-of-networks. Rather, by leveraging new information and communication technologies, al-Qaeda has transformed itself into an organic social movement, making its virulent ideology accessible to anyone with a computer.[6]

Today, almost all terrorist organizations, large or small, have their own web sites.[7] They cooperate with organized crime[8] and use technology to spread propaganda, raise funds and launder money, recruit and train members, communicate and conspire, and launch attacks while governments are trying to counter and catch them using traditional means.[9]

One of the reasons the Internet is such an attractive medium for terrorists, is the technological difficulties associated with tracking and tracing cyber communications in the current environment which, for the most part, is based on Internet Protocol (IP) version 4 (IPv4). Tracking and tracing is particularly difficult using IPv4 because (a) its IP address space is only 32 bits, and (b) the header of the packet is not large enough to hold authenticated tracking and audit information for the entire path of the packet. Therefore, due to very limited header space, IPv4 packet-marking requires tracking data to be split across multiple packets, which allows attackers to insert false information to mask the packet path. These inadequacies with IPv4 significantly hamper tracking and tracing of cyber communications and the investigation of cybercrimes. Internet Protocol version 6 (IPv6) is the next generation Internet Engineering Task Force (IETF) protocol. It has a significantly larger header space and a 128 bit IP address, thereby enabling unique Internet IP addresses to be assigned to all users and the full routing of the packet to be captured.[10] In fact, the implementation of IPv6 will allow for several quadrillion unique IP addresses[11]—enough to assign addresses to all computers, radio frequency identification tags (RFID), sensors, unmanned aerial vehicles, and other devices communicating via networks.[12]

Although the problems associated with IPv4 certainly contribute to problems encountered in the investigation of cyber crimes and suspect communications, another layer of difficulty in tracking and tracing cybercrimes is caused by disparities in the various legal systems around the world. Cyberspace has no borders, but law enforcement, prosecutors and judges, and diplomats do; they must stop at the borders of their sovereign state when investigating cyber activities and proceed according to protocol, which can be complex and time-consuming. The barriers encountered at borders are numerous and include:

- Inadequate and inconsistent legal frameworks governing cybercrimes and government access to information;
- Jurisdictional issues, such as the necessity of letters rogatory, dual criminality requirements, extradition restrictions, conflicts of laws, and inadequate procedural laws;
- Lack of sufficient technical expertise of law enforcement, prosecutors, and judges regarding investigative assistance and the search and seizure of electronic evidence; and
- Inadequate mechanisms and procedures for international cooperation.[13]

The Council of Europe (CoE) Cybercrime Convention attempts to address many of these issues, but, unfortunately, it has been signed by only 43 countries and ratified by only 16 of them, including the recent ratification by the United States Senate.[14] Likewise, the European Union's *Council Framework Decision 2005/222/JHA of 24 February 2005 on attacks against information systems*, which went into force on March 16, 2005, mirrors many of the provisions in the Cybercrime Convention.[15] The EU's 25 Member States have until March 16, 2007, however, to implement the Decision into national law.[16] With 233 nations connected to the Internet, 200 of which are developing countries, the Cybercrime Convention and EU Decision are a step in the right direction, but they are a far cry from a harmonized global framework with any immediate benefits regarding investigations and tracking/tracing of cyber communications. In fact, in a perverse sense, the CoE Cybercrime Convention has created a measure of lethargy on cybercrime issues in the international arena. With a solid multilateral agreement in place open for signature and ratification, multilateral pressure to address the legal gaps regarding cybercrime has actually eased. Terrorists have been quick to take advantage of the glacial pace of multilateral action.

*Thus, when the technical difficulties presented by IPv4 are coupled with the legal and logistical difficulties encountered in tracking and tracing and investigating cyber communications, terrorists have found the almost-perfect medium by which to continue their activities in pursuit of their goals.* Even if law enforcement is lucky enough to be able to track back a packet path, it may not be so fortunate in receiving the cooperation it needs outside its own borders to advance the investigation, seize needed evidence, and prosecute. This is why effective information sharing is so crucial to countering terrorism.

Public and private sector entities must work together to protect their critical infrastructure, facilities, operations, and personnel. Information sharing is a crucial element for the detection, prevention, and mitigation of cyber attacks and terrorist activities. It requires a commitment from the public and private sectors and involves the establishment of systems and networks, security protocols, and well-tested—and trusted—policies and procedures. Information sharing must occur on many levels for it to be effective. This includes:

- Intra-governmental sharing between government agencies and departments;
- Inter-governmental sharing of information between layers of government (local, state/provincial, and national);
- Public-private sharing of information between industry and government at all levels; and
- The sharing of information between foreign government and intelligence agencies.

Many issues work against effective information sharing, including cultural issues, lack of trust, reluctance to engage in the mutual recognition of clearances, jurisdictional and legal/policy issues, and reputational concerns. The U.S. government has spent much of the last decade coaxing and cajoling industry into participating in information sharing programs designed to advance cyber security and protect critical infrastructure. Likewise, local, state, and federal officials have struggled to implement effective information sharing initiatives. All of these efforts have, at best, achieved minimal results. A December 2005 investigative report developed by the U.S. House Committee on Homeland Security Democratic Staff bluntly noted:

> *"Despite numerous directives, exhortations, and invitations to do so, federal policymakers have failed to develop uniform standards for converting classified intelligence into an unclassified or "less classified" format that can be disseminated rapidly to appropriate state, local, and tribal authorities to thwart terrorist attacks. They likewise have failed to create effective mechanisms through which the particular intelligence needs of those authorities can be voiced and met, or where their own information assets can be shared with the Intelligence Community (IC). This distressing lack of leadership has persisted for more than four years."*[17]

The U.S. General Accountability Office (GAO) concurred in a March 2006 report on information sharing, stating that, "More than 4 years after September 11, the nation still lacks government wide policies and processes to help agencies...improve the sharing of terrorism-related information that is critical to protecting our homeland."[18]

Information sharing systems are inherently dependent upon effective information security. When information is exchanged—especially classified and intelligence information—it is imperative it be done in a trusted environment and that authentication, authorization, confidentiality, and integrity of the data can be assured. Although technological tools have been developed to help meet this challenge, the weaknesses associated with IPv4 and the lack of a framework for harmonized global assistance present formidable challenges to establishing such a trusted environment.

*Thus, we have a conundrum; tracking and tracing cyber communications and global cooperation in cyber investigations are dependent upon trusted information sharing; yet trusted information sharing is undercut by weaknesses in tracking and tracing and gaps in global cooperation.* Bottom line: Terrorists are more effective at using our technologies than we are at stopping them and tracking and tracing their online activities.

## TERRORIST USE OF ICTS

### Vision

Al-Qaeda has mastered the art of using the media to advance their own goals. On the heels of 9/11, Ayman al-Zawahiri, Osama bin Laden's second in command, declared:

> *"We must get our message across to the masses of the nation and break the media siege imposed on the jihad movement. This is an independent battle that we must launch side by side with the military battle."* [19]

Experts agree that Zawahiri's vision has been realized. Jihadists around the globe do more than just log on to read al-Qaeda postings, view propaganda, review training materials, and make donations. They are more than consumers of *jihadi* content; they have become virtual foot soldiers and are *producing* it.[20]

That terrorists are producers of persuasive content is a logical consequence. Despite popular belief that those who join jihadi groups are disenfranchised, poverty-stricken youth who are victims of globalization, an extensive analysis of al-Qaeda-linked terrorists by former CIA official Marc Sageman indicated that two-thirds of those in the sample population were middle-class professionals, with university educations and technical capabilities. Terrorism is not due to economic deprivation and lack of societal advancement; terrorism has a bourgeois foundation, not unlike its Christian and Jewish enemies.[21] We must understand our enemy is educated, technically proficient, cunning, and socially adaptable.

## THE REALITY

Since 9/11, Zawahiri has made numerous televised broadcasts, published memoirs, given interviews, and has posted about a dozen speeches on the Internet. Televised broadcasts of his addresses have been subsequently posted to Internet sites.[22] The most famous producer of terrorist content, however, was Abu Musab Zarqawi, the terrorist leader in Iraq who claimed alliance with al-Qaeda and perpetuated the myth put forward by the Bush Administration that al-Qaeda had operations in Iraq.[23] Zarqawi posted beheadings of U.S. contractors and took credit for suicide bombings on behalf of al-Qaeda.[24] He showed his face in an Internet video and accused President Bush of lying to Americans about military successes in Iraq.[25] His postings and media campaign, conducted largely over the Internet, helped attract al-Qaeda followers to Iraq and establish the irony that the U.S. invasion in Iraq, not Saddam Hussein, had enabled al-Qaeda to use Iraq as a breeding ground for terrorist activities. Zarqawi was killed by U.S. forces on June 7, 2006.[26]

Two other jihadi content-producers, whose actions behind-the-scenes significantly contributed to the development of a flat, global and virtual al-Qaeda, are sympathizers Mustafa Setmariam Nasar and Younis Tsoulis. Setmariam, a red-haired, fair complexioned Syrian who leveraged his knowledge of Western culture gleaned from years living in Europe, has been touted by some to be "al-Qaeda's most influential strategist since 9/11."[27] Setmariam was one of al-Qaeda's biggest proponents of the use of weapons of mass destruction (WMD), calling on the Internet for a WMD attack against the United States: "Dirty bombs for a dirty nation."[28] U.S. authorities believe he helped instruct recruits on the use of WMDs at Afghanistan training camps.

Most importantly, Setmariam was a leading theorist behind the diffusion of jihad globally and the transformation of al-Qaeda into a flatter, looser, organization that could be more resilient—and harder to defeat—with jihad waged on a global scale by local cells and supported by a massive influx of recruits from around the world. He also devised recruitment strategies that played on the "Jewish-Crusader oppression of Muslims" and the "degeneracy of the Western world."[29] He developed this strategy at the al-Ghurba camp in Afghanistan during 1998-2001, where he taught al-Qaeda leaders. His lectures were distributed throughout Muslim countries and Europe and were incorporated into the 1,600-page document, "The Call for a Global Islamic Resistance," which was posted on the Internet in December 2004. Some experts believe his work could have influenced the London and Madrid bombings. Setmariam was arrested in Pakistan in fall 2005, but his capture has not diminished his influence among his target audience of jihadists worldwide. As noted by Jordanian journalist Fuad Hussein:

> *"I monitor the Islamist Web sites every day, and every day there are new postings of Setmariam's research, writings, chapters of his books and tapes. He has big credibility because [the jihads] know his history. People read this in Iraq, the Arab world, in Europe and all over the world."*[30]

In November 2005, Scotland Yard arrested Younis Tsouli, who was suspected of being involved in a bomb plot and noticed immediately afterward that postings on Internet message boards by a person known online as "Irhabi 007" had stopped (Irhabi means "terrorist"). Searches of Tsouli's home following his arrest turned up stolen credit card information that was used to pay Internet service providers whose networks Tsouli used to post jihadi propaganda. It was at this point, that the investigators realized they may have bagged the elusive Irhabi 007, the terrorist whose Internet postings the international intelligence community had tried to track and trace for over two years—without even uncovering his identity.

Irhabi was able to join password-protected sites used by al-Qaeda for instruction, propaganda, and recruitment. He secretly and securely disseminated manuals of weaponry, videos of attacks and inflammatory materials, and assisted Zarqawi in releasing his communications thru one of the well-known al-Qaeda sites, Muntada al-Ansar al-Islami (Islam Supporters Forum). Irhabi 007 became the jihadi ICT expert and Internet "help line" of sorts, answering questions and helping other jihadists learn to use the Internet to achieve their objectives. He offered instruction on how to post videos, defeat and enhance security, conduct anonymous browsing, crack server vulnerabilities, and use third-party hosts to disseminate information. He posted a lengthy document entitled "Seminar on Hacking Websites," and it is believed he helped distribute a film produced by Zarqawi containing footage of attacks and Osama bin Laden as well as comments about Abu Ghraib prison and the new Iraqi government.[31] His influence continues; two persons suspected of plotting bombings were arrested recently in Toronto, and they reportedly received advice from Tsouli prior to his arrest.[32] Like Setmariam, Tsouli's capture has done little to quell his impact through the Internet. Prior to his arrest, he released his will over the Internet, which provided links to information that would help jihadists with their computer skills.

Beyond terrorists' use of the Internet to train, recruit, spread propaganda, and fundraise, there is the looming threat of a major cyber attack on critical infrastructure or an attack using a WMD. Setmariam has openly expressed his angst that the planes that crashed into the World Trade Center did not possess a WMD. He argues that the loss of life in the U.S. was too low to justify al-Qaeda's loss of sanctuary in Afghanistan. As a practical matter, he realizes that defeating the U.S. through traditional means would take years and require enormous sacrifice. Thus, he favors defeat more quickly through the use of a WMD, whether biological, chemical, or nuclear.[33] Following his arrest, cohorts posted his final proposal urging jihadis to attack while the U.S. and its allies are bogged down with Iraq and Afghanistan:

> *"I reiterate my call for mujahideen who are spread in Europe and in our enemies' countries or those able to go there, to move fast to hit countries that have a military presence in Iraq, Afghanistan or the Arab peninsula or to hit their interests in our countries and all over the world."*[34]

Setmariam misses the point that an attack on critical infrastructure would be much easier to plan and launch than one using WMD, which would require a physical presence in the targeted country. With stricter immigration controls and increased monitoring of international movements, technology offers an attractive way to conspire and commit acts of terror from foreign lands without the need to apply for visas or set foot on the soil of the targeted country. Indeed, the al-Qaeda training manual, "Military Studies in the Jihad Against the Tyrants," instructs jihadists that 80% of the information needed regarding a target can be obtained from public information.[35] The Internet affords these terrorists the perfect tool.

U.S. government officials have repeatedly warned that terrorists could launch cyber attacks against critical infrastructure[36] or "consequential infrastructure," which includes the ICT systems of private sector companies that, when manipulated, could cause a catastrophic event of enormous consequence, harming masses of civilians or wreaking economic chaos.[37] There is now ample evidence that terrorists have been utilizing sophisticated technologies to communicate, conspire, and plan attacks on such infrastructure utilizing sophisticated technologies.

For example, after September 11, the U.S. Federal Bureau of Investigation (FBI) discovered that online users, whose activity was routed through switches in Saudi Arabia, Pakistan, and Indonesia, were exploring the digital systems of emergency telephone, electrical generation and transmission, water storage and distribution, nuclear power plants, and gas facilities.[38] Computers seized in Pakistan in July 2005 contained material from "casings" of key financial institutions located in New York, Washington, D.C., and Newark, New Jersey, prompting Homeland Security alerts to these organizations and locales.[39] These sorts of attacks are not solely aimed at the U.S., however. France's terrorism chief, Jean-Louis Bruguiere, announced that al-Qaeda was planning an attack on a leading Asian financial center. Tokyo, Sydney, and Singapore were cited as possible locations because countries in the region are considered to be less prepared for this type of attack.[40] Attacks on financial systems are all the more likely to come from terrorists versus nation states because, unlike countries that are all dependent upon the functioning of global financial systems, terrorists specifically reject the global market economy.[41]

ICTs can easily function as weapons for asymmetric attacks. U.S. Department of Defense (DoD) officials have expressed concern that cyber attacks could be directed at disrupting military operations. Military operations are dependent upon a wide range of civilian high technology products and services such as communications and computer networks, technology components of weapons systems, supervisory control and data acquisition (SCADA) systems and operating platforms, and other software and electronic products.[42]

Irrespective of whether terrorists use the Internet for cyber attacks or whether they plot and plan a WMD attack, most certainly ICTs will be involved. From the *offensive* side of the terrorist, cell phones were used to detonate the bombs in Madrid; the Internet was used by the 9/11 hijackers in planning and communicating; and the suspects arrested in Toronto used it for plotting bombings. There are numerous other examples. From the *defensive* side of civilian populations, critical infrastructure and SCADA systems are known to have software vulnerabilities and are almost always connected to the Internet; responders are dependent upon radio, telephone, cellular, satellite, and Internet communications; and business operations are completely dependent upon ICTs.

## WHAT TO DO

To date, Western governments have failed to grasp the full threat and danger of terrorists' use of ICTs—and they have failed to understand how to use these technological means as a strategic tool against al-Qaeda. Bruce Hoffman, RAND's Corporate Chair in Counterterrorism and Counterinsurgency, bluntly stated in testimony before the U.S. House of Representatives Permanent Select Committee on Intelligence that, "To date, at least, the United States…has not effectively contested the critical, virtual battleground that the Internet has become to terrorists and their sympathizers and supporters worldwide."[43] Hoffman further noted that terrorists' propaganda on the Internet has "acquired a veneer of truth and veracity simply because of their unmitigated and unchallenged repetition and circulation throughout the Internet."[44]

It is important that governments—particularly the U.S.—recognize the power behind jihadi content and the role it plays in fueling terrorism around the globe. Jarret Brachman articulately summed up the current situation:

> *"Agencies tasked with monitoring the jihadi movement's use of email, chat rooms, online magazines, cell phone videos, CD-ROMs, and even video games look for immediate intelligence indicators and warnings. However, there has been little directive (or bureaucratic incentive) for these agencies to situate the technological activity they monitor in a broader strategic context. Unfortunately, it is the strategic – not the operational – objectives of the jihadi movement's use of technology that engenders the most enduring and lethal threat to the United States over the long term.*
>
> *If Western governments made reading the online statements posted by al-Qaeda's ideologues a priority, they would better realize how the jihadi movement is not simply using technological tools to recruit members, receive donations, and plan attacks. In actuality, al-Qaeda's use of the Internet and other new technologies has also enabled it to radicalize and empower armies of new recruits by shaping their general worldview."*[45]

This cyber mobilization of jihadists is, in part, due to a "lack of cyber influence/neutralization on the part of governments to counter these recruiting efforts."[46] Nevertheless, terrorists' unfettered ability to engage in these cyber activities could be significantly curtailed if we had the ability to track and trace cyber communications, with global cooperation from law enforcement, courts, governments, and industry. This will require the establishment of trusted environments between the public and private sectors for information sharing that are based upon proven security technologies, legal agreements, diplomatic accords, and operational controls. Quite simply, we must learn to better leverage the ICTs we invented against our enemies and to mobilize our public and private sector resources if we are to hope to restore the rightful role of the rule of law in our societies that terrorists have up-ended.

*Thus, it seems perfectly obvious that actions aimed at increasing cyber security could significantly counter terrorism.* The U.S. Congressional Research Service noted in its report, *Computer Attack and Cyberterrorism: Vulnerabilities and Policy Issues for Congress,* that:

> *"It remains difficult to determine the identity of the initiators of most cyberattacks, while at the same time security organizations continue to report that computer virus attacks are becoming more frequent, causing more economic losses, and affecting larger areas of the globe....The challenge of identifying the source of attacks is complicated by the unwillingness of commercial enterprises to report attacks, owing to potential liability concerns....However, while the number of random Internet cyberattacks has been increasing, the data collected to measure the trends for cyberattacks cannot be used to accurately determine if a terrorist group, or terrorist-sponsoring state, has initiated any of them."*[47]

This is the hard, cold truth, as noted by journalists Rita Katz and Michael Kern:

> *"The unwitting end of the hunt [for Irhabi] comes at a time when al-Qaeda sympathizers ... are making explosive new use of the Internet. Countless Web sites and password-protected forums—most of which have sprung up in the last several years—now cater to would-be jihadists like Irhabi 007. The terrorists who congregate in those cybercommunities are rapidly becoming skilled in hacking, programming, executing online attacks and mastering digital and media design—and Irhabi was a master at all of those arts.*
>
> *But the manner of his arrest demonstrates how challenging it is to combat such online activities and to prevent others from following Irhabi's example....The Internet has presented investigators with an extraordinary challenge.* But our future security is going to depend increasingly on identifying and catching the shadowy figures who exist primarily in the elusive online world."[48]

In the face of this information, governments' failure around the globe to advance cyber security is stunning. The difficulties in tracking and tracing communications are well known and the solutions provided by IPv6 are also well understood, yet little funding or attention has been given toward upgrading public and private sector networks to IPv6, beyond a memorandum from the U.S. Office of Management and Budget (OMB) directing federal agencies and departments to transition their network backbones and interfacing networks to IPv6 by June 8, 2008.[49] The memorandum also directs government entities to "ensure that all new IT procurements are IPv6 compliant."[50] Exceptions will require advance, written approval. The memorandum sets forth a timeline of steps toward IPv6 transition. Government entities were supposed to have completed an inventory of existing IP-compliant devised and technologies and an impact analysis of fiscal and operational impacts and risks.[51]

Little progress has been made. A recent study commissioned by Cisco that surveyed 200 U.S. government decision-makers revealed that only 31 percent said their agency had completed its inventory and only 20 percent had done its impact analysis. Only 2 percent had completed their planning. The reason: Money. OMB issued the mandate but provided no funding to pay for the transition.[52] DoD issued its own directive for moving to IPv6 on June 6, 2003, well ahead of the OMB memorandum, for transition by June 2008. It is farther along,[53] however, DoD efforts alone are not enough to impact the overall state of cyber security in the U.S., much less globally.

Cost is the most significant barrier to IPv6 implementation, but upon examination, this reason is hardly supportable. A 2005 study conducted for the U.S. National Institute of Standards and Technology (NIST) by Research Triangle Institute estimated the conversion cost to be $25 billion over the next 25 years. Participants in the study, however, also identified potential *annual benefits* of $10 billion from IPv6 associated with VOIP, remote access products and services, and improved network efficiencies.[54] Considering the cost of the Global War on Terrorism (GWOT), the costs associated with IPv6 conversion are small even accounting for the fact that they may be off by several fold.

Additionally, the problems associated with international cooperation and information sharing are well recognized by legal, policy, and technical experts, yet little action has taken place in multilateral fora to push these issues forward, other than the CoE Cybercrime Convention and the G8's work through its High Tech Crimes Task Force. The G8's efforts to counter cybercrime began in 1997 with the establishment of a "24-Hour Contacts for International High-Tech Crime" and has slowly advanced through the intervening years to include high-level statements and recommendations on countering organized crime and terrorism.[55] According to the U.S. Department of Justice, the current membership in the 24/7 Points-of-Contact Network is 45 nations[56]—hardly enough to impact the security of an Internet connected to 233 countries when response times have to be immediate and seconds matter.

Likewise, information sharing programs remain mired in distrust and reluctance, largely because: (a) governments have failed to agree upon mechanisms for the mutual recognition of clearances and protection of shared information, (b) industry has remained reluctant to share information with governmental entities absent assurances of protection, (c) technical difficulties remain unresolved regarding the security of various classifications of information shared over a network accessed by multiple levels of users, (d) intelligence agencies do not believe that their sources and methods will be protected, and (e) the public and private sector alike do not believe that their information will be protected outside the jurisdiction where it originated.

Research and development is currently underway to address the technical issues associated with cyber security, but the funding is woefully inadequate. The U.S. Department of Homeland Security's Advanced Research Projects Agency (HSARPA) cyber security program budget for fiscal year 2006 was only $16 million, and its FY 2007 budget is projected to be between $18-22 million—a paltry sum considering that the worldwide economic impact from cyber attacks has been estimated to be as high as $226 billion in 2003, with the average New York Stock Exchange company suffering shareholder losses of $50-200 million in the days following an attack.[57] HSARPA is the primary government entity funding unclassified cyber security R&D—the R&D that will most directly benefit the private sector and most of the communications infrastructure in the U.S.

Non-technical issues require government leadership and multinational participation. The lack of a serious, concentrated effort to mobilize governments globally to share information and cooperate on cyber matters related to terrorism is inexcusable and, in fact, ludicrous. Instances where governments are able to track and trace cyber events find they are stymied by these global inadequacies. *It would surely be much easier to get countries to sign up to a multilateral information sharing and cooperation effort than getting them to commit troops to Iraq or fund nation building efforts.* Moreover, these efforts would have a direct benefit to every citizen and business in the world. The only plausible reason for the failure of the Bush Administration to take up this effort—which would be a natural one for the U.S. since it leads the world in information security efforts—is that it simply continues to believe it can counter terrorism more effectively through traditional means and does not understand, at the highest levels of government, how critical this need is. As Bruce Hoffman noted:

*"Today, Washington has no such program in the war on terrorism. America's counterterrorism strategy appears predominantly weighted towards a "kill or capture" approach targeting individual bad guys. This line of attack assumes that America's contemporary enemies—be they al-Qa'ida or the insurgents in Iraq—have a traditional center of gravity. It also assumes that these enemies simply need to be killed or imprisoned so that global terrorism or the Iraqi insurgency will both end. Accordingly, the attention of the U.S. military and intelligence community is directed almost uniformly towards hunting down militant leaders or protecting U.S. forces—not toward understanding the enemy we now face. This is a monumental failing not only because decapitation strategies have rarely worked in countering mass mobilization terrorist or insurgent campaigns, but also because al-Qai'da's ability to continue this struggle is ineluctably predicated on its capacity to attract new recruits and replenish its resources."*[58]

CONCLUSION

If the U.S. is to hope to win the war against terrorism, it must begin an urgent information security initiative that will create the proper environment for militaries, prosecutors, investigators, governments, and industry to track and trace communications, share information, and cooperate. It is not hard to imagine the difficulty we would have today in countering traditional crime if phone companies were unable to capture the origin and destination of telephone calls and if law enforcement and intelligence communities did not have the legal authority to access telephone records. Essentially, that is the current situation with respect to most Internet communications. Communications from terrorists are flying around the globe and, for the most part, we can not identify where they came from, who sent them, or track them. We cannot get immediate and adequate help from law enforcement in many countries to even trace packet paths. We can not get private sector companies to disclose critical event information regarding their own system activities that could impact millions of people.

We cannot expect to tackle a $21^{st}$ century problem with $20^{th}$ century approaches. If we are to effectively counter terrorism, it is imperative that governments immediately allocate resources—including appropriate funding—and dedicate *continuous, sustained attention at the highest levels* to:

- Transition public and private sector networks to IPv6 to facilitate tracking and tracing of IP communications;
- Increase cyber security R&D to advance tools and capabilities for tracking and tracing and the development of early warning systems;
- Develop trusted platforms and corresponding policies for sharing information from an array of sources and with varying degrees of protection to support tracking and tracing and responding to cyber terrorist activities and/or attacks;
- Launch a multilateral initiative that complements and builds on existing and functioning mechanisms aimed at advancing global capabilities regarding rapid response on cyber investigations, prosecutions, and information sharing;
- Extend the traditional defense capabilities of land, air, and sea to include a specialized joint area of the military for cyber; and

- Change their perception and strategies toward countering terrorism by including, as an important component, an examination of terrorists' use of digital technology.

REFERENCES

1. Jody R. Westby is CEO of Global Cyber Risk LLC, located in Washington, DC and serves as Adjunct Distinguished Fellow to Carnegie Mellon CyLab. She also chairs the American Bar Association's Privacy & Computer Crime Committee and is a member of the World Federation of Scientists' Permanent Monitoring Panel on Information Security.
2. "Internet Usage Statistics—The Big Picture: World Internet Users and Population Stats," Internet World Stats, http://internetworldstats.com/stats.htm.
3. "Internet World Stats: Usage and Populations Statistics," http://www.internetworldstats.com/.
4. "Cells from Hell," *U.S. News & World Report*, *http://www.usnews.com/usnews/news/terror/graphics/cellsofhellmap.pdf.*
5. *See e.g.,* Scott Atran, "The 'Virtual Hand' of Jihad," *Jamestown Terrorist Monitor,* Vol. 3 No. 10, May 19, 2005, http://jamestown.org/terrorism/news/article.php?articleid=2369701; Clay Wilson, *Computer Attack and Cyberterrorism: Vulnerabilities and Policy Issues for Congress,* Congressional Research Service, The Library of Congress, CRS Report for Congress, RL32114, Apr. 1, 2005 at 18, http://www.fas.org/sgp/crs/terror/RL32114.pdf ("Wilson").
6. Jarret M. Brachman, "High-Tech Terror: Al-Qaeda's Use of New Technology," *The Fletcher Forum of World Affairs,* Vol. 30 No. 2, Summer 2006 at 149, http://fletcher.tufts.edu/forum/30-2pdfs/brachman.pdf ("Brachman").
7. Bruce Hoffman, "The Use of the Internet by Islamic Extremists," Testimony before the House Permanent Select Committee on Intelligence, Rand Corporation, CT-262-1, May 4, 2006 at 4, http://www.au.af.mil/au/awc/awcgate/congress/hoffman_testimony4may06.pdf ("Hoffman").
8. David E. Kaplan, "Paying for Terror," *U.S. News & World Report*, Dec. 5, 2005, http://www.usnews.com/usnews/news/articles/051205/5terror_10.htm; Louise I. Shelley, "Organized Crime, Terrorism and Cybercrime," *Security Sector Reform: Institutions, Society and Good Governance,* Nomos Verlagsgesellschaft: Baden-Baden (Alan Bryden and Philipp Fluri, eds.), 2003, http://www.american.edu/traccc/resources/publications/shelle31.pdf; Louise I. Shelley, "The Nexus of Organized International Criminals and Terrorism," Transnational Crime and Corruption Center, 2000, http://usinfo.state.gov/eap/Archive_Index/The_Nexus_of_Organized_International_Criminals_and_Terrorism.html.
9. *See e.g.,* Nadya Labi, "Jihad 2.0," *The Atlantic Monthly*, July/Aug. 2006, http://www.theatlantic.com/doc/prem/200607/online-jihad ("Labi"); Audrey Kurth Cronin, "Cyber-Mobilization: The New *Levee en Masse*," *Parameters,* U.S. Army War College Quarterly, Summer 2006 at 77-87,

http://www.carlisle.army.mil/usawc/Parameters/06summer/cronin.htm; Brachman at 149-164; Steven Donald Smith, "Terrorists Use Internet for Propaganda, Defense Officials Say," American Forces Information Service, May 5, 2006, http://www.au.af.mil/au/awc/awcgate/dod/20060505_5036.htm; Jacqueline S. Porth, "Terrorists Use Cyberspace as Important Communications Tool," United States State Department, May 5, 2006, http://www.au.af.mil/au/awc/awcgate/state/terr_net_5may06.htm; Hoffman; Hanna Rogan, "Jihadism Online: A study of how al-Qaida and radical Islamist groups use the Internet for terrorist purposes," Norwegian Defense Research Establishment, FFI/RAPPORT-2006/00915, http://rapporter.ffi.no/rapporter/2006/00915.pdf; Marc Lynch, "Al-Qaeda's Media Strategies," *The National Interest,* Spring 2006, http://www.nationalinterest.org/ME2/dirmod.asp?sid=&nm=&type=Publishing&mod=Publications%3A%3AArticle&mid=1ABA92EFCD8348688A4EBEB3D69D33EF&tier=4&id=E7689667566548ECB882400570ECFD2D; Gabriel Wiemann, *Terror on the Internet,* United States Institute of Peace, USIP Press Books, 2006, http://bookstore.usip.org/books/BookDetail.aspx?productID=134280; Maura Conway, "Terrorist 'Use' of the Internet and Fighting Back," Trinity College, Dublin, Ireland, Sept. 2005, http://www.oii.ox.ac.uk/microsites/cybersafety/extensions/pdfs/papers/maura_conway.pdf; *Cyber Operations and Cyber Terrorism,* DCSINT Handbook No. 1.02, U.S. Army Training and Doctrine Command, Deputy Chief of Staff for Intelligence and Assistant Deputy Chief of Staff for Intelligence – Threats, Fort Leavenworth, KS, Aug. 15, 2005, http://www.au.af.mil/au/awc/awcgate/army/guidterr/sup2.pdf; Steve Coll and Susan B. Glasser, "Terrorists Turn to the Web as Base of Operations," *The Washington Post,* Aug. 7, 2005 at A01, http://www.washingtonpost.com/wp-dyn/content/article/2005/08/05/AR2005080501138.html; Craig Whitlock, "Briton Used Internet as His Bully Pulpit," *The Washington Post,* Aug. 8, 2005 at A01, http://www.washingtonpost.com/wp-dyn/content/article/2005/08/07/AR2005080700890.html; Susan B. Glasser and Steve Coll, "The Web as a Weapon," *The Washington Post,* Aug. 9, 2005 at A01, http://www.washingtonpost.com/wp-dyn/content/article/2005/08/08/AR2005080801018.html; Wilson; Jon Swartz, "Terrorists' use of Internet spreads," *USA Today,* Feb. 20, 2005, http://www.usatoday.com/money/industries/technology/2005-02-20-cyber-terror-usat_x.htm; David Talbot, "Terror's Server," *Technology Review,* Massachusetts Institute of Technology, Feb. 2005 at 46-52; Gabriel Weimann, *How Modern Terrorism Uses the Internet,* United States Institute of Peace, Special Report 116, Mar. 2004, http://www.usip.org/pubs/specialreports/sr116.pdf; "Dot-Com Terrorism: How Radical Islam Uses the Internet to Fight the West," *The New Atlantis*, Spring 2004 at 91-93, http://www.thenewatlantis.com/archive/5/soa/TNA05-StateOfTheArt-Dot%20Com%20Terrorism.pdf; Timothy L. Thomas, "Al Qaeda and the Internet: The Danger of "Cyberplanning," *Parameters,* Spring 2003 at 112-23, http://carlisle-www.army.mil/usawc/Parameters/03spring/thomas.htm; Maura Conway, "Reality Bytes: Cyberterrorism and Terrorist 'Use' of the Internet," *First Monday,* Nov. 2002, http://www.firstmonday.org/issues/issue7_11/conway/; Michael A. Vatis, *Cyber Attacks During The War on Terrorism: A Predictive Analysis,* Institute for Security Technology Studies, Dartmouth College, Sept. 22, 2001, http://www.ists.dartmouth.edu/analysis/cyber_a1.pdf.

10. Howard F. Lipson, *Tracking and Tracing Cyber-Attacks: Technical Challenges and Global Policy Issues,* Carnegie Mellon University, Software Engineering Institute, CERT® Coordination Center, Special Report CMU/SEI-2002-SR-009, Nov. 2002, http://www.cert.org/archive/pdf/02sr009.pdf.
11. *Linux in the Network,* Chapter 14, "The Next Generation Network," 14.2.1 "Advantages of IPv6", http://nfs-uxsup.csx.cam.ac.uk/pub/doc/suse/suse9.1/adminguide9.1/ch14s02.html.
12. Brad Grimes, "The riddle of IPv6," *Washington Technology,* Aug. 7, 2006 at 31-33 ("Grimes").
13. Jody R. Westby, ed., *International Guide to Combating Cybercrime,* ABA Publishing, 2003 at 41-49, http://www.abanet.org/abastore/index.cfm?section=main&fm=Product.AddToCart&pid=5450030.
14. Council of Europe *Convention on Cybercrime*—Budapest, 23.XI.2001 (ETS No. 185) (2002), http://conventions.coe.int/Treaty/Commun/ChercheSig.asp?NT=185&CM=8&DF=8/5/2006&CL=ENG; the United States Senate ratified the CoE Cybercrime Convention on August 3, 2006.
15. Council of the European Union, *Council Framework Decision 2005/222/JHA of 24 February 2005 on attacks against information systems*, Official Journal of the European Union, OJ L 69 of 16.3.2005, http://europa.eu.int/eur-lex/lex/LexUriServ/site/en/oj/2005/l_069/l_06920050316en00670071.pdf.
16. "Attacks against information systems," European Union, http://europa.eu/scadplus/leg/en/lvb/l33193.htm
17. *Beyond Connecting the Dots: A Vital Framework for Sharing Law Enforcement Intelligence Information,* Investigative Report, U.S. House Committee on Homeland Security Democratic Staff, Dec. 28, 2005 at 1, http://www.fas.org/irp/congress/2005_rpt/vital.pdf.
18. *Information Sharing: The Federal Government Needs to Establish Policies and Processes for Sharing Terrorism-Related and Sensitive but Unclassified Information,* United States Government Accountability Office, Report to Congressional Requesters, GAO-06-385, March 2006 at 4, http://www.gao.gov/new.items/d06385.pdf.
19. Brachman at 149.
20. Id. at 149-150.
21. William Darlymple, "What's Cooking in Madrasas?" *The Week,* Special Report, Dec. 11, 2005 ("Darlymple"). Interestingly, however, reports indicate that few al-Qaeda terrorists seem to have little grasp of Islamic law or teachings, in sharp contrast to the Pakistani jihadists who were educated in the madrasas schools, who are steeped in radical Islamism but from impoverished backgrounds with little formal education and training. Some experts believe Osama bin Laden

actually despises the "juridical approach" of the clerics educated in madrasas schools in Pakistan, preferring his own version of Islamism as the solution to Muslim problems.

22. Craig Whitlock, "Keeping Al-Qaeda in His Grip," *The Washington Post,* Apr. 16, 2006 at A1, http://www.washingtonpost.com/wp-dyn/content/article/2006/04/15/AR2006041501130.html.
23. Karen DeYoung and Walter Pincus, "Zarqawi Helped U.S. Argument That Al-Qaeda Network Was in Iraq," *The Washington Post,* June 10, 2006 at A15, http://www.washingtonpost.com/wp-dyn/content/article/2006/06/09/AR2006060901578.html ("DeYoung and Pincus").
24. Craig Whitlock, "Death Could Shake Al-Qaeda In Iraq and Around the World," *The Washington Post,* June 10, 2006 at A1, http://www.washingtonpost.com/wp-dyn/content/article/2006/06/09/AR2006060902040.html ("Death Could Shake Al-Qaeda"); *see also* Labi.
25. DeYoung and Pincus.
26. Ellen Knickmeyer and Jonathan Finer, "Insurgent Leader Al-Zarqawi Killed in Iraq," *The Washington Post,* June 8, 2006, http://www.washingtonpost.com/wp-dyn/content/article/2006/06/09/AR2006060901972.html.
27. Paul Cruickshank and Mohanad Hage Ali, "Jihadist of Mass Destruction," *The Washington Post,* June 11, 2006 at B2, http://www.washingtonpost.com/wp-dyn/content/article/2006/06/09/AR2006060901972.html ("Cruickshank and Ali").
28. Id. (quoting Setmariam).
29. Id. (quoting Setmariam); *see also* Brachman.
30. Id.
31. Rita Katz and Michael Kern, "Terrorist 007, Exposed," *The Washington Post,* Mar/ 26, 2006 at B1, B4, http://www.washingtonpost.com/wp-dyn/content/article/2006/03/25/AR2006032500020_pf.html ("Katz and Kern").
32. Cruickshank and Ali.
33. Id.
34. Id.
35. Al-Qaeda Surveillance Techniques.
36. Wilson at 1.
37. Jody R. Westby, ed., *International Guide to Cyber Security,* ABA Publishing, 2004 at 18, http://www.abanet.org/abastore/index.cfm?section=main&fm=Product.AddToCart&pid=5450036 ("Westby Cyber Security").
38. Barton Gellman, "Cyber-Attacks by Al Qaeda Feared," *Washington Post*, June 26, 2002, http://www.washingtonpost.com/ac2/wp-dyn/A50765-2002Jun26.
39. "Al-Qaeda surveillance techniques detailed," *USA Today,* Dec. 29, 2004, http://www.usatoday.com/news/washington/2004-12-29-terror-surveillance_x.htm ("Al-Qaeda Surveillance Techniques").
40. Martin Arnold, "Asian terror attack feared," *Financial Times,* Aug. 26, 2005 at 1.
41. Wilson at 8.
42. Id. at 1, 3-4, 8-13.
43. Hoffman at 16.

44. Id. at 17.
45. Brachman at 150.
46. Timothy L. Thomas, Foreign Military Studies Office, Ft. Leavenworth, KS, Aug. 8, 2006, email to author.
47. Wilson at 7.
48. Katz and Kern (emphasis added).
49. Karen S. Evans, "Transition Planning for Internet Protocol Version (IPv6), Memorandum for the Chief Information Officers, Executive Office of the President, Office of Management and Budget, Office of E-Government and Information Technology, M-05-22, Aug. 2, 2005 at 1, http://www.whitehouse.gov/omb/memoranda/fy2005/m05-22.pdf ("OMB IPv6 Memorandum").
50. Id. at 2.
51. Id.
52. Grimes at 31; *see also* Matthew Weigelt, "IPv6 looms on the horizon," *FCW.Com,* July 31, 2006, http://www.fcw.com/article95434-07-31-06-Print. .
53. Captain R.V. Dixon, "IPv6 in the Department of Defense," Defense Information Systems Agency, Joint Interoperability Test Command, http://www.usipv6.com/ppt/IPv6SummitPresentationFinalCaptDixon.pdf.; Grimes at 32.
54. RTI International, *IPv6 Economic Impact Assessment,* U.S. Department of Commerce, National Institute of Standards and Technology, Planning Report 05-2, Oct. 2005, http://www.nist.gov/director/prog-ofc/report05-2.pdf.
55. Westby Cyber Security at 78-82.
56. Tim Spring, "Web of Crime: Who's Catching the Cybercrooks?" *PCWorld.com,* Aug. 26, 2005, http://www.house.gov/list/speech/ca03_lungren/082605webofcrime.html.
57. Brian Cashell, William D. Jackson, Mark Jickling and Baird Webel, *The Economic Impact of Cyber-Attacks,* Congressional Research Service, RL 32331, Apr. 1, 2004 at Summary, 9-12, http://www.cisco.com/web/about/gov/downloads/779/govtaffairs/images/CRS_Cyber_Attacks.pdf.
58. Hoffman at 18-19.

# 8. LIMITS OF DEVELOPMENT
# FOCUS: DEVELOPMENT OF SUSTAINABILITY

# LIMITS TO DEVELOPMENT: SUSTAINABILITY REVIEWED

WOUTER VAN DIEREN
IMSA Amsterdam International Consultants
Amsterdam, The Netherlands

*Progress or Regression: the world is flat (Friedman) versus the world is too small (Meadows).*

THE WORLD IS FLAT

- Ongoing economic growth
- Free trade = proper allocation of capital, labour, resources
- Markets to replace governments
- No environmental limits
- In the end, the poor get their share
- The American Empire

THE WORLD IS TOO SMALL

- Climate change
- The energy gap
- Water and sanitation
- Hunger and poverty
- Desertification and food

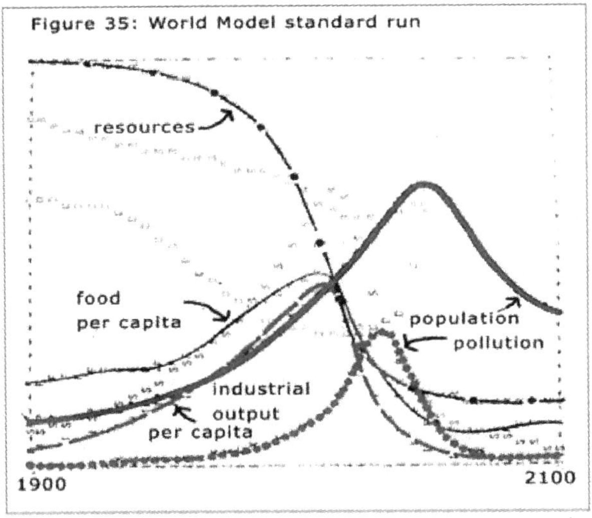


*The Club of Rome report:* Limits to Growth *1972-2006.*

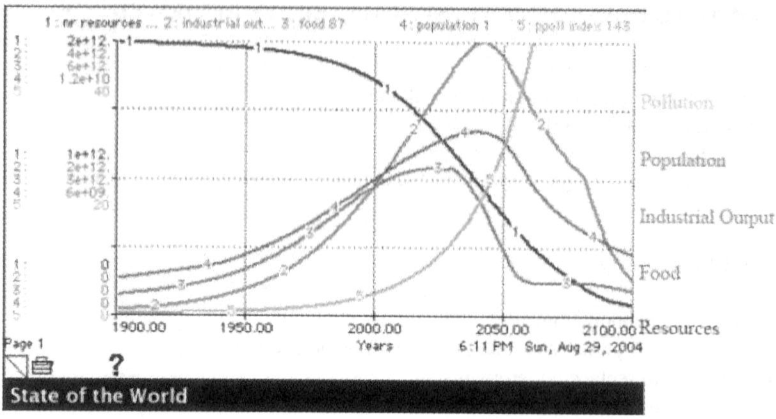


*The 30-year update, 2004: The Reference Scenario.*

LIMITS TO GROWTH 1972-2006

- We did not predict oil depletion in 2000
- We did not forecast any event in 2000
- We did include innovation jumps
- We did not describe one scenario, but many
- Most indicate overshoot & collapse by ca 2050
- Equilibrium is possible

THE NON-SUSTAINABLE REALITY

- Cartel of White House, Pentagon, US Army, Texas Oil, Texas Weapons
- DoD 2005 budget: $ 400 billion
- US Army is largest single fossil fuel burner in the world
- Reborn Christians: *Climate change = the apocalypse is near!*
- "Kyoto etc. is unchristian"
- Loss of farmlands: 40 mln hectare/year
- Pollution: estimated health damage from fine particles € 9 bln per year in the Netherlands
- Deforestation: at the present pace all tropical rain forest will have vanished within 22 years
- Fish supplies: 24% overfished, 52% fished at maximum level
- Growing military grip on resources
- Short-term goals (budget deficit, terrorism, war) prevail over investments in education, care, hunger and poverty alleviation
- Soaring debts (USA nearly 40% GNP)
- Rich/poor: 20% of world population has 83% GNP (so 80% has only 17%)

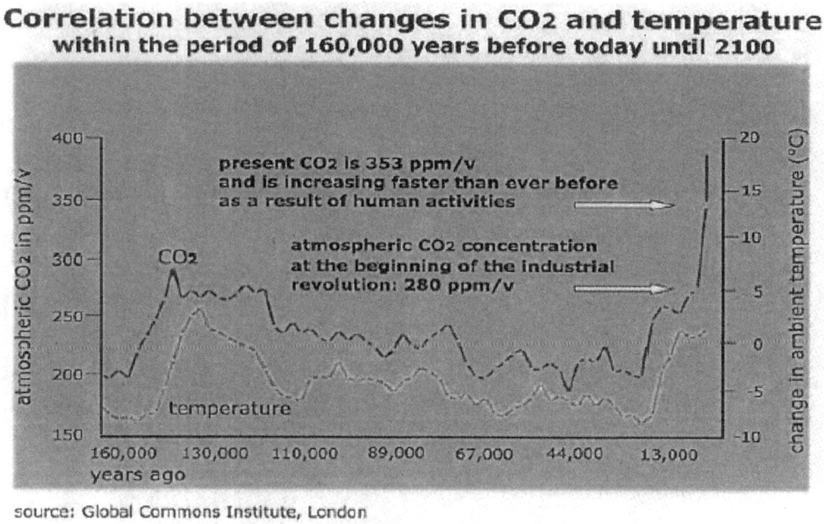


*Global warming is here.*

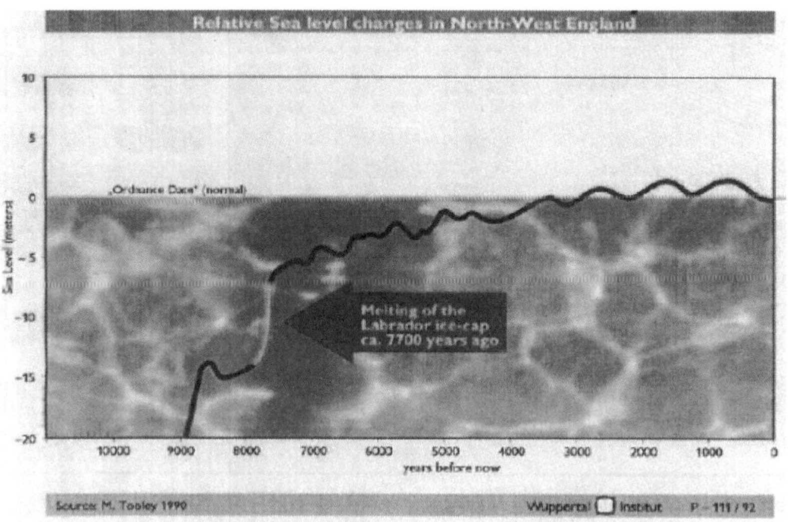


*The flood can arrive suddenly.*

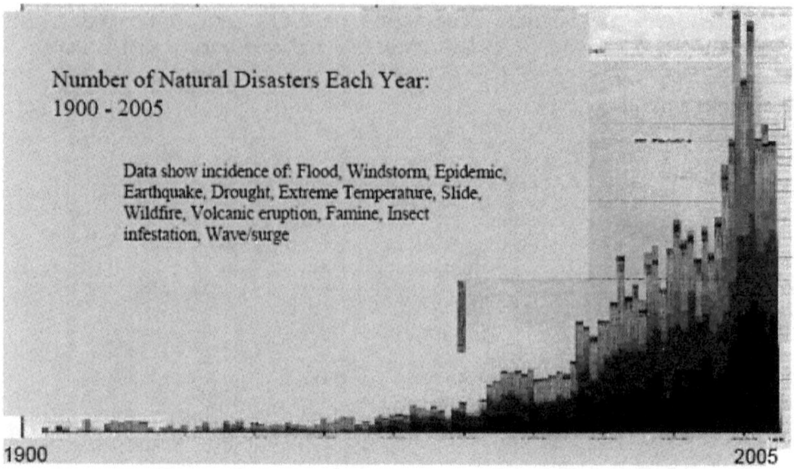


*Natural disasters.*

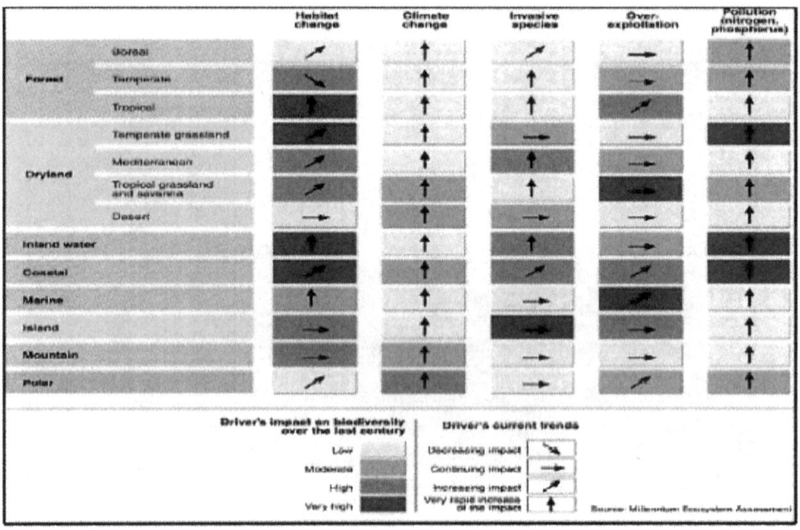


*Growing pressure on ecosystems.*

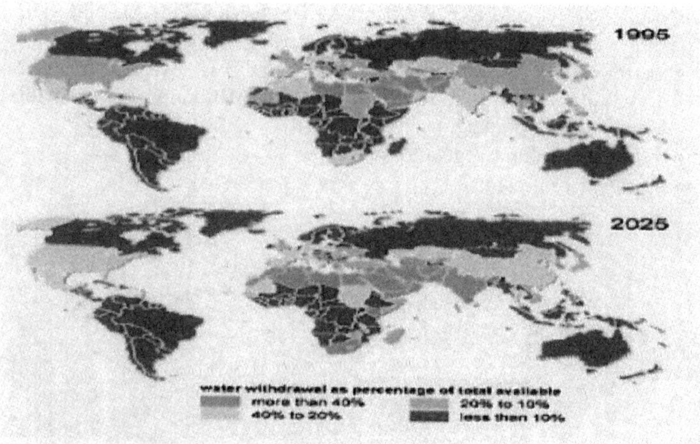


Figure 1: Water stress 1995-2025 (source: GEO-2000 The State of the Environment (UNEP))

*Increasing water stress.*

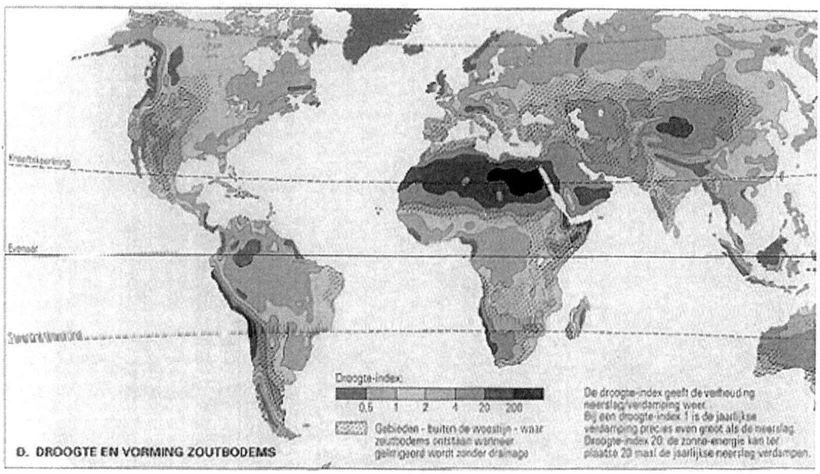


*Global dry areas and salinization.*

Aridity index: precipitation/evaporation
40% of the earth surface is arid: inclined to salinization

INCREASE OF SALINE WASTELANDS IS MOSTLY MANMADE

- Almost 50% of the irrigated area (globally) is severely damaged or affected by salinisation as a result of overexploitation or unsustainable irrigation methods (no drainage)
- Removal natural vegetation
- Seawater intrusion

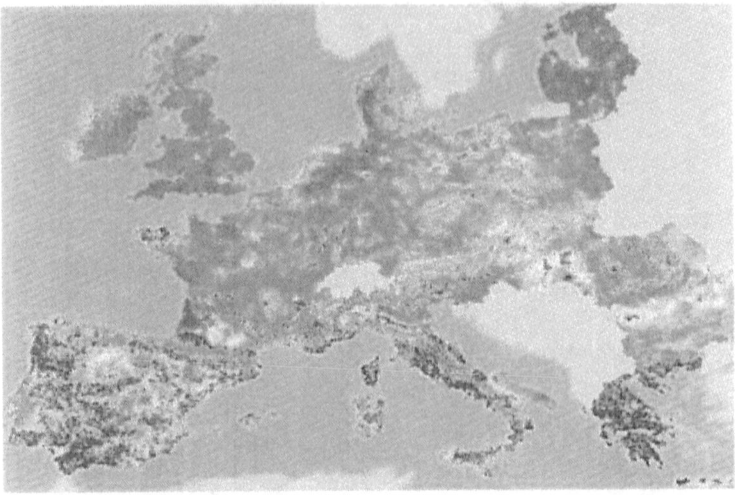

*EU land-degradation risk map combining erosion and salinisation risk as indicators.*

*1.6 Km ... each way!*

POVERTY

- Each day 24.000 people die of hunger, among which 14.000 children <5 years
- Each day 30.000 children die of treatable diseases, because they have insufficient access to medicines
- 115 mln children do not attend school
- More than 1 billion people survive on less than 1 $ per day
- 18% of the world population lack access to (safe) drinking water
- 39% of the world population lack waste-water hygiene

THE MILLENNIUM DEVELOPMENT GOALS (MDG) OF THE UN

- Drinking water
  - By 2015: half the proportion of 1990
  - Access to drinking water within 1.6 km
- Sanitation
  - By 2015: half the proportion of 1990
  - Access to toilets
- Slums
  - "Improved lives" for 100 million slum dwellers

*IF* THE GOALS ARE MET... ...IN 2015, THERE WILL STILL BE:

- almost 800 million people without drinking water
- some 1,600 million people without basic sanitation

POVERTY AND MDGS

And how about slums?
MDG #7, Target #11 states:

*"By 2020, to have achieved a significant improvement in the lives of at least 100 million slum dwellers".*

HOW MANY SLUMS ARE THERE?

According to the Executive Director of UN Habitat, if we continue with business as usual:

- One billion in slums in 2003
- Two billion in slums in 2030
- Three billion in slums in 2050

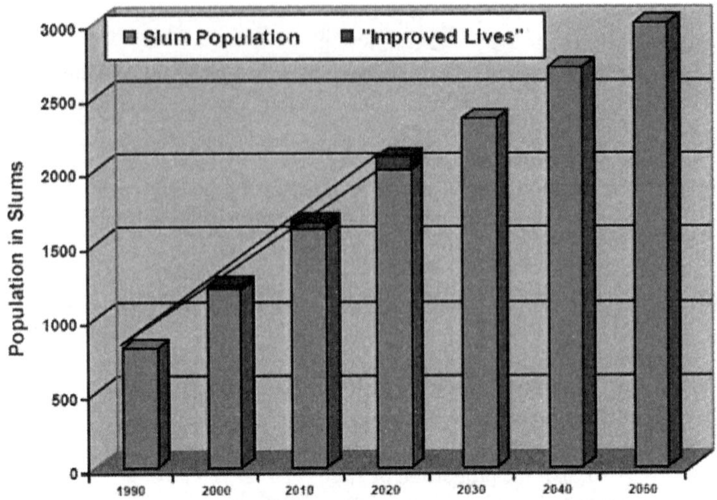


(source: Development Alternatives)

*Slum Dwellers and MGD 7 (From UN Habitat Data).*

THE OTHER ENERGY CRISIS

- 174 countries cannot afford nuclear power
- 132 countries politically unsafe for nuclear power
- In 2050, 3 billion in slums: what energy?
- In 2050, 3 billion rural poor: what energy?

ENERGY FOR THE POOR

- A solar water device per village
- A solar panel per family
- A smokeless cook-stove per house
- A hand winded 40W lamp at $2

ELECTRICITY EXTRAPOLATIONS?

- Forrester 1970: "Nuclear Power is the answer, but what was the question?"
- Reduction of energy-demand by Factor 10 in 50 years
- Stopping the Hummerisation of the economy
- No economy can survive if it sells out resources at low prices, which have to be renewed at high prices

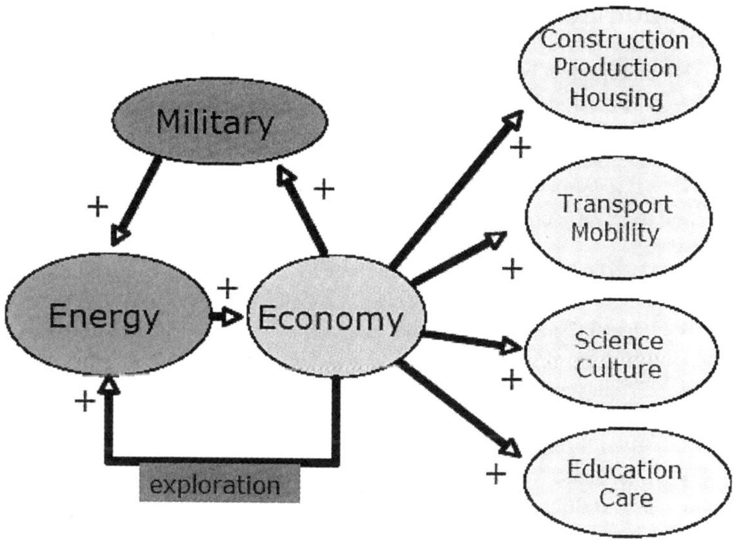


*Added Value + Capital Stocks.*

CONCLUSIONS

- The world runs out of breath
- Ideologies dominate, not science (free market, Islamic, Christian)
- With Business as Usual (BAU) we need two worlds soon
- New solutions needed: think out of the box

ACHIEVEMENTS SINCE 1970

- Global environmental science structure
- From zero to multiple legislation
- Technology advancement:
  - Energy conservation
  - Diversification of energy generation
  - Pollution control
  - Process redesign
  - Water quality improvement
  - Waste handling

MAJOR INCENTIVES

- Global treaties; regional & federal unities
- Legislation, regulation
- Fiscalities
- Internal Engineers Fun
- Dow Jones Sustainability Group Index
- (Not yet) consumer behaviour

SOME SOLUTIONS

- The CO2-neutral economy
- From fossil to solar
- Driving eco-efficiency (Factor 10)
- Rural technologies (Development Alternatives)
- Biosaline agroforestry
- Sustainable national income
- Triple Bottom Line investments
- Corporate Social Responsibility (CSR)

COLORADO DELTA PROJECT

- Unconventional resources:
  - Re-use drainage water: waste water
  - Brackish groundwater where available
  - Saline wasteland: 2x seawater salinity
- Biosaline forestry research:
  - Long-term objectives renewable energy and $CO_2$ sequestration

*Spring 2004: Starting up nursery in Mexico.*

*1 year later: Spring 2005.*

*Autumn 2005.*

UNDERSTANDING TRENDS

- Dutch EPA has created a quadrant representing four mental models of the world
- In an opinion poll, the public voted on its preferences
- The given percentages are the outcome
- The dominant political and corporate mental model of the world (quadrant 1) received only 8% support.

WHAT PEOPLE WANT AND WHAT POWER DECIDES

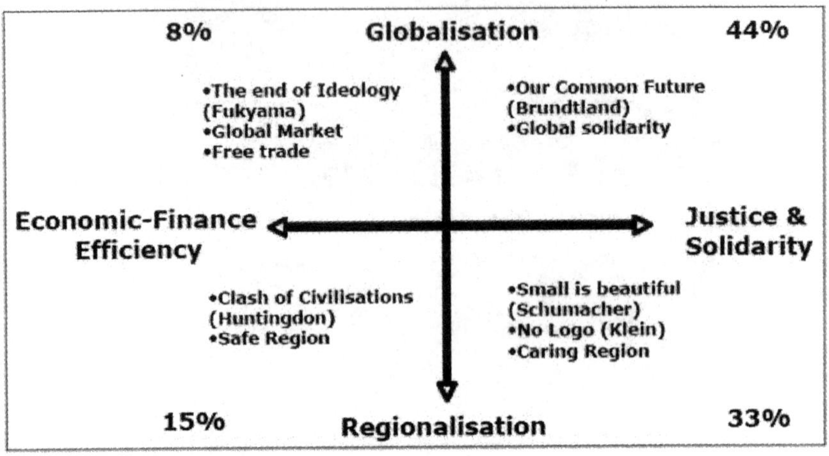


*The world is diverse, not flat.*

# NEW CONCEPTS ON SUSTAINABLE DEVELOPMENT

GERALDO G. SERRA
University of São Paulo, São Paulo, Brazil

## FROM LIMITS TO SUSTAINABILITY

From the time of Malthus,[1] the idea that development has limits has been accepted, rejected and modified by many authors both through scientific discoveries or by technological progress. On the one hand, it is easy to see that "green revolution" has increased food production at unimaginable levels since Malthus' times, but on the other hand many people are starving all over the world, particularly in certain parts of Africa.

In 1970, "The Club of Rome" published its "The Predicament of Mankind," a "quest for structured responses to growing world-wide complexities and uncertainties." The research project commissioned by the "Club of Rome" at the time resulted in a model proposed by Jay Forrester and a book[2] by Dennis L. Meadows and others. The main conclusion of this book is that if the growth trends in world population, industrialization, pollution, food production and resources depletion do not change, in less than a century we will reach the limits of growth. It is important to note that they are dealing with "growth," which is not the same as "development."

However, in 1992, the Rio summit postulated that sustainable development is possible. At the time, Agenda 21 was considered to be an effective program to assure sustainable development for humanity at the new millennium. Being a meeting mostly concerned with the environment, Agenda 21 directives referred to a development, sustainable in terms of environment, not considering other aspects such as human movements and economic structures.

A review of the "The Limits of Growth" carried out mainly by Donella H. Meadows,[3] co-author of the first book, defines the characteristics of sustainable development as follows:

- It should use only renewable resources;
- If this is absolutely impossible, it should use non-renewable resources at a rate that allows enough time to create a substitute.

Although the book says that we have already passed beyond the limits of growth, it tries to envision a sustainable future by means of science and technology. So, developing substitutes for non-renewable resources is nowadays an important task for science and technology.

In 1997, the Brundtland commission defined sustainable development as "development that meets the needs of the present without compromising the ability of future generations to meet their own needs."[4] This definition doesn't enter into conflict with the above-mentioned, because, as Meadows referred to growth, his concept is more operational when analysing a concrete case. It is clear that growth without development is possible, however, the inverse is very difficult. The conclusion is that limits to growth also can be seen as limits to development.

## FROM GROWTH TO DEVELOPMENT

Robert Malthus was not concerned with development but with growth of population and of food production. It seemed to him that population growth could be expressed as a geometric progression and that food production, being only possible by addition of new cultivated land, could only be expressed as an arithmetic progression.

Therefore as, regardless of their ratios, values of geometric progressions inevitably surpass values of arithmetic ones, he concluded that famine was inevitable and could only be avoided by disease and wars. Such a terrible prospect for human kind makes Malthus a target for strong criticism.

In modern times, economic texts refers to production growth, Gross National Product (GNP)[5] growth, prices growth, etc., avoiding the more complex concept of development that seems to have other social or political meanings. Even the Club of Rome report refers to limits of growth and not of development. Indeed, development seems to be a more qualitative concept than mere growth and more difficult to define.

The UN considers that development "is about creating an environment in which people can develop their full potential and lead productive, creative lives in accord with their needs and interests." This means to enlarge peoples' choices by building human potential.

A correlation between development and product growth can be established. The assumption is that development implies distribution of production outcome. So, development is associated with the relationship between production and population and can be achieved not only through increases in production but also through population growth control. Anyway, projections of the World Bank[6] say that "in the next 35 years 2.5 billion people will be added to current population of 6 billion."

## FROM GDPPC TO HDI

Gross Domestic Product per capita (GDPpc), the quotient of GDP and population of an economy, is a good step in the direction of a more correct approach to development evaluation. Indeed, Malthus was already thinking in terms of resources divided by population.

However, a high GDPpc doesn't necessarily mean that product is indeed being distributed among inhabitants and in most countries we observe a considerable concentration of income.

Human Development Index (HDI) is a comparative index and was created by the economist Mahbub ul Haq. It consists of an average between an index of life expectancy, an education index and a GDPpc index and maintains a significant correlation with GDPpc index.

The table below presents the 25 higher GDPpc and the corresponding HDI.

25 Countries with higher GDPpc

| Country | GDPpc U.S.$ 2005 | HDI 2003 |
|---|---|---|
| Luxembourg | 75130 | 0.949 |
| Norway | 64268 | 0.963 |
| Iceland | 53472 | 0.956 |
| Switzerland | 50524 | 0.947 |
| Ireland | 48351 | 0.946 |
| Denmark | 48000 | 0.941 |
| Qatar | 47519 | 0.849 |
| United States | 42101 | 0.944 |
| Sweden | 39658 | 0.949 |
| Netherlands | 38333 | 0.943 |
| Austria | 37528 | 0.936 |
| Finland | 37014 | 0.941 |
| United Kingdom | 36599 | 0.939 |
| Japan | 35787 | 0.943 |
| Belgium | 35750 | 0.945 |
| Canada | 35064 | 0.949 |
| Australia | 34714 | 0.955 |
| Germany | 33922 | 0.930 |
| France | 33734 | 0.938 |
| Italy | 30450 | 0.934 |
| United Arab Emirates | 28582 | 0.849 |
| Spain | 27226 | 0.928 |
| Singapore | 26835 | 0.907 |
| New Zealand | 26441 | 0.933 |
| Kuwait | 26020 | 0.844 |

25 Countries with lower GDPpc

| Country | GDPpc U.S.$ 2005 | HDI 2003 |
|---|---|---|
| São Tomé & Príncipe | 430 | 0.604 |
| Mali | 421 | 0.333 |
| Bangladesh | 03 | 0.520 |
| Zimbabwe | 383 | 0.50 |
| Togo | 378 | 0.512 |
| Cambodia | 375 | 0.571 |
| Tajikistan | 364 | 0.652 |
| Guinea | 355 | 0.466 |
| Mozambique | 46 | 0.379 |
| Central African Rep. | 336 | 0.355 |
| Uganda | 326 | 0.508 |
| Tanzania | 324 | 0.418 |
| Nepal | 322 | 0.526 |
| Gambia. The | 304 | 0.470 |
| Niger | 278 | 0.281 |
| Madagascar | 263 | 0.499 |
| Rwanda | 242 | 0.450 |
| Sierra Leone | 219 | 0.298 |
| Eritrea | 206 | 0.444 |
| Guinea-Bissau | 181 | 0.348 |
| Malawi | 161 | 0.404 |
| Ethiopia | 153 | 0.367 |
| Congo, Dem. Rep | 119 | 0.385 |
| Burundi | 107 | 0.378 |
| Myanmar | 97 | 0.578 |

Comparing these tables, we see that the GDPpc of the top economy is 774.5 times that of the last economy, which, together with job opportunities, constitutes the main attraction for international migration. A second conclusion is that if somebody is "beyond the limits" it must surely be the advanced economies of the first table and not the ones of the second table, which need urgent help to grow. On the other hand, considering that planetary population continues to grow, the world economy is "condemned" to growth. The problem, therefore, is what sort of development we need and what scientific and technological innovations are necessary to make this development sustainable.

Working in 166 countries, the UNDP concentrates on the following issues:

- Democratic Governance: challenge scepticism around the ability of the poorest countries to progress towards stable democracies.
- Poverty Reduction: besides income, poverty reduction implies equity, social inclusion, women's empowerment, and respect for human rights.
- Crisis: prevention and recovery: implies both natural disasters and armed violence
- Energy and Environment: both aspects are considered essential to a sustainable development.
- HIV/AIDS: not only HIV/AIDS but also other diseases can be considered huge obstacles to a sustainable development.

Therefore, the UN considers that besides being an end in itself, democratic governance is very important to fight poverty, overcome crises, provide energy, protect the environment and control diseases, bringing these issues together as priorities in the path to development.

In agreement with these directives, the UN Millennium Project intends to invest in development to achieve a set of goals. Jeffrey Sachs presents the project saying: "This triumph of the human spirit gives us the hope and confidence that extreme poverty can be cut by half by the year 2015, and indeed ended altogether within the coming years. The world community has at its disposal the proven technologies, policies, financial resources, and most importantly, the human courage and compassion to make it happen."

The goals of the Millennium Project consubstantiate what the world understands as being development priorities:

- Eradicate extreme poverty and hunger
- Achieve universal primary education
- Promote gender equality and empower women
- Reduce child mortality
- Improve maternal health
- Combat HIV/AIDS, malaria, and other diseases
- Ensure environmental sustainability
- A global partnership for development

Of course, these goals refer mainly to undeveloped or developing countries and even this case should be understood differently in different regions. For instance, the question of gender equality sounds much more important in certain countries than in others. The document does not seem concerned with possible limits to development and in goal 7, "environmental sustainability" is understood as sanitation and availability of drinking water. The only reference to conservationism is the directive "Reverse loss of forests."

FROM ECOLOGY TO SOCIOLOGY

From the above, we see that both environmental and socio-economic obstacles can limit development in a given region or country. On one hand, natural resources are not distributed equally along the Earth's surface: on the contrary, irregularities are the rule. These differences refer to energy sources, water availability, raw materials and so on. On the other hand, socio-economic discrepancies are also profound, due to historical, cultural or political causes.

However, among the 25 countries with highest GDPpc and HDI, some are poorly gifted in terms of energy sources or raw materials availability and among the other 25 countries some have a fertile soil, good sources of energy and plenty of water. Indeed, the UN seems to concentrate its efforts to promote development in aspects related to human resources, like health and education.

THE DEVELOPMENT CONUNDRUM

As we saw, the development conundrum bifurcates in two different questions:

- Should we continue to grow the same way we did until today?
- How could underdeveloped countries develop but growing?

Considering the enormous differences among countries and regions, the first thing to explain is which countries, groups of countries or regions we are talking about when discussing growth and development. Indeed, developing countries seem to grow more than developed countries as we can see hereafter.

| Real GDP Growth $_a$ | 2003 | 2004$_a$ | 2005$_b$ | 2006$_b$ | 2007$_b$ |
|---|---|---|---|---|---|
| World | 2.5 | 3.8 | 3.2 | 3.2 | 3.3 |
| High Income | 1.8 | 3.1 | 2.5 | 2.5 | 2.7 |
| Developing Countries | 5.5 | 6.8 | 5.9 | 5.7 | 5.5 |

PPP = purchasing power parity; e = estimate; f = forecast.
a GDP in 1995 constant dollars; 1995 prices and market exchange rates.
b GDP measured at 1995 purchasing power parity (PPP) weights.
Source: http://www.finfacts.com/biz10/globalworldincomepercapita.htm

As expected, annual growth rates of developing countries are much higher than developed ones. However, "developing countries" in this table aggregates very poor countries and emergent economies like China, India and Brazil.

The two questions are interconnected because planetary resources can't be enough to maintain all humanity in high levels of consumption, but only to a part of it. Indeed, as Furtado[7] says, based on the Club of Rome report, if through a magical word, all humanity achieves the same level and variety of consumption of developed countries, world resources would be exhausted in few months.

The first of the two questions presented above can be answered in the following ways:

- Yes
- No. We have to stop.
- No. We have to change.

The first of these answers—business as usual—presents obvious shortcomings, considering the rising prices of certain resources, particularly oil and metals, which seem signalling scarcity of them, besides growing $CO_2$ levels in the atmosphere and other evidences of environment exhaustion.

The second answer is the "equilibrium" alternative proposed by Jay Forrester. Unfortunately, economic equilibrium is very unstable and history shows countries' GDPs growing or shrinking. Besides, that answer means that under-developed countries should resign themselves to poverty without any hope, and this seems highly improbable.

The third answer implies that science and technology can produce significant innovations and discoveries both in substitution materials and in new processes. The search is for more productive processes, both in terms of energy as in terms of use of materials, for resources recycling processes and very low pollutants output in atmosphere, watersheds and landscape. That answer allows some space to poverty reduction and human development in underdeveloped countries. However, what is the effort size and how much time it would take?

## QUANTIFYING THE EFFORT

Let us consider simply the first target of the first goal of the UN Millennium Project: "Halve, between 1990 and 2015, the proportion of people whose income is less than $1 a day." The number of people living on less than U.S. $1 can be roughly something around 1 billion people. To halve this number by 2015 maintaining per capita incomes of other people the effort is not that much considering that in 2004 world GDP is $41,290 billion dollars (The World Bank) and growing at 5% per annum, which means roughly $2,000 billion dollars per annum.

World population was 6.3 billion people and was growing at 1.2% per annum. To maintain today's conditions, the world GDP needs to grow at the same rate, which means 495 billion U.S.$ per annum. However, while the GDP of the advanced economies[8] is growing at a rate around 3% per annum, the GDP of emerging economies and developing countries is growing at a rate around 7%. Even Sub-Saharan Africa is growing at rates around 5.5%. The conclusion is that the world economy is growing at an average rate that is 4 times higher than that needed to maintain world per capita GDP, even in the poorest regions of Africa. Indeed, in 2005 only in 18 countries GDP grew less than 1.2% and the list includes some poor countries but also some of the most advanced economies of Europe:

**GDP - Gross domestic product**

2005vConstant prices, annual percent change

| | |
|---|---|
| Netherlands | 1,1 |
| Lebanon | 1,0 |
| Malta | 1,0 |
| Burundi | 0,9 |
| Germany | 0,9 |
| Togo | 0,8 |
| Jamaica | 0,7 |
| Netherlands Antilles | 0,7 |
| Côte d'Ivoire | 0,5 |
| Kiribati | 0,3 |
| Portugal | 0,3 |
| Italy | 0,1 |
| Kyrgyz Republic | -0,6 |
| Lesotho | -0,7 |
| Seychelles | -2,3 |
| Guyana | -2,8 |
| Maldives | -3,6 |
| Zimbabwe | -6,5 |

Therefore, the world does not need to grow faster than it grows today to develop poor regions and countries. Of course, accountability and governance are key issues in international aid, but better trade conditions are crucial. It requires new life to trade reforms, which is precisely the contrary of what is going on in World Trade Organization meetings nowadays where the most ruthless selfishness prevails.

Reports[9] on the effort completed, suggest that there are some progresses in poverty reduction due mainly to a favourable growth in world economy, but asks for strengthening infrastructure and national investment climates.

The Millennium Goal Reports, prepared by the World Bank presents the following graphics to resume the fight against poverty:

FIGURE 1 Poverty headcount by region, 1990–2002, and forecasts to 2015

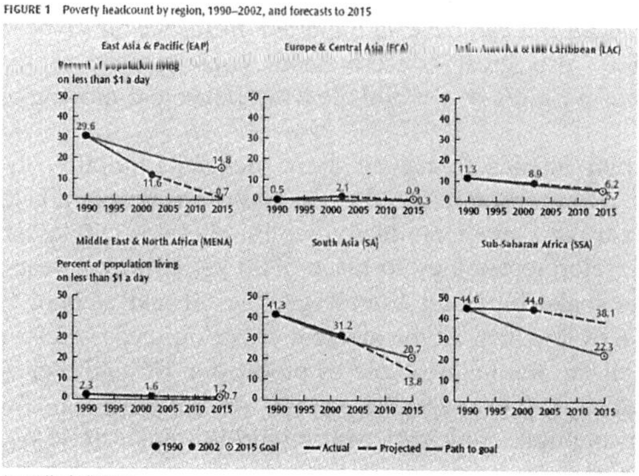


Source: World Bank staff estimates.

But how would the world maintain the rates of growth and development it experiments today? The answer takes us back to sustainable development priorities.

SUSTAINABLE DEVELOPMENT PRIORITIES

Sustainable development has three main priorities:

- Substitutes for non-renewable resources
- Recycling technologies
- Solution for undesirable outputs of human activity.

<u>Substitutes for non-renewable resources</u>

A sustainable development, as we saw, can only be achieved by avoiding the depletion of non-renewable resources and discovering substitutes for them. These are important tasks for science and technology nowadays. Fossil fuels are among the most important non-renewable resources that desperately need a substitute. Indeed, scarcity is already reflected in rampant prices increases these last years. On the other hand, discharging pollutants into the environment—earth, water or atmosphere—began to produce its deleterious results.

The case of Brazilians "ProAlcool" and "Biodiesel" illustrates this priority. This program was created in 1975 to attempt to reduce oil importation, after OPEC decided to increase prices. The first goal was to reach 3 million cubic meters of alcohol by 1980, which happened one year before. Together with internal oil production it succeeded at least in maintaining importation levels:

From 1975 on, automakers produced more than 6 million vehicles fuelled only by hydrated alcohol and more than 10 million burning a mixture of gasoline and around 20% anhydrous alcohol (200 proof), reducing $CO_2$ emissions by more than 110 million tons. It estimated that oil economy is around 550 million barrels. Nowadays the production of flex fuel motors associated with rampant oil prices is once again encouraging ProAlcool. Around 75% of all cars sold in 2006 incorporate the new technology.

By 2010 Brazil needs to cultivate 2.5 million hectares of sugar cane to meet the demand, and this is a real concern, because it can have an important influence on soybean and coffee prices. Besides, there is also great concern about consequences to the environment in certain regions where we already see birds leaving fields and moving to towns and cities.

Now, the "biodiesel" program intends to reduce diesel demand. Initially, the biodiesel is being added to diesel in a proportion of 2%, but the aim of the program is to raise this proportion to 40%. This year the average productivity of biodiesel is 600 kg/ha, but the aim of the program is to develop technology to reach 5000 kg/ha in few years. The most common process used to make biodiesel from vegetable oil and alcohol is transesterification, with glycerin as a by-product. At the moment castor oil seeds (*Ricinus communis*), soybean oil and palm oil are the main source of production of biodiesel in Brazil.[10] The Brazilian Government says that renewable sources currently account for 43.8% of Brazil's total energy consumption, compared with a world average of 13.6% and less than 6% in developed world.

Recycling technologies

Most metals, as well as plastics, glass and paper can be recycled.[11] But also buildings can be recycled to new functions, avoiding demolition construction waste and need of new materials and energy. Great successes are obtained with aluminium beer and soft drinks cans, paper and other components of municipal wastes. Today, more than 50% of all aluminium cans are recycled in the world. The U.S. recuperates 51% of all aluminium cans and in Brazil 95% of all new cans are made of recycled material.

Several materials can be recycled like tires, plastics, steel, batteries and many others.

Solution for undesirable outputs of human activities

The path to a sustainable development needs more: it is not enough to create substitutes for non-renewable resources, because today undesirable outputs of human activity are thrown in the atmosphere, in watersheds and over the landscape. This includes solid, liquid and gaseous waste. Solid waste both from industrial plants and from large urban agglomerations constitutes today as one of the more intractable problems confronting growth.

For instance, there is great expectation on technologies such as "clean coal," by far the most abundant fossil fuel source.[12] However, burning coal produces around 9 billion tons of $CO_2$ each year. "Clean coal" technologies are addressing this problem. This involves many different approaches such as: "washing," electrostatic precipitation, filters, desulphurisation, re-burning techniques and some advanced technologies as Integrated Gasification Combined Cycle (IGCC) and Pressurized Fluidised Bed Combustion (PFBC). At any rate, "zero" emission technologies are the goal.

## CONCLUSIONS

Hereafter a number of conclusions are now presented:

- The world population[13] will continue to growing the next decades.
- Differences between the upper and lower levels of GDPpc and HDI are enormous resulting in the largest international migration wave in world history.
- GDP growth doesn't mean automatic human development as defined by the UN HDI, but it is a very important factor.
- Therefore, instead of imposing limits on development, what is important is to discover the path to sustainable development, which seems to be one of the main tasks science and technology face today.
- We already have very successful programs of substitution of non-renewable resources, of recycling technologies and development of cleaner industrial processes.
- However, it seems that certain economies are facing limits that are not of a physical nature, but rather social and political.
- Sustainable development needs much more research in many scientific and technological areas.

REFERENCES

1. Forrester, Jay W. "World Dynamics," Wright-Allen Press, Massachusetts, 1973.
2. Meadows, Dennis L. et al. "The Limits of Growth," Universe Books, New York, 1972.
3. Meadows, Donella H. et al. "Beyond the limits: confronting global collapse, envisioning a sustainable future."
4. World Commission on Environment and Developments (the Brundtland Commission) report Our Common Future (Oxford: Oxford University Press, 1987).
5. Gross National Product differs from Gross Domestic Product because it includes profits of capital held abroad.
6. http://www.worldbank.org/depweb/english/modules/social/pgr/
7. Furatado, Celso. "O mito do desenvolvimento econôico". Círculo do Livro, São Paulo, 1974.
8. 29 countries with highest GDPpc.
9. Global Monitoring Report 2006—Millennium Development Goals: Strengthening Mutual Accountability, Aid, Trade and Governance. The World Bank, Washington DC, 2006.
10. Biodiesel—The new fuel from Brazil. National Biodiesel Production and Use Program.
11. Brazil is recycling more than 10 billion aluminium cans of bier and soft drinks per annum and U.S. around 54 billion.
12. http://www.uic.com.au/nip83.htm
13. 2005 World Population Data Sheet. Population Reference Bureau.

# 9. DEFENCE AGAINST COSMIC OBJECTS

# OVERVIEW OF RECENT RESEARCH ACTIVITIES ON COSMIC OBJECTS

WALTER F. HUEBNER
Southwest Research Institute, San Antonio, TX, USA

Three spacecraft missions have explored asteroids and comets. NASA's Deep Impact Mission flew by periodic Comet 9P/Tempel 1 and launched a 370 kg copper impactor at its nucleus on 4 July 2005. Dr. M J.S. Belton will report on this. NASA's Stardust mission flew by periodic Comet 81P/Wild 2 and collected samples of dust particles. The dust particles were returned to Earth in a capsule in January 2006 for laboratory analysis. JAXA's Hayabusa spacecraft landed in November 2005 on Asteroid 25143 Itokawa to sample its surface. Dr. H. Yano will report on the exciting findings from this Japanese mission. All of these missions added knowledge and data about the composition and structure of Near-Earth Objects - data that will be needed if we have to mitigate a possible collision of one such object with Earth. In the spring of this year, a conference titled, "Near-Earth Objects Hazard: Knowledge and Action" was held in Belgirate, Italy. I will review discussions at the conference with some emphasis on Asteroid 99942 Apophis that may have the potential of colliding with Earth. Finally, I will report on a NASA Workshop on Near-Earth Objects: Search, Characterization, and Mitigation, conducted in response to a request by the U.S. Congress.

## INTRODUCTION

This has been an active year for Near-Earth Object (NEO) discoveries, missions, data analysis, and plans for potential Earth impact mitigation. The plan to find and catalogue 90% of NEOs larger than 1 km in size is on track for completion by 2008. However, because of the fast rate of discovery, the rate of characterizing these objects cannot keep pace. Even characterizing by remote sensing, such as size, spin state, mineralogy, etc., using ground-based telescopes has fallen behind the desired rates, not to mention *in situ* characterization using rendezvous and flyby missions to determine mass, mass distribution, and shape. Each object in its class (asteroid or comet) looks different from the previously investigated object, suggesting that different processes are at work. Mike Belton and Hajime Yano will illustrate these differences in separate presentations.

Fireballs have been observed over New Mexico. These observations relate to the entry of small objects into the earth's atmosphere and will be presented in a separate presentation by John Zinn. Such entry phenomena are useful for studying the physics of NEOs.

Two conferences were held on the issues of Potentially Hazardous Objects (PHOs) at Belgirate, Italy and at Vail, USA. These conferences are summarized below.

## MEETING ON NEAR-EARTH OBJECTS HAZARD: KNOWLEDGE AND ACTION

The meeting took place 26–28 April in Belgirate, Italy, and had international participation from Finland, France, Germany, Italy, Russia, Ukraine, and USA, including representation from ESA, NASA, and the Russian Planetary Defense Center.

Topics covered included the record of past impacts on Earth, new ground- and space-based telescopes, new methods for analysis, thermal detections of asteroids, taxonomy, mineralogic identifications, light curves (spin periods), and albedo and size determinations. Of special interest was ESA's Don Quixote mission.

In July 2002, the ESA General Studies Programme provided funding for preliminary studies of six space missions, proposed by European groups, that could make significant contributions to our knowledge of NEOs. Following the completion and presentation of the six studies, NEOMAP was established in January 2004 and charged with revising the scientific rationale for the six missions in light of current knowledge and international initiatives and producing a set of prioritised recommendations for observatory and rendezvous missions in an international context. NEOMAP's recommendation was that ESA should give highest priority to the Don Quixote concept.

The Don Quixote concept is a test of a mitigation pre-cursor mission. As such, two spacecraft will be used: Sancho (an orbiter) and Hidalgo (an impactor). Sancho is intended to arrive at the target NEO a few months ahead of Hidalgo to perform reconnaissance and measure size, shape, mass distribution, and bulk density. Hidalgo will impact the NEO. The impact will be observed by Sancho, which will remain with the NEO for several months to provide information on the regolith properties, interior structure, and mechanical properties, in addition to making a direct measurement of the dynamical response ($\Delta v$ of the asteroid) to an impact. The results will provide crucial information for all further development of mitigation strategies, including numerical modeling. The time frame for launch is 2011 to 2017.

About 20% of all discovered NEOs pass within 0.05 AU of the Earth's orbit and are considered Potentially Hazardous Objects (PHOs). The number of discovered NEOs larger than 1 km in size stands now at about 830. Of these about 160 are considered PHOs. However, the total number of NEOs of all sizes discovered so far is almost 4000, of which nearly 800 are PHOs. It is estimated that between about 920 and 1280 NEOs larger than 1 km in size exist. One of the objects that is being followed very closely is Asteroid 99942 Apophis, which was discovered on June 19 2004. It is about 320 m in size, is in an Aten type orbit, and will make a close approach to Earth on 13 April 2029. If it goes through an about 600 m wide "keyhole" the close approach will alter its orbit that could lead to a collision with Earth in 2036 (however, the current probability of 1/5900 is less than the general "background" risk). Four other PHOs of significant size will pass Earth within lunar orbit in the next 150 years.

# WORKSHOP ON NEAR-EARTH OBJECTS: SEARCH, CHARACTERIZATION, AND MITIGATION, VAIL, USA

The workshop was part of a response to a U.S. Congressional request to NASA for new initiatives on NEOs. It was held 26–29 June in Vail, Colorado, USA.

NASA authorization act of 2005 proclaims: "The Congress declares that the general welfare and security of the United States require that the unique competence of the National Aeronautics and Space Administration be directed to detecting, tracking, cataloguing, and characterizing near-Earth asteroids and comets in order to provide warning and mitigation of the potential hazard of such near-Earth objects to the Earth."

"The Administrator shall plan, develop, and implement a Near-Earth Object Survey program to detect, track, catalogue, and characterize the physical characteristics of near-Earth objects equal to or greater than 140 meters in diameter in order to assess the threat of such near-Earth objects to the Earth. It shall be the goal of the Survey program to achieve 90 percent completion of its near-Earth object catalogue (based on statistically predicted populations of near-Earth objects) within 15 years after the date of enactment of this Act."

"The Administrator shall transmit to Congress not later than 1 year after the date of enactment of this Act an initial report that provides the following: (A) An analysis of possible alternatives that NASA may employ to carry out the Survey program, including ground-based and space-based alternatives with technical descriptions. (B) A recommended option and proposed budget to carry out the Survey program pursuant to the recommended option. (C) Analysis of possible alternatives that NASA could employ to divert an object on a likely collision course with Earth."

In May 2006, abstracts for White Papers were requested to be submitted by 25 June, 2006. From the list of abstracts, about forty were chosen for presentation at the Vail conference. The presentations were divided into three categories: Detection, Characterization, and Deflection and Mitigation. Each of these areas has its own working group that will analyze abstracts and presentations to contribute to a White Paper to be presented to Congress by 28 December 2006. There is no official plan to publish the abstracts or the presentations.

Much emphasis was placed on long (years) warning times. However, it is well to remember that Comet C/1983 H1 (IRAS-Araki-Alcock), which has an orbital period of 963.22 years, was discovered 27 April 1983. It passed the Earth only two weeks later, on 11 May 1983, at a distance of 0.0312 AU, thus, falling into the group of PHOs. Two other comets passed Earth at even closer distances: Comet D/1770 L1 Lexell passed Earth at a distance of 0.0151 AU and 55P/1366 U1 Tempel-Tuttle at a distance of 0.0229 AU.A key issue was to extend the search capabilities to smaller sized objects: 140 m in size and to identify 90% of these within 15 years. Long-period comets, in spite of their size and high velocity, and therefore high kinetic energy, in the Earth neighborhood were considered less urgent than asteroids because of their unpredictability (orbital periods of more than 200 years). While seismometers were mentioned in a few presentations, there was almost no discussion about asteroid seismology. Asteroid seismology is the key for determining interior structure and strength of materials. Radio tomography, more useful for the determination of interior structure of comets, also was not discussed besides an occasional mentioning. Detection using infrared technology should greatly enhance the discovery rate, because asteroids emit more photons in the thermal infrared in the inner solar system than they are able to reflect solar photons in the visible range of the spectrum. One of the conclusions of the meeting was that detection is cheaper from ground, but faster from space.

# METEOR IMPACT HAZARDS AND SOME METEOR PHENOMENA

JOHN ZINN
Los Alamos National Laboratory, Los Alamos, USA

My own work in recent years has included some studies of small Leonid meteors and their interactions with the upper atmosphere. But for the context of these seminars I will try first to review existing knowledge about impacts of large meteors (or asteroids or comets) with the earth.

There has been a lot of work in recent years on impacts of large meteors, and on questions about the probabilities of more large impacts in the future. One product of these studies, from a symposium in Turin, Italy, in 1999, has been the establishment of a scale of hazards known as the Torino Scale. This classifies the probable future meteor encounters on a scale ranging from 0 to 10. A zero on this scale implies either a zero probability of an earth encounter or a zero expected hazard from such an encounter. On the other hand, a 10 implies certain collision with disastrous consequences on a global scale. Figures 1 and 2, from the NASA Ames web site (impact.arc.nasa.gov), represent a quick summary of the Torino Scale. They include a color code superimposed on the numerical scale, where 'red' includes categories 8 to 10—representing collisions that are certain but which will produce disasters either over a small region or a large region, or the entire earth, respectively. They also indicate the probable time interval between such collisions, where category 10 events will occur about once every 100,000 years, or longer. Categories 5 to 7 (orange) represent probable close encounters with objects which, if they did hit the earth, would have serious consequences.

The Torino scale is the product of numerous studies including geological studies of craters from past impacts and astronomical monitoring studies of asteroid and comet populations and orbits.

# THE TORINO SCALE

## Assessing Asteroid/Comet Impact Predictions

| | | |
|---|---|---|
| No Hazard | 0 | The likelihood of collision is zero, or is so low as to be effectively zero. Also applies to small objects such as meteors and bolides that burn up in the atmosphere as well as infrequent meteorite falls that rarely cause damage. |
| Normal | 1 | A routine discovery in which a pass near the Earth is predicted that poses no unusual level of danger. Current calculations show the chance of collision is extremely unlikely with no cause for public attention or public concern. New telescopic observations very likely will lead to re-assignment to Level 0. |
| Meriting Attention by Astronomers | 2 | A discovery, which may become routine with expanded searches, of an object making a somewhat close but not highly unusual pass near the Earth. While meriting attention by astronomers, there is no cause for public attention or public concern as an actual collision is very unlikely. New telescopic observations very likely will lead to re-assignment to Level 0. |
| | 3 | A close encounter, meriting attention by astronomers. Current calculations give a 1% or greater chance of collision capable of localized destruction. Most likely, new telescopic observations will lead to re-assignment to Level 0. Attention by the public and by public officials is merited if the encounter is less than a decade away. |
| | 4 | A close encounter, meriting attention by astronomers. Current calculations give a 1% or greater chance of collision capable of regional devastation. Most likely, new telescopic observations will lead to re-assignment to Level 0. Attention by the public and by public officials is merited if the encounter is less than a decade away. |
| Threatening | 5 | A close encounter posing a serious, but still uncertain threat of regional devastation. Critical attention by astronomers is needed to determine conclusively whether or not a collision will occur. If the encounter is less than a decade away, governmental contingency planning may be warranted. |
| | 6 | A close encounter by a large object posing a serious, but still uncertain threat of a global catastrophe. Critical attention by astronomers is needed to determine conclusively whether or not a collision will occur. If the encounter is less than three decades away, governmental contingency planning may be warranted. |
| | 7 | A very close encounter by a large object, which if occurring this century, poses an unprecedented but still uncertain threat of a global catastrophe. For such a threat in this century, international contingency planning is warranted, especially to determine urgently and conclusively whether or not a collision will occur. |
| Certain Collisions | 8 | A collision is certain, capable of causing localized destruction for an impact over land or possibly a tsunami if close offshore. Such events occur on average between once per 50 years and once per several 1000 years. |
| | 9 | A collision is certain, capable of causing unprecedented regional devastation for a land impact or the threat of a major tsunami for an ocean impact. Such events occur on average between once per 10,000 years and once per 100,000 years. |
| | 10 | A collision is certain, capable of causing a global climatic catastrophe that may threaten the future of civilization as we know it, whether impacting land or ocean. Such events occur on average once per 100,000 years, or less often. |

*Figure 1. The Torino scale of asteroid/comet impact predictions – summary chart.*

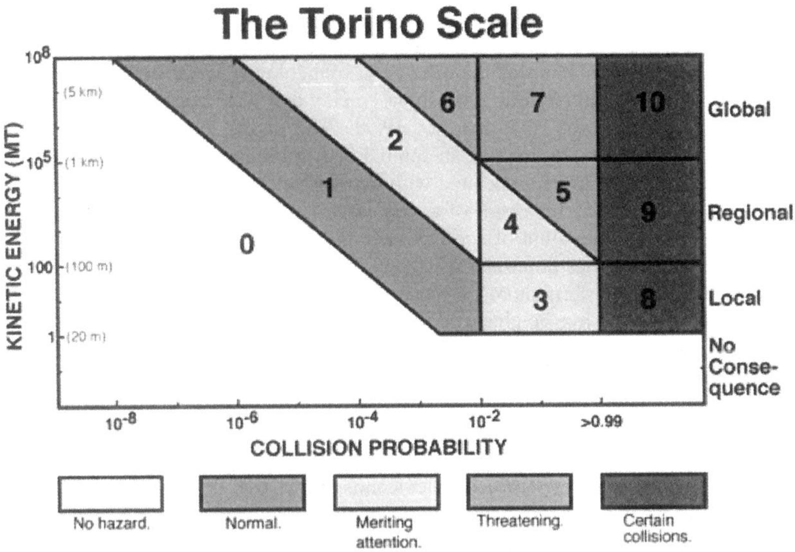


*Figure 2. The Torino scale of asteroid/comet impact hazards— a summary diagram*

Another useful table, shown below, gives estimated time intervals between impact events of various sizes. (http://www.seds.org/nineplanets/nineplanets/meteorites.html)

| Impactor Diameter (meters) | Yield (megatons) | Interval (years) | Consequences |
|---|---|---|---|
| < 50 | < 10 | < 1 | meteors in upper atmosphere most don't reach surface |
| 75 | 10-100 | 1000 | irons make craters like Meteor Crater; stones produce airbursts like Tunguska; land impacts destroy area size of city |
| 160 | 100-1000 | 5000 | irons, stones hit ground; comets produce airbursts; land impacts destroy area size of large urban area (New York, Tokyo) |
| 350 | 1000-10,000 | 15,000 | land impacts destroy area size of small state; ocean impact produces mild tsunamis |
| 700 | 10,000-100,000 | 63,000 | land impacts destroy area size of moderate state (Virginia); ocean impact makes big tsunamis |
| 1700 | 100,000-1,000,000 | 250,000 | land impact raises dust with global implication; destroys area size of large state (California, France) |

Data from 'The Impact Hazard', by Morrison, Chapman and Slovic, published in *Hazards due to Comets and Asteroids*

The most recent category 10 occurrence appears to have been the "K-T Extinction" event, which occurred about 65 million years ago, and which killed off the dinosaurs and almost all land animals, and produced forest fires over most of the earth landmasses. The trail of evidence about this event is quite interesting. Various bits of data were collected over a period of several decades, and gradually assembled into a coherent picture. One of the earlier bits was the discovery, by geologist W. Alvarez in 1980, of a thin (approx 5-mm thick) layer of a grayish-green clay that was found to exist over most of the earth and dated to 65 million years BP, the end of the cretaceous era (Alvarez, 1980). The layer was found to include an anomalously high concentration of the element iridium, which is known to be associated with extraterrestrial material, and therefore led to the inference that it was the result of a very large meteor impact. The layer was also found to contain a large amount of soot, as well as large numbers of previously molten andesite spherules. Another important discovery, by geologist A.R. Hildebrand in 1990, was evidence on the western shore of Haiti of the ancient occurrence of an enormous tsunami, with probable wave heights of several kilometers. This was followed by similar discoveries in western Cuba. The tsunami evidence appeared to point to a source about 1000 km to the west. Somewhat earlier, in 1978, there had been a discovery, by G. Penfield, of geomagnetic anomalies in the Caribbean off the northern Yucatan peninsula that were aligned along an arc with its southern ends pointing toward the Yucatan coast. Intrigued by this discovery, Penfield managed to obtain a gravity map, made in the 1960's under the auspices of Petroleos Mexicanos (PEMEX), that showed another arc, this one on land on the Yucatan peninsula itself, which connected up with the offshore magnetic anomaly arc, forming a nearly perfect circle with a diameter of about 180 kilometers, and centered at the village of Puerto Chicxulub. This led to a reexamination of archived geological drill-cores from studies conducted by PEMEX in 1951, which revealed a hard igneous andesite layer about 1.3 km below the surface. Still later, a study of NASA satellite photographic images showed a partial ring of sinkholes that aligned perfectly with the ring of magnetic and gravitational anomalies, and was interpreted as associated with the ancient subsidence of a crater rim (Pope et al. 1996). Some further evidence has suggested that the original crater was actually 300 km wide, and the 180 km ring was just the inner wall. The totality of these observations have led to the conclusion that they were the result of an impact of an extremely large object, probably a stony meteoroid or asteroid of 10 to 15 km diameter, with a mass of 1-4 $\times 10^{15}$ kilograms and delivering an impact energy of the order of $10^8$ megatons.

These observations have led to numerous theoretical studies and computer simulations. I will show a movie from simulations done by Galen Gisler at Los Alamos. The computer simulations show that the impact would have resulted in an explosion of vaporized meteoritic and surrounding terrestrial material so powerful as to propel a fraction of the material vertically at more than the earth's gravitational escape velocity, and propel a larger fraction into ballistic orbits that would have resulted in later precipitation of re-condensing material into the atmosphere over the entire earth's surface. The aerodynamic heating associated with the reentering molten material would have produced an infrared radiant heat flux on the earth surface large enough to kill all exposed land animals and to initiate forest fires over the entire land surface. Observational evidence in support of these simulation results includes the fact that essentially all terrestrial animal species became extinct at about that time, with the exception of a few burrowing species. Further evidence of the reentering ballistic particles and the heating was in the observation of large concentrations of soot in the Alvarez global clay layer, as well as the large numbers of andesite spherules.

After the heating episode from the explosion and the subsequent reentering material there would have been a very long period, probably lasting for years, of extreme global cooling due to high concentrations of atmospheric aerosols, largely sulfates and soot, deposited in the stratosphere. The impact occurred in a geologically unique sulfur-rich region, kicking up billions of tons of sulfur and other materials into the atmosphere, much of it in the form of vapor.

A composite gravity anomaly gradient and photographic survey image of the Yucatan crater, named "Chicxulub" after the contemporary village at its center, is shown in Figure 3. This one is from a web site http://miac.uqac.ca/MIAC/chixculub.htm, which includes several other images derived from several kinds of data. See also http://en.wikipedia.org/wiki/Chicxulub_Crater.

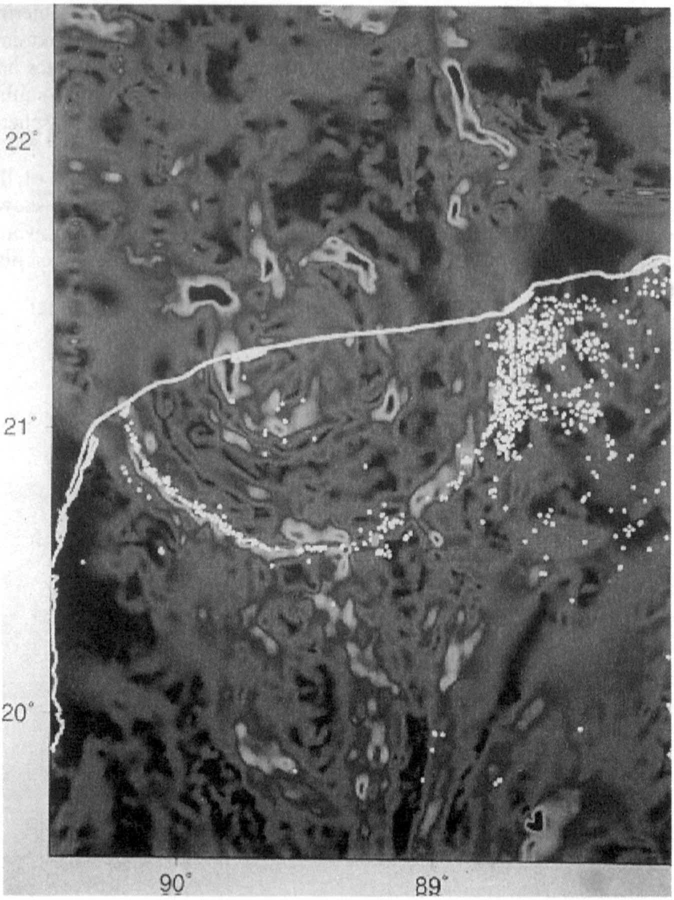


*Figure 3. A composite image of the Chicxulub crater, constructed from gravity anomaly gradient data, together with satellite imagery showing sinkhole distributions.*

Several other craters about the same age as Chicxulub have been discovered—all between latitudes 20°N and 70°N. This has led to the hypothesis that the Chicxulub crater may have been only one of several impacts that occurred at about the same time. The fragmentation of meteors before impact appears to be a common phenomenon, leading to formation of extended crater chains. (See http://science.nasa.gov/headlines/y2006/12may_craterchains.htm?list4750. It is estimated that an event of this scale will occur about once every one hundred million years.

Altogether about 150 terrestrial impact craters have been identified. One of the better known ones is the Barringer Meteor Crater in Arizona (USA). A photo of the crater is shown in Figure 4. (From www.barringercrater.com/science/). It is about 1.2 km wide and 570 feet deep, and about 49,000 years of age. The crater is believed to have been produced by the impact of a nickel-iron meteoroid about 150 ft in diameter, weighing 300,000 tons and traveling at about 12 km/s, with an estimated kinetic energy of 2.5 to 5 Megatons. Small balls of meteoritic iron were found randomly mixed with other ejecta debris over a wide area surrounding the crater. An impact of this size is estimated to occur about once every 1000 years.

*Figure 4. A photograph of the Barringer Meteor Crater in Arizona.*

The most recent known very large terrestrial encounter is the Tunguska event, which occurred in Siberia at 7:14 AM on 30 June, 1908. It is described in the web site www.s-d-g.freesave.co.uk/tunguska.html. It is currently believed to have been associated with a comet or stony meteor about 50 m in diameter that exploded about 6 km above the earth's surface with an explosive force of 15 to 30 Megatons. It leveled 2150 square kilometers of Siberian forest. The force of the blast knocked people off their feet at a distance of 70 km, and they described feeling intense heat. Seismic tremors were measured at the Irkutsk Magnetic and Meteorological station 893 km away, and the blast wave was detected about 45 minutes later. The blast wave was later picked up in Germany and in England. Over the next few weeks there was a night sky glow, reportedly such that one could read in their light (presumably from noctilucent clouds). In the United States the Smithsonian Astrophysical Observatory and the Mount Wilson Observatory observed a decrease in atmospheric transparency that lasted for several months. Several scientific expeditions to the Tunguska region, beginning in the 1920's, were unable to find any crater, but only a vast area of scorched and fallen trees. The fallen trees all seemed to be aligned in a radial direction away from the blast center. Siftings of the soil showed large numbers of microscopic glass spheres containing high concentrations of nickel and iridium, indicating that they were of extraterrestrial origin. Greenland ice core data from the 1908 period also show elevated concentrations of iridium. The actual cause of the Tunguska event is still being debated, but it was almost certainly the result of the impact of a large extraterrestrial object. Whether it was a comet or asteroid is still an open question. An event of Tunguska size is expected to occur about once every 300 to 1000 years.

A very large number of small meteoroids enter the earth's atmosphere every day, adding up to a total daily influx of about 100 tons of material. Most of these meteoroids are very small, just a few milligrams each. The tiny meteoroids are no danger to the earth, but they are a constant danger to spacecraft. Meteors with masses of less than 10 kilograms can be expected to burn up (i.e., ablate and vaporize) high in the upper atmosphere. Meteor velocities cover a range from 10 to 70 km/s, and the associated kinetic energies are proportional to the velocity squared. A 1-kilogram meteor traveling at 70 km/s carries an energy equivalent of 0.6 tons of TNT, and emits a very bright visible flash. Military satellites have been observing such explosions for decades.

While the Torino scale and the geological observations provide interesting statistical information on the likelihood of large meteoroid encounters, they give no definite information about whether a disastrous event will occur next week or next year. There are several ongoing studies of the populations and orbits of observable asteroids and comets. The Lincoln Near-Earth Asteroid Research program (LINEAR) in Socorro, New Mexico has produced orbital data on over 300 asteroids. Another program, by the NASA/Jet Propulsion Laboratory, called Near-Earth Asteroid Tracking (NEAT), has facilities on Haleakala, Maui, Hawaii and at Palomar Mountain in California. So far none of the objects tracked appear to be on collision courses with the earth. However, it is believed that there are as many as 2000 asteroids with diameters larger than one kilometer that have earth-crossing orbits. Also, it is believed that there are some 12 billion comets in the Oort cloud and another billion in the Kuiper belt. The orbits of a large number of periodic comets are known, and they do not present a foreseeable danger. However, an aperiodic comet could appear at any time with little or no warning.

An interesting event occurred in May of 1996 when the Asteroid $JA_1$ was discovered to be passing within 280,000 miles of the earth, and the information became available only 4 days earlier. Figure 5 is a diagram of the orbit of 1996 $JA_1$ (from www.cfa.harvard.edu/cfa/ps/lists/OrbitDiagrams.html). That web site includes orbit diagrams of 22 other near-earth objects, long-period comets and periodic comets. More orbit data can be found at http://neo.jpl.nasa.gov/ca/. The orbit of another large asteroid, 2004 XP14, that passed the earth on 3 July 2006 at 1.1 times the earth-moon distance is shown in Figure 6.

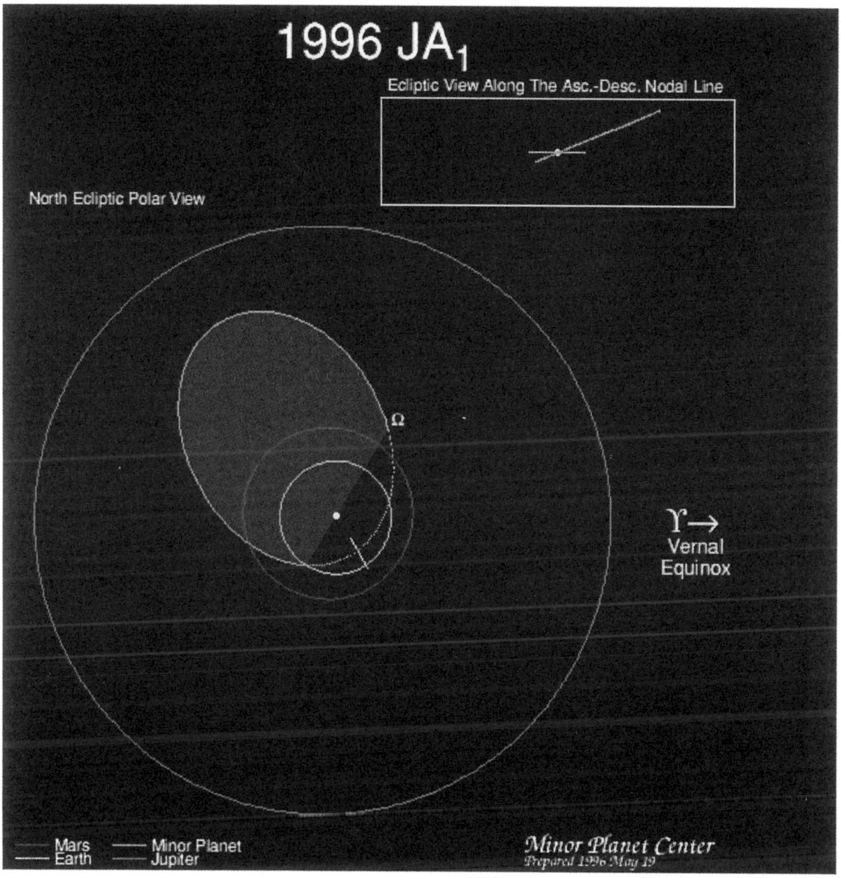


*Figure 5. Orbit diagram of asteroid 1996 $JA_1$, together with the orbits of Earth, Mars and Jupiter.*

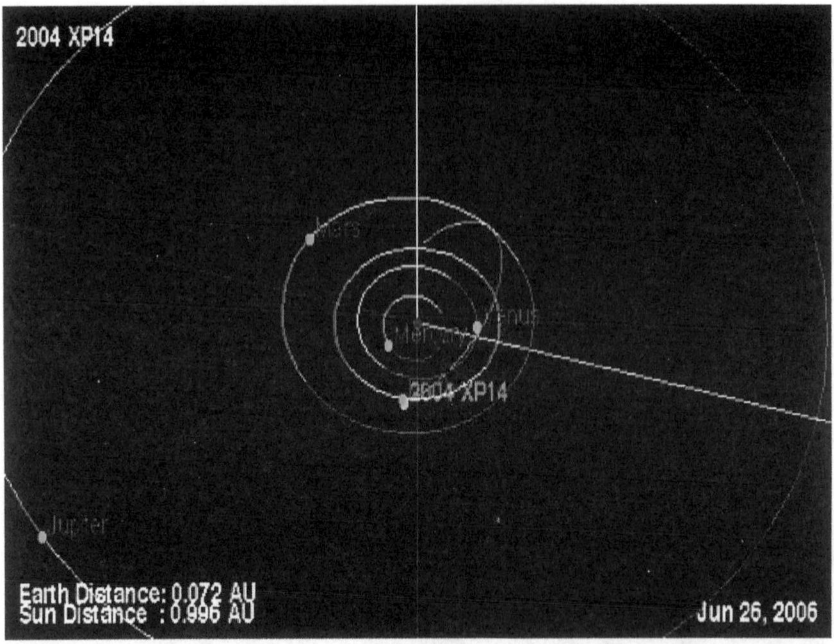


*Figure 6. Orbit diagram of asteroid 2004 XP14, together with orbits of Earth, Venus, Mars, and Jupiter. This object, with an estimated diameter of 370 to 820 m, passed the earth at a distance of .0029 AU on 3 July 2006.*

It has been proposed that there should be an organized effort to explore the possibility of intercepting a threatening object before it reaches the earth. Further discussion of this subject is mostly outside the scope of this report. However, since most meteors seem to be quite fragile, it does seem worth considering the possibility of intercepting and explosively fragmenting one, if it was positively known to be on a collision course with the earth, and known to be very large. The intercept would have to occur at a very great distance from the earth.

Our own work in this area has been focused entirely on Leonid meteors and observational data from the Leonid meteor storm period 1998 to 2002. A Leonid shower occurs each November when the earth passes through the orbit of the comet Temple Tuttle, a periodic comet with a 33-year orbital period. Many of the Leonid storm observations were the product of a joint NASA and U.S. Air Force campaign that included coordinated flights to two scientific aircraft, and participation of scientists from several countries. Other data were obtained from smaller campaigns including optical measurements by the USAF Optical Range in Albuquerque, New Mexico, and from collaborations between Cornell University, the Los Alamos National Laboratory, and Sandier National Laboratories. The meteors observed were all fairly small, with masses of 500 grams or less and energies below 1.3 gigajoules (equivalent to 0.3 tons of TNT), and all of them burned up at altitudes above 80 km. My own participation has been in the analysis of these data and construction of computer models to simulate the phenomena.

An interesting effect observed in several of these cases, and previously described in the literature, was the splitting of the meteor trails into pairs of parallel trails. Three photographs showing these double trails are shown in Figures 7, 8 and 9 (recorded by the USAF Starfire Optical Range). The double trails develop within about 15 seconds to several minutes after the meteor arrival. The photographs all show distortions of the trails produced by the complicated upper atmospheric winds, but the trail doubling effect can be seen over parts of each of them.

The Arch

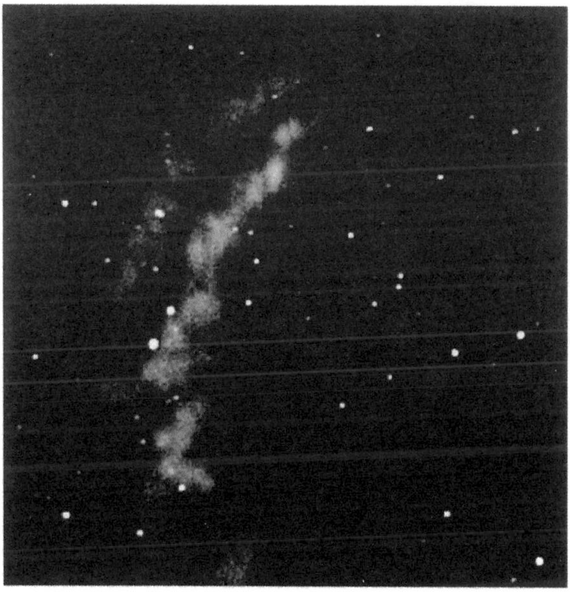

Nov 20, 2002 8:54:51 UT

*Figure 7. A telescopic photo of the meteor train from the "Arch" event from the Leonid meteor shower of November 2002. The photo was taken about 90 seconds after meteor arrival. (Much detail in these photos has been lost in converting from color to grayscale.)*

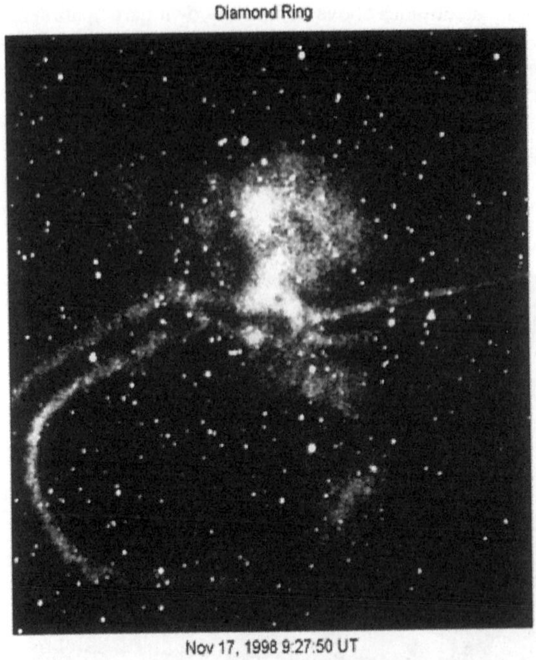


*Figure 8. A telescopic photo of the meteor train from the "Diamond Ring" event, from the Leonid shower of 1998. The time of the image was about 90 seconds after meteor arrival.*

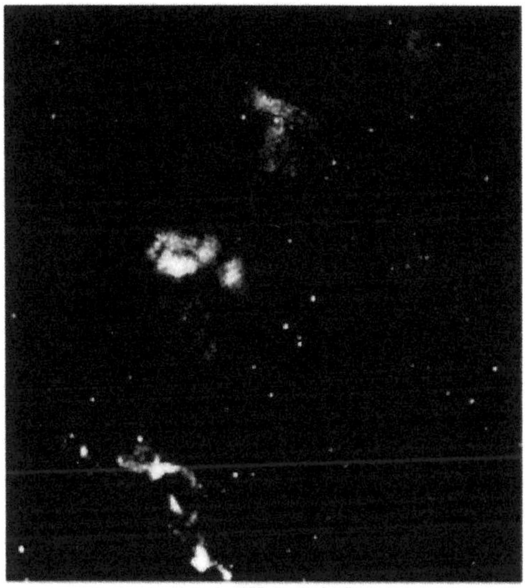


*Figure 9. A telescopic photo the meteor train from the "Puff Daddy" event from the Leonid meteor shower of November 1999—about 90 seconds after the meteor arrival.*

Figure 10 shows some results from a computer simulation of the trail evolution in the "Diamond Ring" case shown in Figure 8. The trail doubling is produced in the buoyant rise of the very hot and expanding quasi-cylindrical trail. The buoyant force is strongest in the vertical central plane through the cylindrical trail and least along the edges, resulting in the splitting of the trail into a pair of counter-rotating linear vortices. The computer simulation treats the hydrodynamic motions in 2-D Cartesian coordinates, and also includes radiation transport and chemistry.

The physical separation of the pairs of vortices, inferred from the computer model results, agrees quite well with the separations of the meteor trail pairs measured from the photographs. Moreover, the measured luminosities of the trails are in good agreement with the results of the chemistry computations that are a part in the computer model. The optical emissions at these late times are primarily airglow in the red to near-infrared $O_2$ "atmospheric" band system, from populations of metastable $O_2(b^1\Sigma_g^+)$ molecules. There is also somewhat weaker emission in the atomic sodium 589.3 nm yellow line doublet, produced in an autocatalytic reaction cycle involving reactions of Na and NaO with ozone and atomic oxygen. From combination of the optical data and the computer model results we infer that the sodium content in the meteor was about .024%.

The 2-D hydrodynamics model that produced the results in Figure 10 was initialized from the results of a more detailed radiation transport, hydrodynamics and chemistry computation in 1-D cylindrical coordinates. This latter simulation model treats the early-time explosive quasi-cylindrical expansion of the air and ablated meteor vapor at cross-sections along the meteor track, as the very large kinetic energy of the ablated vapor is converted to heat.

This 1-D cylindrical model is initialized in turn from the results of a meteor ablation and energy deposition model that treats the rate of ablation and deceleration of the meteor as it enters the atmosphere, and the rate of deposition of the ablated vapor kinetic energy. Some results of this model and early-time cylindrical expansion model are shown in Figs 11 and 12. Figure 11 is a composite plot of energy deposited by the meteor vapors per linear centimeter along the meteor trajectory, for two different meteors including the "Diamond Ring" and a slightly larger meteor known as the "Glowworm". The Diamond Ring reached a terminal altitude of 87.5 km before ablating totally, while the Glowworm penetrated to a slightly lower altitude of 85 km. In both cases the deceleration of the meteor body was negligible out to the point where the meteor was totally consumed. Figure 12 shows computed temperature contours along the trail behind the Diamond Ring meteor, as obtained from several sets of 1-D cylindrical computations for different cross sections along the trajectory when the meteor was at 91 km altitude.

These Leonid data and the computer models are described in more detail in references 10 and 11.

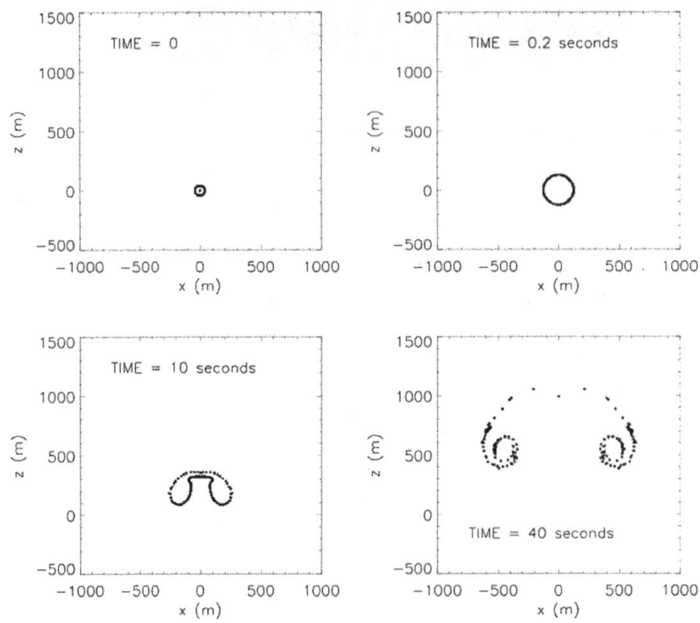


*Figure 10. Results of a 2-D computer simulation of the Diamond Ring event, showing the evolution of the pair of line vortices produced in the buoyant rise of the hot meteor train.*

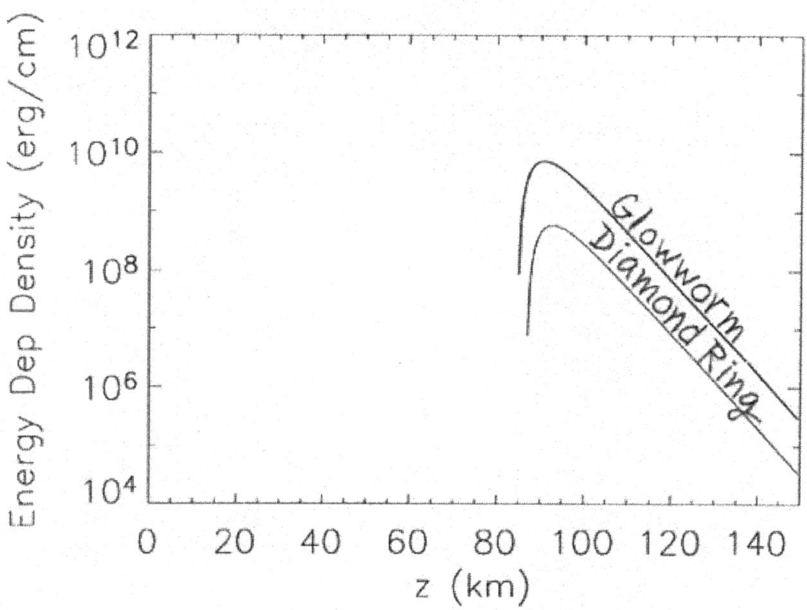


*Figure 11. Computed energy deposition profiles (ergs/cm) vs. altitude along the trajectories of the Glowworm and the Diamond Ring.*

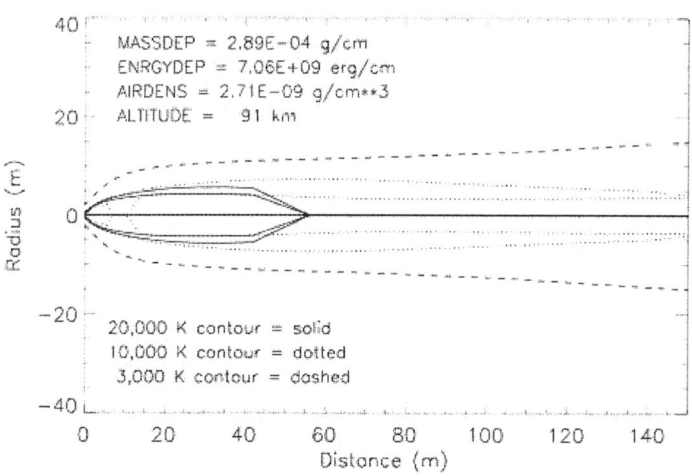


*Figure 12. Computed temperature contours in the Diamond Ring meteor trail for the instant when the meteor was at 91 km altitude.*

REFERENCES


1. Alvarez, L.W., Alvarez, W., Asaro, F., Michel, H.V. (1980) *Science* **208**:1095-1108.
2. Alvarez, W. (1997) "T. rex and the crater of doom." Princeton University Press.
3. Atkinson, A. (1999) "Impact Earth." Virgin Pub Ltd., Great Britain.
4. Gisler, Galen. Personal communication, June 2006.
5. Hildebrand, A.R. & Wolbach, W.S. (1989) *Lunar Planet. Sci. Conf. XX* (abstr) 414-415.
6. Jones, E.M. & Kodis, J.W. (1982) in *Geological Implications of Impacts of Large Asteroids and Comets on the Earth*, Geol. Soc. Am. Spec. Pap. 175-186.
7. Melosh, H.J., Schneider, N.M., Zahnle, K.J., Latham, D. (1990) "Ignition of global wildfires in the Cretaceous/Tertiary boundary." *Nature* **343**:251-254.
8. Pope, K.O., Ocampo, A.C., Kinsland, G.L., Smith, R. (1996). "Surface expression of the Chicxulub crater." *Geology* **24**(6): 527-30. PMID 11539331.
9. Robertson, D.S., McKenna, M.C., Toon, O.B., Hope, S., Lillegraven, J.A. (2004) "Survival in the first hours of the Cenezoic." *Geological Society of America Bulletin*, May/June 2004, 760-768.
10. Zinn, J. & Drummond, J. (2005) "Observations of persistent Leonid meteor trails: 4 Buoyant rise/vortex formation as mechanism for creation of parallel meteor train pairs." *J. of Geophys. Res.* **110**:A04306.
11. Zinn, J., Judd, O., ReVelle, D.O. (2004) "Leonid meteor ablation, energy exchange, and trail morphology." *Advances in Space Research* **33**:1466-1474.

# SCIENTIFIC RESULTS FROM THE DEEP IMPACT MISSION

MICHAEL J.S. BELTON
Belton Space Exploration Initiatives, LLC,
Tucson, USA

## INTRODUCTION

### Goals of the mission

NASA's Deep Impact mission, which was successfully carried out on July 4$^{th}$ 2005 at comet 9P/Tempel 1, was conceived as a scientific exploration of the deep subsurface properties of a typical Jupiter family (short period) comet nucleus to depths of 20-30 m. This depth was expected to extend below the mantle of the comet formed by prolonged solar exposure, to possibly primordial material below. It achieved this by means of a hypersonic (10.3 km/sec) collision with a largely copper impactor spacecraft released from a flyby "mother" spacecraft. The "mother" spacecraft was instrumented with two high performance cameras and a 1–5μ, moderate resolution, near-infrared spectrometer. These made rapid remote observations of the results of the impact and provided high spatial resolution contextual physical and compositional measurements of the comets surface. The mission was 100% successful in achieving its technical goals.

### Why target comets?

Cometary nuclei are of interest in NASA's program of solar system exploration from two points of view. First, they may have a lot to tell us about conditions in the earliest phases (~$10^6$ y) of the formation of the solar system and how macroscopic (i.e., pre-planetary) objects were initially accumulated. Secondly, as one of the two categories of large objects that impact the earth (asteroids and comet nuclei) with frequencies of interest to human society and with the potential for causing catastrophic disruptions to human civilization, we need to gain a firm understanding of their physical and compositional structure in order to evaluate the nature of the threat and to be able to design operational systems that can avert any threat with the highest probability of success. A decade ago quantitative knowledge was severely lacking about both kinds of objects. Now the situation is rapidly changing as a result of six successful missions to comets, including Deep Impact, and three major missions to asteroids mounted by the U.S., European, Soviet, and Japanese space agencies and a robust earth-based research program. These advances, plus the potential results of ESA's Rosetta mission, now en route to comet Churyumov-Gerasimenko, should transform our understanding of these objects.

### The mission profile

The Deep Impact spacecraft arrived in the vicinity of comet Tempel 1 co-joined. The 364 kg impactor spacecraft separated one day before impact at a range of 8.9 x $10^5$ km on a direct trajectory to the comet with a relative speed of 10.3 km/s. The 601 kg mother spacecraft diverted to a trajectory that would miss the comet by 500 km and would arrive some ~10 min post-impact. The mother spacecraft was able to observe the impact and the development of the ejecta cloud for about this length of time. Later the mother spacecraft turned to look back at the collapsing cloud of ejecta and make further observations. The energy delivered on impact was 19 gigajoules (~4.5 tons of TNT equivalent). A crater some 100-200 m in diameter and 20-30 m deep was expected to have been formed.

Observation strategy

The instrumentation consisted of a fast-frame rate camera on the impactor that provided contextual information on the location of the impact to resolutions as high as ~20 cm/pixel. The mother spacecraft sported two cameras which provided a global view of the nucleus and were able to follow the development of the ejecta cloud. These cameras attained resolutions of about 1 and 10m/pixel respectively. The high resolution camera also fed a 1-5μ slit spectrometer with a minimum resolution $\lambda/\Delta/\lambda$ of 216 that provided compositional information of the surface and the impact ejecta.

## OVERVIEW OF SCIENTIFIC RESULTS

In setting out the science, I have taken the point-of-view of a reader who is primarily interested in the impact threat mitigation problem and what they could learn from a fast flyby/impactor mission and what it might contribute to the design of a robust operational mitigation system. And so I place emphasis on what we have learned of the global physical properties of the nucleus (shape, spin, mass, topography of the surface) and the composition and texture of near surface materials. Because of the large amount and fineness of the material released as ejecta the impact crater itself was not observed. It was expected that crater formation and dispersal of the ejecta cloud would have been complete by ~7 minutes post impact, but unfortunately this was not the case. The STARDUST mission spacecraft is still operational in interplanetary space and it has sufficient resources to be targeted to Tempel 1 (J. Veverka, *Priv. Comm.*). Such a mission of opportunity has been proposed to NASA with one of the prime goals to observe the Deep Impact crater.

The results reported below are the work of many members of the Deep Impact science team as well as the author. The results are taken from papers presented at the 2005 Lunar and Planetary Conference and from papers currently submitted to a special issue of *Icarus* devoted to the Deep Impact mission. Goals are discussed in Belton and A'Hearn (1999. *Adv.Space Res.* 24:1167-1173) and an initial appraisal of results is in A'Hearn et al. (2005. *Science* 310:258-264).

Size, shape and spin

These properties were determined from the mother spacecraft during the last 60 days on approach to the comet. Spin determination is based on photometric time series taken while the nucleus is unresolved. The amplitude and shape of the light curve are also used to constrain the overall shape of the nucleus. The spin period was determined to high accuracy at 1.6976±0.00006 days. There is some evidence that the spin period changes noticeably over orbital timescales but no evidence for excited spin. The size and shape of the nucleus and the space orientation of the rotation axis are determined from spatially resolved observation of the nucleus during the last day of observations before impact. These observations, plus photometric surveys of the surface, also yield the albedo,

**Figure 1. Shape Model** (P.C. Thomas):

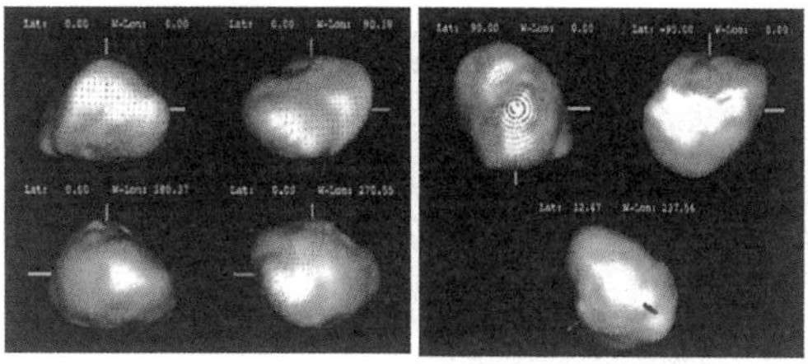

**Parameters:**

Mean radius: 3.04 ± 0.1 km — Effective a/b: ~ 1.5

Surface area: 123.83 km² — Volume: 117.80 km³

Range of radii: 2.36 – 3.95 km — Fraction of surface imaged: ~ 1/3 at <10m/pxl

thermal inertia, and scattering properties of the surface material. Figure 1 illustrates the derived shape. The topographic irregularity is not unexpected for objects of this size. The single scattering albedo is typically 0.036 with a variation of ±20%. Small bright areas (see below are about 2 times the average brightness. The color of the nucleus is slightly red (spectral gradient ~12%) in the visible. The thermal inertia is uniformly very low indicating a very porous surface. Figure 2 illustrates the orientation of the rotation axis. The dawn terminator is to the top of the image.

**Figure 2. Spin Pole direction:**

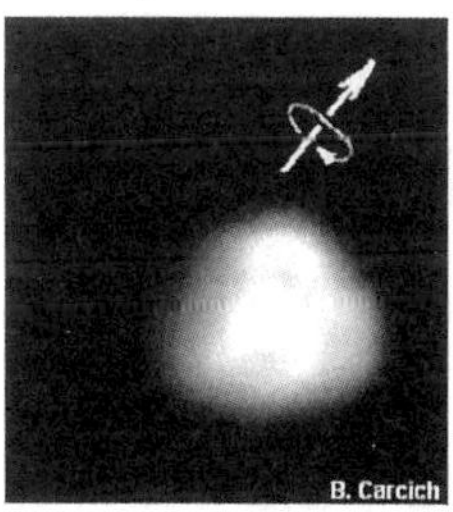


**Two sources of information:**

- Time lapse of resolved images
- Light curves under different geometries

Thomas Pole (RA, Dec J2000): 293.8, 72.6 (± 5) deg.

Light curve Pole (RA, Dec J2000): : 317, 81 (± 8) deg.

Angular separation: 9.8 deg    Obliquity: 11.5 deg

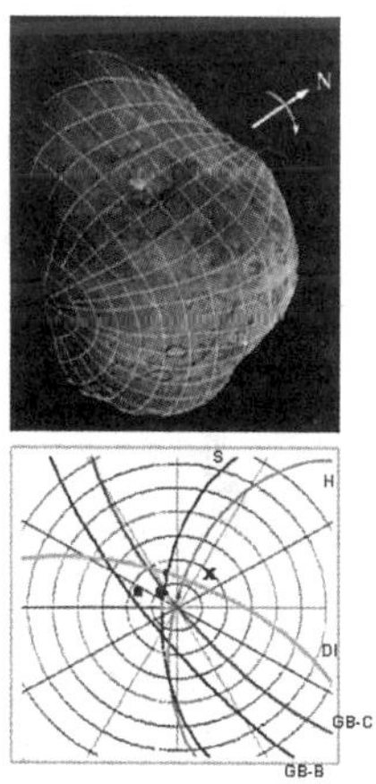


The surface of the nucleus

The surface of the nucleus in cylindrical projection is shown in Figure 3. Approximately 30% was imaged on approach and addition information of the shape of the nucleus was obtained in silhouette during the post-impact look back phase. The spatial resolution is highly variable over the surface after images from all three cameras have been combined to give this view.

**Figure 3. General characterization of the Nucleus Surface**

Obvious from this view are features that appear to be layered, a few impact craters, and features that have appeared to flowed in the past. There are also many highly localized bright spots. Figure 4 shows a close up of the layered areas in the better resolved region of the picture. Some of the layers are bounded by low scarps that appear to be backwasting; others appears as linear outcroppings on a sloping surface. Different layers have different topographic expression: one is cratered, another sports bright spots, another is characterized by very rough topography. Some layers appear to have been "exhumed."

**Figure 4: The Layered Nucleus :**

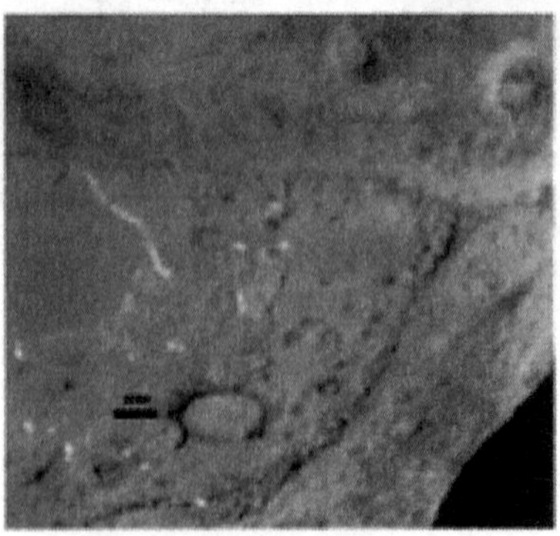

Figures 5 and 6 show how the smooth layers are located in topographic lows and the evidence for a unique source region and flow dynamics.

Figure 5: Smooth areas and Topography

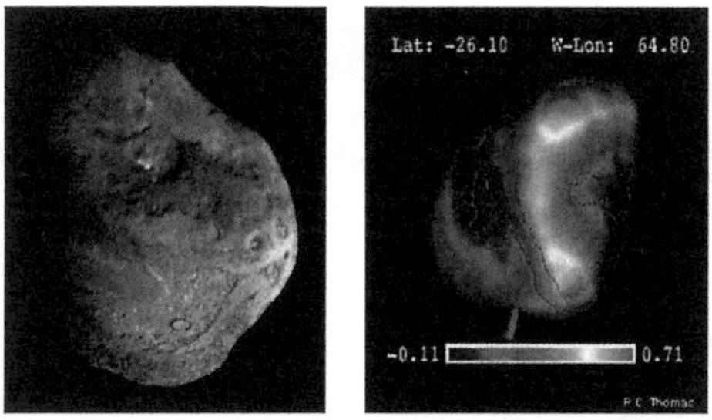


Smooth areas in low spots    Note prominent ridge (Height in km)

Figure 6: Origin and flow in smooth areas

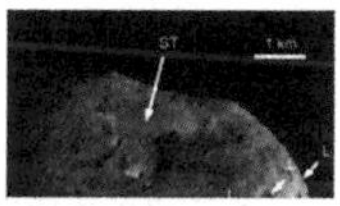

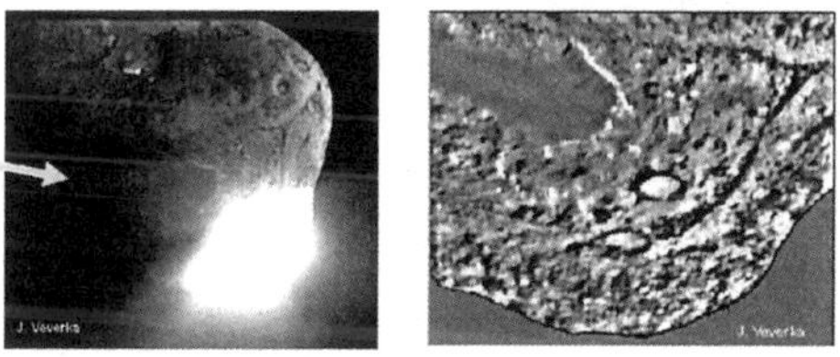


Smooth – geologically young – evidence for flow - evidence for a source region – endogenic activity? Or, could they be primordial?

Spectroscopy of the surface

The near-IR spectrometer was able to scan the surface to build up a spectral-image cube. The primary scientific result was the location of regions of enhanced ice content on the surface. These regions coincide with regions that have an anomalous blue color in images as is shown in Figure 7. The implications of this have recently appeared in *Science* (J. Sunshine et al. 2006, *Science* 311:1453-1455). The regions have ~5% water ice in the form of ~10–50 micron diameter particles. No water ice is detected on the surface outside of these regions.

The impact and its ejecta

The location of the impact is shown in Figure 8 and the complex phenomena that occurred subsequently occur over a region of ~200 m on the surface. The approach altitude was about 30 deg from the local surface and as far as can be determined the following events occurred (after P. Schultz):

Figure 7: Spectroscopy of the surface

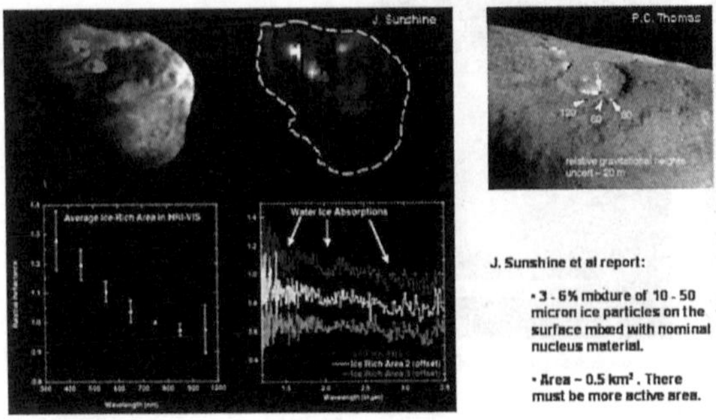


| | |
|---|---|
| 0 sec: | initial faint flash |
| 0 - 0.13 sec: | fading moving source (100 - 170m) |
| 0.25 sec: | bright flash |
| 0.25 – 0.37 sec: | bright, fast moving, plume emerges (moving downrange with 4.8 km/sec projected velocity) |
| 1 – 10 sec: | ejecta plume emerges (shadow gives estimate of crater diameter) |
| 10 – 200 sec: | high-angle diffuse plume emerges (can see shadow) |
| 200 – 800 sec: | broad ejecta cone develops |

Fallback of ejecta plume seen as late as 48 min to 75 min after impact (J. Richardson).

Figure 8: The impact Location

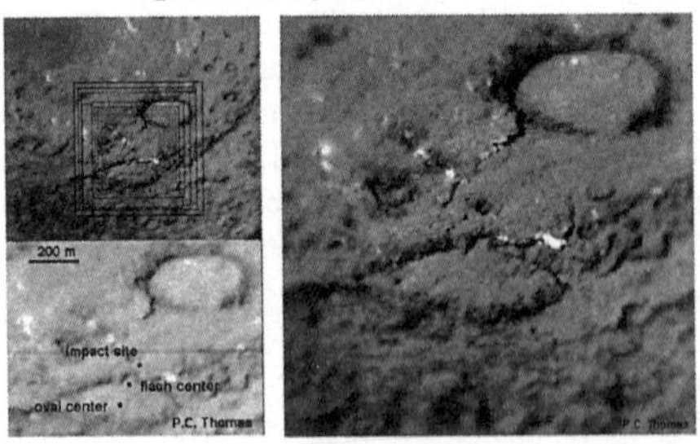


The vapor plume which moved downrange at a projected velocity of ~4.8 km/sec is thought to be self-luminous (for about 0.3 sec) and composed of liquid silicate droplets that have condensed. The ejecta plume, for which estimates of $10^7$ kg total mass released have been made, stays attached to the surface as expected for a weak target with shear strength < 65 Pa. Modeling of the dynamics of the ejecta, particularly in the look back phase, implies a local gravity field of ~30 mgal, which in turn implies a bulk density of ~400 kg/m$^3$ for the nucleus. The development of the initial events is shown in Figure 9, which also shows the location of the spectrometer slit.

**Figure 9: Events around the time of impact**

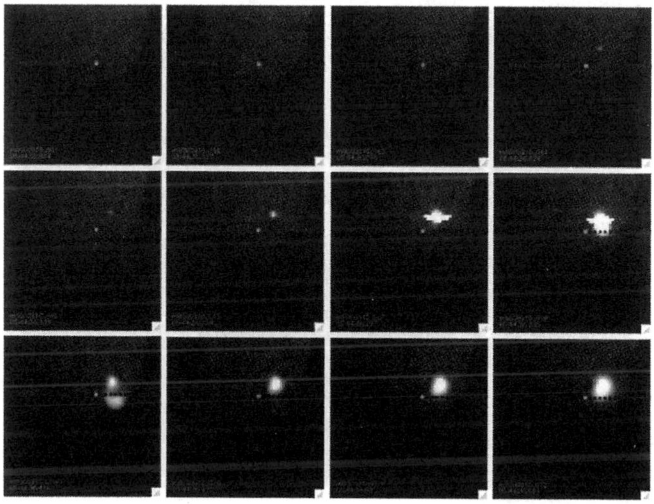

The spectral observations in the 3 micron region taken through the impact period are shown in Figure 10. The ice particles must be very small in size, ~1-10μ in diameter, and the vapor plume contains water vapor and organics neither of which were seen prior to the impact. The composition of the ejecta was well observed remotely from the Spitzer telescope yielding considerable information on the detailed composition (Lisse et al. 2006 *Science,* in press) of the ejecta (Figure 11).

The cometary atmosphere, inhomgeneity and activity

On approach to impact, the near-IR spectrometer was able to scan over the inner coma. Evidence of large variations in the $CO_2/H_2O$ ratio were found indicating compositional inhomogeneity in the source areas of activity. On approach, the photometric time series also showed ample evidence for localized areas to produce abrupt and short lifetime outbursts. The results of such outburst were also seen in images taken with the Hubble space telescope.

These outbursts appear only with approximate periodicity but seem to be connected to the onset of illumination in loosely defined areas by the sun. The mechanism that originates the outburst remains obscure however.

Figure 10: Spectroscopy of the impact

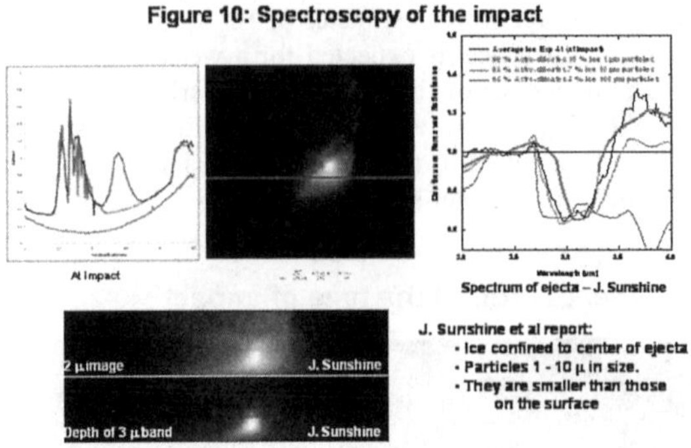


Figure 11. Composition of the ejecta. Spitzer Observations)

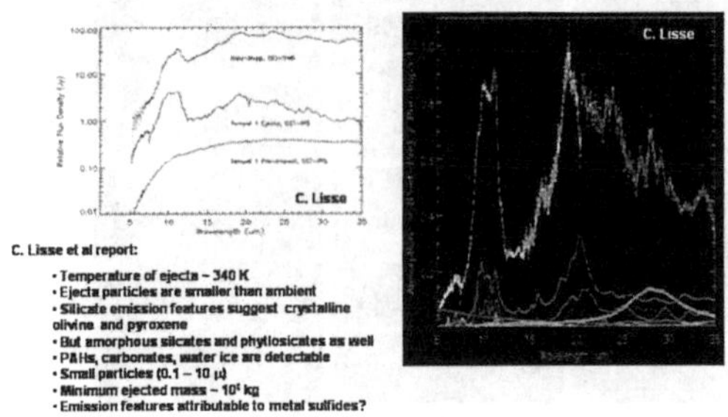


<u>Cosmogony of cometary nuclei</u>

The pervasive presence of layers with differing properties on the surface of Tempel1 is an area of active research stimulated by the Deep Impact results. The smooth areas may be geologically young and indicate the presence of contemporary endogenic activity spilling onto the surface: a relatively new concept in cometary physics. Three have been found on the 30% of the surface observed and so this phenomenon may be common. The other layers may be primitive and we are investigating the hypothesis that they are ubiquitous on Jupiter family comets and are an essential element of their interior structure. Weaker evidence for layering has been found on the two other comets that have been visited by spacecraft–Borrelly and Wild 2. If this hypothesis is true then the current view of collisional evolution in the Kuiper belt, which leads to a picture of comet interiors as in a "rubble" pile, will need a re-evaluation.

# 10. WFS GENERAL MEETING
# CULTURAL EMERGENCY
# FOCUS: TERRORISM

# REPORT OF THE PERMANENT MONITORING PANEL ON TERRORISM (PMPT)

AHMAD KAMAL
Senior Fellow, United Nations Institute for Training and Research
New York, USA

The Permanent Monitoring Panel on Terrorism (PMPT) held its Fourth Meeting in Erice in May this year, at a session that was significant for two major changes in its consideration of the scourge of terrorism, and in the search for do-able solutions to this scourge.

The first change was that the division of the PMPT into sub-groups was simplified, from four to just two: one on Motivation, and a second on Mitigation. The first group was focused on an examination of those underlying factors and root causes which motivate terrorist acts in the first instance, while the second group focused its attention on actual concrete measures aimed at mitigating the consequences of terrorist acts.

The second major change lie in the expansion of the inputs into the Fourth Meeting, with detailed written contributions being received from His Royal Highness Prince Hassan bin Talal of Jordan, and from His Excellency Boutros Boutros-Ghali, former Secretary General of the United Nations, both of whom thus participated in the discussions in this manner. In addition, new participants from Iran and Pakistan, with an intimate knowledge of the "other" point of view, lent greater depth to the debate and discussions, and hopefully enabled the rest of the participants to get a better understanding of the complexity of the problem that we face.

The Final Report of the Fourth Meeting of the PMPT is a voluminous document, but copies can be made available to any members who might be interested.

Many important elements emerged during this Fourth Meeting of the PMPT, and they all need to be highlighted before this distinguished audience.

There can be no doubt that the Mitigation efforts have produced quantifiable results. Many terrorist acts have been foiled, and more will be foiled as vigilance and preparedness increase. With every passing day, a deep analysis of different vulnerabilities has enabled the latter to be plugged, one by one. As that happens, the statistics about culminated terrorist acts show a considerable decline, even if the attempted terrorist acts still show a constant increase in numbers and geographical spread.

The real question, then, is whether all these Mitigation efforts have succeeded in dampening the threat. In order to answer that question, we have to first try to understand the basic objectives of the terrorist, and the manner in which he measures the consequences of his actions. His is a political objective, not a military one. He does not even have to actually commit the act in order to achieve his objective. Just the threat of an act does the trick. The objective is not death and damage, but rather the creation of fear and panic to make a point, and as an ancillary, to bleed the economy and confidence of the target. It would appear that from this point of view, the perpetrator of terrorist acts has a comparative advantage.

It therefore behooves us to judge whether our Mitigation efforts have been cost-effective. The enormous amounts spent in the post 11 September 2001 world, estimated by some to have exceeded one trillion dollars, and by others at over two trillion dollars or double that figure. The terrorists themselves have only had to spend a minute fraction of this amount in order to draw that astronomical response. They can thus see this as great success.

The enormous expenditure in the target states does not appear to have stemmed the endless supply of nascent terrorists, or to have made the world any safer. Uncertainty has grown exponentially, and every time a hole in the dyke is plugged, others appear elsewhere in the most unexpected places. National economies are under increasing pressure everywhere, investor confidence is shaken and volatile, and frustrations are boiling over into conflicts with devastating consequences for innocent civilians.

An unfortunate consequence of this is that the cancer has spread through the whole of the global body politic. While most Western countries appear to believe that they alone are under attack, they forget the enormous burden that is now being borne by the developing countries, all of which have become front-line states in their own regional environments. In comparative terms, it is the latter which are really paying the price in a world of terrorism.

That is why it is all the more important now to address Motivation frontally and seriously. It is not normal for educated young married middle-class individuals to fly planes into buildings, or to blow themselves up in the middle of civilian crowds. Basic science tells us that there is an intimate link between cause and effect, and that all incidents must therefore have underlying reasons. The level of motivation that we are witnessing can only come from very deep frustrations, and these have to be identified and addressed. Many of them are known, such as the perceptions of continuing political and economic injustices in the regions in which they live, frequently as a result of external forces and interventions. Many continue to live under the heavy yoke of non-democratic governments, supported from abroad despite all the glib talk of democracy. Turmoil in the Middle East appears to be central in this situation, even though the consequences of that turmoil have now spread throughout a large swath of territory from North Africa all the way to South East Asia. All of us in the whole world have thus become front-line individuals and front-line states: under constant threat. It is quite possible that we are witnessing a new Hundred Years War. After all, the Middle East situation has now existed for 60 years already, and no light is yet visible at the end of this tunnel. Could it be that this protracted conflict is the World War III that we have all been trying so hard to avoid?

It would appear that never in its entire history has the world known this level of uncertainty and fear and panic. This may perhaps be the result of the fact that no society is now insulated from events in other parts of the world. Feeding on each other, reactions everywhere are becoming increasingly illogical. Centuries of legal norms are being jettisoned at the altar of a "war on terror." Day by day, extremism is gaining strength and even respectability, everywhere. Individuals, who would have been ostracized for their outrageous ideas just a few years ago, are now being carried around on the shoulders of adoring crowds, egged on by the visual media. Even educated and developed societies are beginning to act irresponsibly under a cloud of collective mob psychology. In just a short period of time, the entire fabric of social and legal norms, built up over centuries of thought and action, has been frayed into tatters. It is as if the whole structure of society as we have known it, or wanted it to be, has started slipping out of our fingers as loose sand, leaving us to wonder when and where this will all end. *Quo vadis* indeed.

What complicates the problem even further is globalisation. As the world shrinks in time and space, we all become part of a global village. That in turn, sets off flows of information and migration, moving like a tsunami, carrying with them a flood of ideas, some constructive and others highly destructive. It is becoming virtually impossible to filter the good from the bad, without giving up all the liberalism and open door policies that we have now nurtured for a couple of hundred years at least. Information flows are an essential part of our world today. Many have tried to censor them, and failed. Truth can no longer be concealed from the public. Awareness is globalised, and all can now see and judge for themselves.

Globalisation has also spotlighted the problems of migration in the world today. Migration flows are part of the basic history of the world. All the populations of the world have common origins in Africa. All have reached where they are now settled only through migrations. Efforts to stifle this fundamental force cannot therefore succeed. The search for a better life is a human right. The best that we can do then is to keep migrations within manageable proportions. The danger, of course, is that shortsighted protectionist immigration policies become a case of double jeopardy for the sending countries, in which the disadvantaged populations of the world cannot survive at home, and cannot leave home either. It is double jeopardy for the receiving countries also; if they allow unfettered immigration, they face deep economic and social problems, and if they close their borders to immigration, then they face aging problems at home and add on to the frustrations in the sending states, thus leading to even more serious security problems for themselves.

By far, the most difficult part of any exercise then lies in the difficulties in drafting a consensus on the need to address Motivation as an essential complement to Mitigation. Mitigation appears to need no more than money and contracts and trigger-happy guns, of which there seems to be no shortage anywhere. The Motivation exercise, on the other hand, needs the courage to enter into open and empathetic dialogue with the opposition, in an effort to understand and address the root causes of frustrations, and a willingness to accept that we ourselves might be a contributing factor, or even the starting point of the slide into the abyss of death and destruction. That requires political foresight, intellectual honesty, and long-term vision and statesmanship, all of which are scarce resources in our modern world.

It was our hope that the World Federation of Scientists, with its intellectual rigor and scientific neutrality would be able to provide the environment in which better sense could be made to prevail. That remains our hope and effort, despite the many shoals through which we have to navigate. The Mitigation portion of our effort has resulted in many concrete suggestions, some of which have probably been useful to our decision-makers. The Motivation exercise, on the other hand, has remained mired in a dialogue of the deaf. Nobody seems to be interested in analysing and addressing root causes, most of which, if not all of which, are known. In fact, even scientists seem to clam up whenever the words "root causes" come up in the discussions. It is as if the missionary convictions of the political decision-makers have nibbled into their intellectual impartiality. They have also become highly politicised, and in the process, have lost their true scientific credentials. No experimentation, no doubts, no dialectics, no dialogue, no discourse, as they continue to parade in their scientific garb.

Despite all these difficulties, the Permanent Monitoring Panel on Terrorism (PMPT) will continue its search for do-able solutions. It has no option but to slog on. Terrorism is such a pernicious disease, and the innocents of the world have suffered so much from it, that we just cannot relax in our efforts to find do-able and durable solutions. The PMPT will meet again next May, once again with the participation of as many exponents of different and differing points of view, in the hope that we will all be able to learn from each other, and identify what are the most promising avenues still to be pursued. We cannot give up. We shall not give up.

# 11. PERMANENT MONITORING PANEL REPORTS

# ITALY'S APPROACH TO NUCLEAR NON-PROLIFERATION

MR. VITTORIO CRAXI
Italian Deputy Minister for Foreign Affairs

The growing number of weapons of mass destruction is a serious threat to global peace and security, with the added risk that WMDs could fall into the hands of terrorist groups. Limiting proliferation, also given that traditional deterrence criteria no longer apply, has become a global priority and is an issue of growing relevance for public opinion.

The Italian government is actively committed in several ways: within the United Nations, in the European Union, in the G8, and in bilateral relations with our main partners.

I would like to start by illustrating Italy's action within the European Union, especially during our Presidency in the second half of 2003, when a European Non-Proliferation Strategy was adopted—which we strongly advocated—and the basis for a consistent European policy was consolidated, hence making the EU a key player in this area.

The European strategy is based on enhancing the global non-proliferation system; promoting the universal nature of international agreements and their enforcement; and consolidating and strengthening cooperation with the United States and other key partners.

The Strategy makes special reference to the Nuclear Non-Proliferation Treaty (NPT), the Chemical Weapons Convention (CWC) and Biological Weapons Convention (BWC), the Hague Code of Conduct Against Ballistic Missiles Proliferation (HCOC) and the Nuclear Test Ban Treaty—the implementation of which the European Union has urgently called for.

The EU cannot overlook the risks that the proliferation of WMDs represents for its Member States, its population and the interests of Europe. The EU countries and institutions are aware that they share the collective responsibility of addressing such risks and contributing to the fight against proliferation.

Diplomatic and political preventive measures and the action of international organizations represent the first line of defence against proliferation. If such measures fail, coercive measures under Chapter VII of the United Nations Charter may be considered.

The Strategy's approach is rooted in the belief that multilateralism is the most effective instrument in reaching the objectives it sets out and that international cooperation must be the reference framework. Non-proliferation is the test bench for the effectiveness of multilateralism.

In Europe's view, the system established by Multilateral Treaties lays the groundwork for all efforts in the area of non-proliferation. However, for such a system to preserve its credibility, its effectiveness must be enhanced.

To this end, the strategy pursues full compliance of the obligations spelled out in the Treaties, by means of existing verification mechanisms and, possibly, new ones.

Within this context, the role of the United Nations Security Council must be strengthened and new forms of partnership with the UN and other international organizations must be sought. In any event, strengthening the role of the UNSC is seen as complementary to the action of verification agencies, to enhance their political, financial and technical capabilities.

Furthermore, the European Union intends to focus on the root causes of instability and insecurity, consolidating its commitment in favour of solving political conflicts, providing development assistance, fighting poverty and safeguarding human rights. Achieving regional and global stability is essential in order to succeed in the fight against the proliferation of weapons of mass destruction.

At an operational level, the Strategy indicates a number of concrete measures for the future: financial support for the projects of the International Atomic Energy Agency, enhancement of export control regimes, enforcement of sanctions against the trafficking of materials used in weapons of mass destruction and more stringent rules for monitoring the transit and transfer of sensitive materials. Within this framework, the EU supports initiatives aimed at identifying and stopping illicit trafficking, especially the Proliferation Security Initiative (PSI) launched by the United States, which Italy has participated in since its establishment.

With regard to nuclear non-proliferation in particular, we consider the Nuclear Non-Proliferation Treaty as the cornerstone of the non-proliferation system. It has played a key role: in the 35 years since its implementation, the NPT has made a fundamental contribution to global peace and security. Thanks to the NPT, countries with nuclear weapons have not increased, as was feared, whereas over the years the number of countries with the intention of developing nuclear weapons has decreased. The NPT has been joined by almost all countries, with the exception of India, Pakistan and Israel, while North Korea has withdrawn. Italy joined the NPT in 1975 after an extensive debate. Since then, that decision has consistently inspired our foreign policy, with the complete consensus of all the subsequent governments and political forces.

The international nuclear non-proliferation system is, however, under constant pressure, with the ensuing risk of considerable erosion. There are several reasons for this: non-compliance with international obligations; the spread of nuclear technology, which makes it possible to develop military capabilities in the guise of civilian initiatives; and a vast black market for nuclear materials.

The NPT Review Conference, held in May 2005, did not reach a satisfactory conclusion, notwithstanding our efforts. In New York we called for progress in the three main areas covered by the Treaty: non-proliferation, disarmament and peaceful uses of nuclear energy. We requested a Common Position of the EU based on a balanced approach liable to safeguard the integrity of the Treaty. We indicated the need to comply with non-proliferation obligations and to implement more effective monitoring and safeguards.

Furthermore, in addition to the priority of complying with nuclear non-proliferation obligations, we also voiced the need to reduce existing arsenals in a verifiable manner. This reiterated the commitment of the G8 in favour of international cooperation initiatives for the elimination of weapons of mass destruction and relative materials through the *Global Partnership,* to which Italy makes a significant contribution. With regard to the right to use nuclear energy for civilian purposes, we claimed the need for a reasonable equilibrium between such a right and the existence of increasingly effective verification mechanisms and safeguards. In other words, the basic requirement must be the compliance with non-proliferation obligations.

The United Nations Summit in September 2005 also failed to reach significant conclusions in the area of disarmament and non-proliferation.

Despite the understandable disappointment with the outcome of both these summits, which we had looked to with great expectations, the positive aspects that did emerge must not be overlooked.

The central role of the NPT was never questioned—indeed all parties confirmed its importance and the need to safeguard it.

Furthermore, the debate on the NPT which began in 2005 has shed light on a number of encouraging signs: firstly, the key role played by the European Union, particularly during the Review Conference. In addition, it is significant that the extensive debate that took place covered a number of fundamental issues, particularly thanks to the EU, which if adequately pursued, could contribute to enhancing the non-proliferation system. The most significant matters include: a more restrictive interpretation of the possibility of withdrawing from the NPT, under Art. IX, international cooperation for the elimination of weapons of mass destruction (Italy had drafted a document on this point that received the support of the EU), the need for new rules on the nuclear combustion cycle and access to these facilities, in compliance with the principles of non-proliferation and the right to use nuclear energy for civilian purposes. A report by the expert group established by the IAEA Director General Dr. El Baradei was submitted on this subject, giving indications on the possibility of creating international consortia providing these services. This issue is currently being followed up by the G8 and the IAEA.

One of the reasons for the failure of the International Community to reach an agreement on how to fight the proliferation of weapons of mass destruction is without doubt the strong disagreement between those who claim that non-proliferation is the top priority and those who deplore the lack of progress in the field of nuclear disarmament.

This is a vicious circle that must be broken. The threat posed by weapons of mass destruction demands immediate action.

We believe that the process must be reinvigorated by seeking to establish as vast a consensus as possible on practical, uncontroversial and mutually-agreeable measures liable to relaunch the disarmament and non-proliferation agenda. To this end we have consulted extensively with our main partners and have been active within the main multilateral fora. Our main objective is to continue enhancing the role of the European Union through the full implementation of the Non-Proliferation Strategy.

I would like to illustrate two approaches that we have undertaken to pursue: the enhancement of the inspection capabilities of the International Atomic Energy Agency (IAEA), through the generalized enforcement of the Additional Protocols and the resumption of effective initiatives within the Geneva-based Conference on Disarmament, through a new negotiation process leading to an agreement limiting the production of fissile material for nuclear weapons (Fissile Material Cut-Off Treaty—FMCT)

Italy, together with the EU and the G8 has been supporting the Additional Protocol: the campaign for its wider enforcement must continue.

The FMCT, for its part, limits the possibility of stockpiling new fissile material for the production of bombs and hence lays the groundwork for the eventual reduction of nuclear armaments. At the Geneva-based Conference on Disarmament, we have urgently called for negotiations to this end and we are actively committed to ensuring that the

FMCT become a priority for the European Union. The possibility of cooperation in the area of nuclear technology for civilian uses with India, launched by the "Bush-Singh Declaration" in July 2005, would certainly benefit from the adoption of a FMCT, as would the entire nuclear non-proliferation system.

The G8 Summit held in July in St. Petersburg, also thanks to Italy's commitment, underscored the importance of the Additional Protocol and of launching negotiations for the FMCT. These two instruments enjoy widespread support in the international community: it would be extremely important if progress could be made in the enforcement of the former and the adoption of the latter.

Italy is also committed to seeking a negotiated solution to the Iranian nuclear question. The NPT recognizes the right to develop nuclear energy for peaceful purposes; the enjoyment of such a right must however comply with non-proliferation criteria, particularly with regard to access to sensitive technologies in the fuel cycle. On July 31, the UN Security Council, with Resolution 1696, invited Iran to suspend its uranium enrichment and spent fuel reprocessing, including research and development. The IAEA Director General will submit a report on this matter by August 31. We very much hope that Iran will fully comply with these requirements.

# PERMANENT MONITORING PANEL ON INFORMATION SECURITY

HENNING WEGENER
Chairman
Ambassador of Germany (ret.), Madrid, Spain

My report on the activities of the Permanent Monitoring Panel since I last reported in August 2005 can be brief, since many aspects of our work have been dealt with in the Plenary Meeting on Information Security held as recently as yesterday. Still, a brief résumé is in order, and some additional facts and achievements need to be recorded.

In the first place, I should report that, in the aftermath of our August meeting last year, the Panel finalized its comprehensive Report and Recommendations entitled "Information Security in the Context of the Digital Divide." As planned, that document was submitted to the World Summit on the Information Society at its Tunis phase (16-18 November 2005) and figures prominently in the document list of the summit under reference WSIS-05/TUNIS/CONTR/01.

Owing to the generous support of the World Federation, three members of our Panel were able to attend the Summit to follow the proceedings, but also to advocate and explain our document. For better effect, an executive summary that also contained information about the World Federation and its activities was widely distributed in the Conference Halls. Attending a good number of the collateral conference events and making multiple acquaintances, we were able to broaden our international network. The main purpose of these efforts at dialogue was to bring across our argument that in a context of global development and of broadening the information society, information security needed to be an ever more important element, given the growing relevance of threats looming in cyberspace. It is exactly in, as yet, fragile, nascent information structures that information security had to be built in from the outset. We insistently made the point that capacity building in such societies and security building had to go hand in hand. In Tunis, we also participated in the organizing sessions of what is now the (UN) Global Alliance for ICT and Development.

Since the Tunis Summit, the Panel has attempted to foster its ties with the follow-up organs of the World Summit. We belong to the Advisory Group of the Global Alliance and have recently offered, as a contribution to the incipient work program of the Alliance, to be recognized as a Community of Expertise—a mechanism the Global Alliance has instituted—for matters of information security. The Panel also follows actively the cybersecurity work of the ITU which has been named, in the Tunis final documents, as the central facilitator/moderator for future work on information security ("Action Line 5: Building confidence and security in the use of ICTs") and has taken steps to register as a future participant in the work and meetings on cybersecurity of the ITU World Telecommunication Development branch.

Our substantive work program for the current year is also geared to the post-WSIS tasks. The analysis and recommendations the Panel is presently working upon will be aimed at the Global Alliance; the new Internet Governance Forum, also created at Tunis; the ITU; and possibly UNESCO and the new UN Human Rights Council.

I referred to the subject areas of our current work assignment when I introduced yesterday's Plenary Meeting. Carrying on from the various presentations at the Plenary, we are working on the following problem areas: (1) an analysis of, and recommendations on, the security challenges emanating from new digital networks; (2) the challenges presented by cyber conflict, in both its essential variants, cyberwar and cyber terrorism; (3) the dramatic rise in Internet censorship by Governments which we refer to as cyber-repression; (4) further contributions to the promotion of cybersecurity in the context of the Digital Divide; (5) new challenges for the necessary balance between security and privacy in the face of terrorist threats; and, finally, (6) the safeguarding of information security in processes of trans-frontier and transcontinental outsourcing, in an effort to bridge the existing legal divides. Not all of these issues can be taken in hand with equal depth, or within identical time frames, but we are confident to produce a valid work product this year as in earlier work periods.

In addition, I would like to recall that the Panel, in addition to its presence on the web site of the World Federation, maintains a more complete open web site of its own (www.itis-ev.de/infosecur) where its collective work products, but also individual contributions and supporting papers can be found.

# PERMANENT MONITORING PANEL ON POLLUTION

LORNE EVERETT
Chancellor, Lakehead University
Thunder Bay, Canada

## MEMBERS OF THE PMP ON POLLUTION

Members of the WFS who are current active members of the Pollution Permanent Monitoring Panel include the following:

Chairman
Richard Ragaini (USA) (ragaini1@llnl.gov)

Permanent Members
Lorne Everett (USA) (leverett@haleyaldrich.com),
Sergio Martelluci (Italy) ("Tor Vergata" <Smart@ing.UniRoma2.It>),
Gina Calderone (USA) (gcalderone@eaest.com),
Paolo Ricci (USA) (apricci@earthlink.net), and
Frank Parker (USA) (frank.l.parker @vanderbilt.edu).

Associate Members
Robert Clark (USA), William Sprigg (USA), Albert Tavkhelidze (Georgia), Vittorio Ragaini (Italy), Majid Hassanizadeh (Netherlands), Joseph Chahoud (Italy), Stephen Kowall (USA), Zenonas Rudzikas (Lithuania), Andrew Thompson (USA), Aurielio Aureli (Italy), Massimo Civita (Italy), Giovanni Barrocu (Italy), and Salvatore Carrubba (Italy).

## SUMMARY OF THE EMERGENCY

The continuing environmental pollution of Earth and the degradation of its ecosystems both constitute one of the most significant planetary emergencies today. This emergency is so overwhelming and so encompassing that it requires the greatest possible international East-West and North-South cooperation to implement effective ongoing remedies.

It is useful to itemize the environmental issues addressed by the Pollution Permanent Monitoring Panel (PMP) since several PMPs are dealing with other environmental issues. Global pollution, for example, including ozone depletion and the greenhouse effect, is being addressed by other PMPs at the World Federation of Scientists.

The Pollution PMP has been involved in addressing the following environmental emergencies:

- Degradation of surface and groundwater quality,
- Degradation of marine and freshwater ecosystems,
- Degradation of urban air quality,

- Impact of air pollution on ecosystems.

The Pollution PMP monitors the following priority issues:

- Degradation and cleanup of existing surface and groundwater supplies from industrial and municipal wastewater pollution, agricultural run-off, deforestation, and military operations.
- Reduction of existing air pollution and resultant health and ecosystem impacts from long-range transport of pollutants and trans-boundary pollution.
- Development of technologies for prevention and/or minimization of future air and water pollution.
- Training scientists and engineers from developing countries to identify, monitor, and clean up pollution.
- Provide an informal channel for experts to exchange views and make recommendations regarding environmental pollution.

In the process of monitoring the priority issues, the Pollution PMP has launched the following initiatives:

- Vulnerability of Groundwater to Pollution in Sicily.
- Containment of Nuclear and Hazardous Wastes in the Subsurface Regime.
- Laser Drilling as a New Technique for Drilling Subsurface Wells.
- Evaluation of the Ecological Impacts of Endocrine-Disruptor Chemicals on Surface Water and Groundwater.

WORKSHOPS AND SEMINAR SESSIONS

The following workshops and seminar sessions have been sponsored by the Pollution PMP and held in Erice, Sicily since its beginning in 1998. These workshops and seminar sessions highlight the global and regional impacts of pollution-related issues in developing countries:

- 1998: Workshop on Impacts of Pharmaceuticals and Disinfectant By-Products in Sewage Treatment Wastewater Used for Irrigation
- 1999: Seminar Session on Contamination of Groundwater by Hydrocarbons
- 1999: Workshop on Black Sea Pollution
- 2000: Seminar Session on Contamination of Groundwater by MTBE
- 2000: Workshop on Black Sea Pollution by Petroleum Hydrocarbons
- 2001: Workshop on Caspian Sea Pollution
- 2001: Seminar Session on Trans-Boundary Water Conflicts
- 2001: Workshop on Water and Air Impacts of Automotive Emissions in Mega-Cities
- 2003: Seminar Session on Water Management Issues in the Middle East
- 2003: Workshop on Monitoring and Stewardship of Legacy Nuclear and Hazardous Waste Sites

- 2005: Task Force Meeting on Groundwater Pollution Vulnerability Mapping of Sicily
- 2006: Workshop on Plastic Contaminants in Water

2006 REVISED PROPOSAL FOR VULNERABILITY MAPPING AND ENVIRONMENTAL DATABASE MANAGEMENT FOR SICILY

Since 2003, the members of the Pollution PMP, in collaboration with Sicilian and Italian universities and agencies, have developed a proposal for safeguarding the drinking water and groundwater resources in Sicily. The proposal for Groundwater Pollution Vulnerability Mapping and Environmental Database Management for Sicily was originally developed in September 2003 and revised in conjunction with Sicilian University and Italian agencies in August 2005 and August 2006. There are three main components of the proposal, as follows:

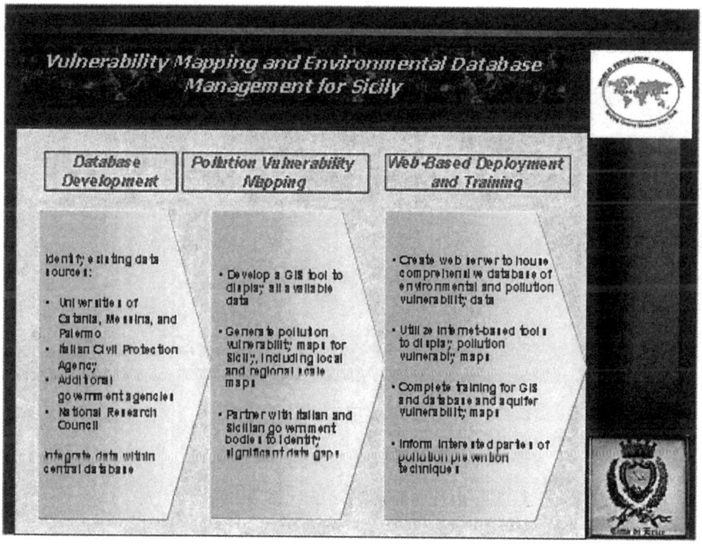


PROJECT PHASES

- Create an up-to-date comprehensive database:
  - Hydrogeologic zones,
  - Site-specific pollution sources.
- Develop/update aquifer vulnerability maps of Sicily with local University collaboration.
- Deploy maps on website with Geographic Information System tools.
- Train the potential end users (i.e., government agencies, local land use planners, developers, etc.).

GROUNDWATER VULNERABILITY MAPPING: DRASTIC

A methodology that allows the pollution potential of any hydrogeologic setting to be systematically ranked.

The system has two major parts: (1) the designation of mapable units—"hydrogeologic settings," and (2) the superposition of a relative rating system called "DRASTIC."

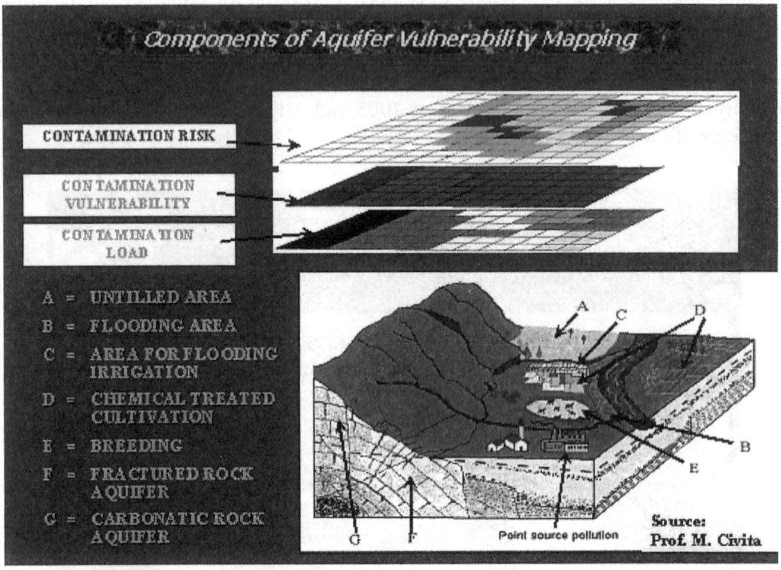


PROJECTED FUTURE ACTIVITIES

Advancements in Laser Drilling as a New Technique

Up to 1.6 megawatts of power: able to drill an 8-in. diameter hole in 4 seconds. Used in environmental drilling, subsurface tunneling, unexploded ordnance removal missions, and structural foundation investigations.

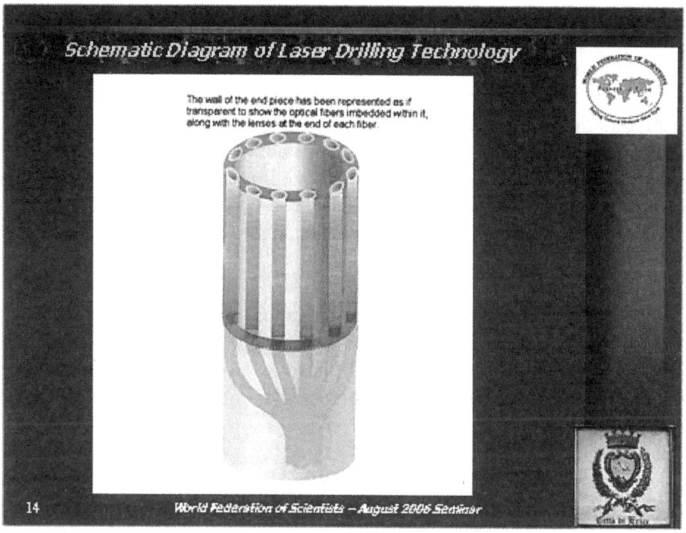


Advancements on Contamination Related to Vapor Intrusion

Atmospheric/indoor air quality issues related to vadose zone soil gas migration. Source material is located in subsurface or residual soil as a non-aqueous phase liquid.

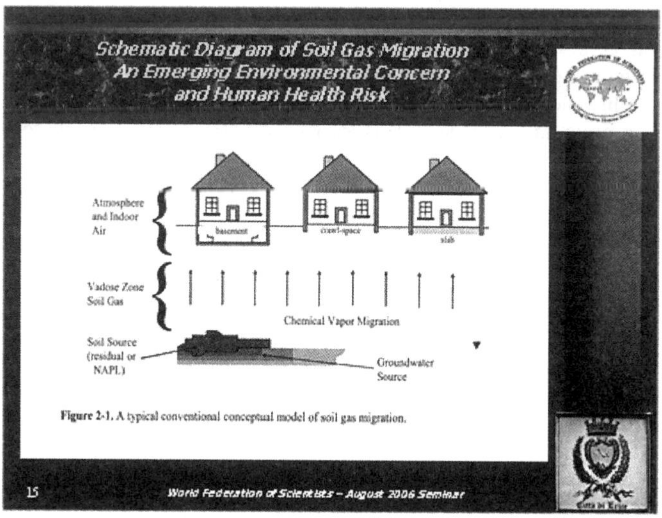

Impact of Endocrine-Disruptor Chemicals on Oceans, Surface Water, and Groundwater

Endocrine-disruptor chemicals are chemicals that can disrupt the endocrine processes by interfering with the hormonal signals that control normal development of the brain and other organ systems.

Chemicals that are known human endocrine disruptors include dioxin, polychlorinated biphenyls, dichlorodiphenyltrichloroethane, and other pesticides. Environmental contamination associated with these types of organochlorine compounds is a global problem resulting in serious consequences for both human and ecological health.

Chlorinated Volatile Organic Compounds in Groundwater

A world-wide issue associated with the increase in industrialization. These toxic compounds tend to remain in the dissolved phase in groundwater, and can migrate deep into bedrock aquifers as dense, non-aqueous phase liquids, and impact drinking water supplies.

Long-Term Stewardship of Nuclear/Hazardous Wastes

This issue impacts nearly every country in the world. There are no universal regulatory criteria for proper disposal practices for nuclear and hazardous wastes. Panel has worked with other participating countries and governmental agencies with the responsibility for addressing this issue. An international proposed Memorandum of Agreement between the World Federation of Scientists and five participating countries was developed to further understand this issue.

Perchlorate Contamination in Groundwater

Used as a strong oxidizing agent. Slow to react under normal environmental conditions. Its lack of reactivity and high solubility in water makes it very mobile in the subsurface. This chemical is an emerging surface water contaminant of concern in the states of California and Arizona, and is associated with adverse impacts to agricultural land.

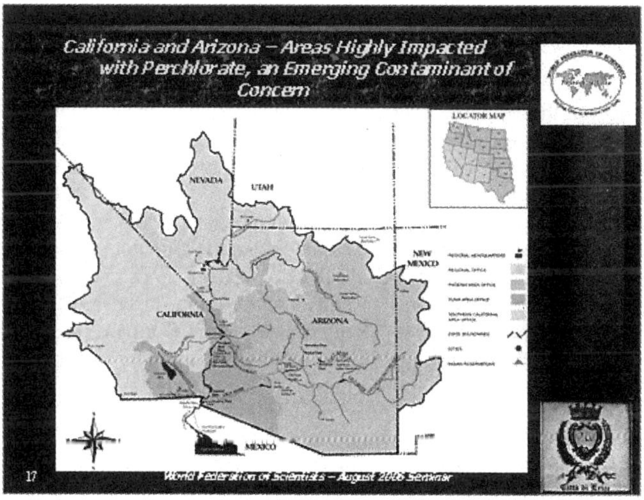
California and Arizona – Areas Highly Impacted with Perchlorate, an Emerging Contaminant of Concern
LOCATOR MAP
NEVADA
UTAH
CALIFORNIA
ARIZONA
NEW MEXICO
MEXICO
17
World Federation of Scientists – August 2006 Seminar

# PERMANENT MONITORING PANEL: LIMITS OF DEVELOPMENT

GERALDO G. SERRA
University of São Paulo, Brazil
Chairman

From 1999 to 2003, the panel examined several aspects of megacities' sustainable development. The main conclusion of the papers and discussions is that megacities concentrate most of today's problems of sustainability because they are huge agglomeration of people and many human activities. Therefore they are, on the one hand, the main centres of demand for energy, water and all sorts of goods and, on the other hand, large generators of pollutants in soil, water and atmosphere.

The last two years were dedicated to monitoring the intense process of interregional and international migration from the viewpoint of its advantages both for reception and emission regions and countries, but also from the viewpoint of social, economical and cultural problems this process generates.

This year scope was the limits of development priorities and conceptual changes required to face new challenges presented by globalisation and threats coming from possible resources exhaustion and environment contamination.

Throughout the year, members were encouraged to discuss this theme through e-mails exchanging their viewpoints on it. In Erice, besides the normal PMP meetings, it proposes and organizes a special session on new concepts on limits of development.

To help the PMP Members to succeed in this task, it invited Wouter van Dieren, one of the founders of the Dutch environmental movement, to this year's meetings and to present a paper at the mentioned special session. He is the Director of IMSA Amsterdam, Institute for Environment and Systems Analysis. Between 1978 and 1988 he was vice-president of Ecoropa, the European Ecological Association. From 1992 to 1997 he was vice-president of the International Advisory Board of the German "Wuppertal Institute for Climate, Environment and Energy." Currently he presides over the Advisory Board of the HKB/SNS-Bank's Environmental Re-Fund. He is a member of the Club of Rome and of the World Academy of Art and Science. He is author or editor of twelve books, among them the 1995 Club of Rome report "Taking nature into account."

The attendance at the PMP main meeting was the following:

- Alberto Gonzalez Pozo, Mexico
- Bertil Galland, Switzerland
- Christopher Ellis, USA
- Geraldo G. Serra, Brazil
- Hiltmar Schubert, Germany
- Juan Manuel Borthagaray, Argentina
- Leonardas Kairiukstis, Lithuania
- Mbareck Diop, Senegal
- Wouter van Dieren, The Netherlands

At the main meeting some members presented papers to stress different aspects of the theme.

Hiltmar Schubert proposed a dynamic approach to limits of development, considering that most assumptions and projections are based on ever changing technologies and economic conditions. He recommends that limits of development should be considered in separate as well as in its mutual effects.

Juan Manuel Borthagaray presented the example of Argentina "a country that during the 90s was perceived as a model of growth in a high index of human development in Latin America, and which became a routine negative reference for development." He considers that human development and democracy are inseparable and examines the recent historical evolution of Argentina from the military regime to hyperinflation and "peso" convertibility until the system breakdown and bankruptcy.

Leonardas Kairiukstis presented a land-use plan for Lithuania and showed how the country's economy is recovering quickly after the end of the Soviet Union.

Alberto Gonzalez Pozo contributed with an analysis of the last years PMP activities and proposed to give more attention to sudden climate changes and the mitigation of damages and other consequences.

Wouter van Dieren presented his views on the evolution of development concepts since the publication of "The Predicament of Mankind" and the works of Dennis Meadows and Jay Forrester. He stressed the non-sustainability of military expenditure, particularly in the USA, and the fact that pressures over the environment are growing. He also shows the importance of international regulations and fiscalities to control over-consumption and pollution, particularly $CO_2$ emissions.

Geraldo G. Serra stressed the difference between mere growth and real development and showed how the evaluation criteria is evolving from Gross Domestic Product, passing by GDP per capita and Purchase Power Parity to HDI, the Human Development Index.

The discussion of this presentation by all members reached the following main conclusions:

- Development is a concept very different from mere growth, though dependent on it.
- Development should be understood at a local, regional and global scales and it is conditioned by many territorial and human aspects.
- 35 years after the Club of Rome statement, we need a more dynamic consideration of limits of development relevant to the changes in resources and environmental stresses.
- A political paradigm of development could soon reach its limits if it does not bring satisfaction to human development expectations.
- Excellence of governance is essential to sustainability.
- Consider the cultural dimension of development: cultural diversity is as important as biodiversity.

The PMP invited William A. Sprigg from the PMP on "Climatology" to verify the advantages of a joint meeting of both PMPs at the 2007 meeting. There is unanimity about the convenience of a joint workshop and as a first approach participants proposed the following themes for it:

- Effects of climate changes in territorial activities.
- Climate changes and international migration.

At the end, members proposed as a preliminary thematic for the 2007 meeting the following items:

- Globalisation: natural and human resources
- Destruction of local economies and cultural diversity.
- Environmental consequences of international outsourcing.
- Human consequences of international outsourcing.

The meeting was considered by most participants as very interesting and productive.

# FLOODS AND EXTREME WEATHER EVENTS PERMANENT MONITORING PANEL

ROBERT A. CLARK
Hydrology and Water Resources
University of Arizona, Tucson, USA

Activities in 2005-2006 of the Permanent Monitoring Panel (PMP) on Defense Against Floods and Unexpected Meteorological Events have centered on two main areas:

1. Studies related to sustainable development of water and related resources in Sicily.
2. Updating of the PMP web pages.

The following PMP members, international consultants, and representatives of Sicilian universities have participated in these activities.

Philip H. Burgi, Bureau of Reclamation, USA
Robert A. Clark, University of Arizona, USA
A. Curtis Elmore, University of Missouri, Rolla, USA
Munther Haddadin, Ministry of Water & Irrig, Jordan
Margaret S. Petersen, University of Arizona, USA
William A. Sprigg, University of Arizona, USA
Giuseppe Rossi, University of Catania
Antonio Cancelliere, University of Catania
Bartolomeo Reitano, University of Catania
Dr. Antonio Boccafoschi, University of Catania
Antonio Cancelliere, University of Catania
Giuseppe Aronica, University of Messina
Angela Candela, University of Palermo

## DEVELOPMENT OF WATER AND RELATED RESOURCES IN SICILY:

Preliminary proposals were prepared during the August 2005 PMP meeting. A follow-up meeting was held in Taormina in October 2005, convened by professors from the University of Catania, to discuss further content of the proposals.

Revised drafts of the proposals were prepared in 2006 under the leadership of Professor G. Rossi (Water Supply) and Professor B. Reitano (Flash Floods). These drafts were discussed at the August 19$^{th}$, 2006, meeting in Erice. It is expected that final revised drafts will be completed by early October 2006.

Work has concentrated on preparing drafts of two proposals, as follows:

1. Improve reliability of municipal water supply in Sicily.

The primary objectives of this research program are to identify critical factors relating to improving reliability of municipal, industrial, and agricultural water supplies in Sicily.

A number of work programs are envisioned in connection with the implementation of this program. All are critical to improving the water infrastructure in Sicily. Unfortunately, the available Sicilian domestic water supply is such that probably more than 50 percent of the population has an insufficient and intermittent water supply. Many cities, towns, and villages experience water shortages for prolonged periods almost every year. The water problem is exacerbated by the fact that 70 percent of the Sicilian water supply is consumed by agriculture which is critical to the Sicilian domestic economy.

The proposed budget for this study involves funding for at least six Sicilian senior and junior scientists. Preliminary estimates of costs of engineering work, meetings, workshops, and operational costs including scientists, total €1.1 million over a period of three years

2. Reduce risk and damages from flash floods in small drainage basins in Sicily.

Sicily is a semi-arid country where climate and geography produce intense short-duration rainfall events that—although infrequent—produce severe flooding over small drainage basins. Such floods frequently result in extensive damage and loss of life. The primary objective of this project is to develop appropriate methodology for a comprehensive flood warning system. The program methodology includes a study utilizing three experimental pilot basins to collect data by radar and ground-based rain and stream gages.

A flood forecasting/warning response system will be developed that utilizes not only observed meteorological data, but quantitative precipitation forecasting, including assessment of flood hazard and flood risk.

This study will include scientists from three Sicilian universities and both national and regional departments of civil protection. The proposed project would last three years and involve at least eight Sicilian senior and junior scientists. The total budget involving personnel, equipment, and other operational costs is estimated to total €1.4 million.

REVISE PMP WEB PAGES

The website for the PMP on Defence Against Floods and Unexpected Meteorological Events is currently under revision to include the following:

1. Home page.
   a. Summary of the emergency.
   b. Priorities in dealing with the emergency.
   c. Reports of meetings and other activities.
      – Initial meeting reports.
      – More recent PMP meeting reports
      – Other related reports.
   d. Special recommendations.
2. Links.
   a. Complete Definition of the Emergency.
   b. PMP reports for 2003, 2004, and 2005.
   c. Water – A Global Emergency.

d. Dust and Sand Storms in Arid Countries.
e. Optimal Use of Environmental Resources in the Kalahari.
f. The World Lab Yellow River Project.
g. Developing a National Master Program for Water and Related Resources.

# 12. LIMITS OF DEVELOPMENT
# PERMANENT MONITORING PANEL MEETING

# THE DYNAMIC CONSIDERATION OF LIMITS OF DEVELOPMENT

DR. HILTMAR SCHUBERT
Fraunhofer Institut for Chemical Technology
Pfinztal, Germany

## INTRODUCTION

If we observe the "Club of Rome" activities since the first publication in 1972 "Limits of Growth," the "New Limits of Growth" in 1992, the "30 Years Up-to-date report" of 2004 and other contributions to this item in optimist and pessimist models, we shall find noticeable variations and consequences for the human being. The prediction of the original statements was based on an instantaneous situation without future variations caused by influences of mankind and reciprocal influences of economic-, social- and political developments and of the report itself.

The following assumptions were made:

1. If present trends continue unchanged, a limit to the growth would be reached sometime within the next 100 years. This would then result in a sudden and un-controllable decline in both population and industrial capacity.
2. These trends can be altered. Moreover, if proper alternatives were made, the world could establish a condition of "ecological stability" that would be sustainable far into the future.
3. The world could embark on this second path, but the sooner this effect started, the greater the chance would be of achieving this ecological success.

Many predictions did not happen, have changed in their meaning or were shifted into the future. Nevertheless, the importance of the Club of Rome results should be estimated on a high level concerning the consequences of human behaviour and the future. All contributions to this item "Limits of Growth" lead to the conclusion that, on principle, most of the statements are right, if consequences are neglected.

The difference of the optimistic or pessimistic model will be only the date, when limits of the quite different resources would take place in 50 or 100 years or beyond.

There will be also the question as to whether we can substitute resources in the long-term or if new developed technologies change the way of living. This may change the behaviour of people, the consumption of resources and other limits of development may occur.

The consequence will be:

*"We need a dynamic consideration of "Limits of Development" and appropriate activities."*

8 years ago (1998), the International Seminar of Erice began with a paper "Limits of Development in a Global View" and one year later, one of the Permanent Monitoring Panels (PMP) was named "Limits of development" (No. 6) and a working group was established. Every year the working group has discussed at Erice one specific problem in

this field. (Examples were: Solid Urban Waste, Water, Migration, Urban- Mobility, Megacities etc.)

To update the problems this year in a general discussion of all our experiences is a very good idea.

THE TEMPORAL VARIATIONS

If we consider the different "Limits of Development", we observe a dynamic background, because these limits are dependent on resource-consumption, behaviour of people, progress of technologies, growth of population, social- and political-development etc.

For instance, 35 years ago, scientists predicted the oil supply would come to an end in the year 2000. Now, some people estimate the year 2070 or beyond.

The UNESCO tables do not show other mineral resources, which could become scarce if their consumption increases in larger amounts. There may be some shortages in the future, such as mercury. But as in the past, there would be substitutes, if the cost of production would rise over an acceptable level. The behaviour of an open market in a globalised world will be a good vehicle to avoid shortage over a long-term. Therefore, we may be able to state—in contradiction to the first statement of the Club of Rome— that nothing will run out on a long-term because:

1. Extrapolation and isolation of non-renewable material out of natural resources is still increasing.
2. Requirements of resources are changing because of the dynamic industrial supply of industrial production by energy- and material-saving processes.
3. Substitutes will replace non-renewable resources.
4. An open market in a globalised world regulates the demand of resources by increasing its value and causes the use of substitutes.

The environmental stress of industrial production, traffic, good consumption etc. and the contamination of the environment will be of importance. We observe the economic "Kuznetcov curve," which means the following: First the environmental stress will grow faster with economic growth, then we observe a neutralization, that means no increase of environmental stress with economic growth and then at the end, the environmental stress decreases with increasing economic growth thanks to the increase of environmental protection activities.

There is a connection between growth of economy and environmental behaviour:

*"The surest way to improve your environment is to become rich."*

Therefore, in many European countries, the extrapolation of resources and industrial production have grown to a large extent without environmental stress in the last 20 years.

From a theoretical point of view, many "Limits of Developments" can be temporally shifted into the future or may be completely avoided, if financial help is available to solve these problems. Precondition is the political will of the responsible people. But these financial preconditions are not realistic, therefore we do have limits.

To decrease the contamination of the environment the following well-known activities should be executed:

1. The dependence of consumption and production of goods from environmental pollution has to be avoided by increasing environmental protection (Kuznetsov Curve).
2. The carbon dioxide output has to be reduced by avoiding Carbon-Combustion (Energy- and Goods production, Traffic etc.).
3. Private and industrial production should avoid toxins in products and waste.
4. Waste-Disposal has to be regulated worldwide.
5. The future decrease of population should be supported in overpopulated regions.

Limits of Development are also influenced by the mental level of the people who are confronted with these limits. Successful training and education of these people may improve the situation, but to a large extent, the success is dependent on the capability and/or ability of the special human being. There is no doubt in the short-term view that we have to exempt distinct mental properties of people who are not willing to improve the situation by own activity. The reasons may be mental inability, indolence or ignorance.

## EXAMPLES OF "LIMITS OF DEVELOPMENT"

<u>Carbon dioxide contamination of the atmosphere has a temporal and long-term priority</u>

In opposition to limits of development caused by lack of non-renewable resources, the global contamination of the environment is of high priority. Countermeasures are very expensive and burden the economic output of industry and government. The saying, "Act local, think global" has, in this connection its special importance, especially in the carbon dioxide business. We shall have no change in the situation, if the large producers such as the USA and China neglect their carbon dioxide output.

<u>Contamination of food products by environmental toxins</u>

As long as the influence of toxins (in quantity and quality) on the human being are not better known, toxins must be avoided as much as possible in environmental contamination. In view of avoidance of epidemics, priorities have to be introduced.

<u>Increasing differences of economic growth caused by globalisation</u>

The general increase of worldwide economic growth by globalisation caused a lower economic increase in the poor- or undeveloped countries compared to industrial countries.

The reasons are that infrastructure, political circumstances and education are lacking.

Very often the ability and will of the home governments are not able to overtake the organization and management of the economic growth of their country.

There are many examples of economic investigations of the poor or undeveloped countries by the "World Bank," which are starving after the World Bank Management drew back its specialists.

Over fishing the oceans

Up to a distinct percentage, over fishing will be not reversible by order of nature, therefore fishing should be done only visibly below this limit.

Contamination of the environment

The growth of population, industrial manufacture, and traffic, increase the contamination of ground, water and air. It is very important to avoid the sources of contamination first, because to clean up the environment later on needs much more financial power which will be not available everywhere.

Environmentally friendly waste management of Megacities will be a real limit of the development. Most of the problems in future will rise in developing countries. Industrial countries have solved the problems to a large extent (Kuznetsov Curve). Additional the influence of environmental toxins burden the health of the human being and causes the die out of rare animals and plants.

Overpopulation of developing countries

The overpopulation of distinct developing countries (e.g., Near East, North Africa, etc.) causes emigration to industrial countries and political pressure. The problem will increase extraordinarily in the future.

## CONCLUSION

Since the first statements of the Club of Rome, we have discovered changes in "Limits of Development." Some limits have increased, some limits have decreased in their meaning.

Nevertheless, some "Limits of Development" in realistic circumstances are foreseeable and of constant danger.

The item of "Limits of Development" is more complicated than we thought at the beginning, because we have to consider the interaction with limits of growth of economy and limits of economic efficiency and the interaction between Limits of Development and Globalisation.

What are the consequences of sustainability for the economic, ecologic and social aims? We have to consider the largest uncertainty in the limits of development: the reactions of the different governments and their political pressure groups.

In conclusion, we have to consider:

1. A forecast of the future situation will be difficult because the influence of measures or counter-measures are changing in their efficiency over time.
2. The networks and their interactions of technical-,social-, economic-, political- and psychological influences are overseeable only for some years

3. As we have learned from the last 35 years after the start of the "Club of Rome" activities: in general we should not neglect that time goes on. The only thing is, when we are confronted with severe problems out of the large palette of the limits: In 50, 100 years of beyond?
4. We must consider each limit of development separate and in its mutual effects.
5. The decoupling of economic growth, environmental protection and prosperity should be an aim for sustainability.

REFERENCES

Wachstum und Ungleichheit, Oschinski, Weber: Die Volkswirtschaft, Magazin für Wirtschaftspolitik 1-2002 page 19ff.

# LIMITS OF DEVELOPMENT REVISITED: THE SOCIAL AND POLITICAL IMPLICATIONS

JUAN MANUEL BORTHAGARAY
Universidad de Buenos Aires
Buenos Aires, Argentina

## A QUESTION

How is it possible that Argentina, a country that during the 90s was perceived as a model of growth and a high index of human development in Latin America, had turned into a routine negative reference for development?

## A DEFINITION

*Human Development.* Conception of development aimed to the expansion of individuals, and of society as a whole to attain a quality of life according to their own values.

## A HYPOTHESIS

"The crisis Argentina is going through, and which all social actors are suffering in various degrees, is a result of wrong and dogmatic visions, as well as an institutional and political system that proved unable to avoid the ensuing collapse. But this incompetence was compounded by international responsibilities that through error or omission did not warn in due time about the need to make changes in the implemented policies."

Human development and democracy are inseparable.

## LIMITS OF DEVELOPMENT REVISITED

Professors Serra and Schubert have rightly redefined the question of **LIMITS**, from Malthus to Bruntland, as something that is historically determined, in the measure that man, through science and technology has displaced the fence once and again. Therefore, we must no longer speak about limits, but about sustainability instead. Much in the way the bard said, *All's well that ends well,* we can work with the assumption that *"All that is sustainable falls within limits."* As to **DEVELOPMENT**, in our present environment, in which positivism is so deeply imbedded, development implies a category that must be measured, and therefore, expressed, in numbers. In the most primitive and roughest of approaches, Gross Domestic Product per capita is a first option. The World Bank and UNDP offer a reliable and consistent source of yearly tables for most countries. But is it that scientific to compare amounts in U.S. dollars that have so different meanings from country to country and from year to year? Economists work miracles to adjust amounts in function of Purchasing Power Parity and to U.S.$ of a particular year. The result is far from scientific, so much so, that *The Economist* magazine has elaborated a *Big MAC* index around the price, in each country, of this globalised commodity. The United Nations Development Programme ought to know what they are talking about when they say **development**. In their own words ***Human Development*** : conception of development aimed to the expansion of individuals, and of society as a whole to attain a quality of life according to their own values.

Even if we remain positivists and want to express this elusive conception of human development through numbers, we should pay attention to a wider *basket* of indicators such as:

- Life expectancy at birth
- Literacy
- Gender inequity
- Territorial inequity, the cause of massive internal migrations
- Social inequity: Ratio between richest 20% over the poorest 20%
- Social inequity: Ratio between richest 10% over the poorest 20%
- Child/mother mortality per 1000 births
- Average weight at different, telltale ages
- Access to safe drinking water
- Access to sanitation
- Gender equality
- Telephones
- Internet connections
- Access to health attention

The list is far from complete, and anyone is entitled to add or eliminate indicators so as to conform perfected wish list menus.

But again, countries do not hold regular censuses using the same methods, and even within a particular country, methods may vary between censuses, so the determination and comparison of trends may be quite biased.

But the conclusion is that we are not dealing with development as such, in terms of U.S.$ and tons, KW and such, but of Human Development.

But in another twist of deconstruction we may find that this development issue is strongly tainted with World's Bank and UNDP universally accepted categorization of:

- Developed Countries: 1st to 57th or High Human Dev. Ctries
- Countries of Intermediate: Dev. 58th to 145th or Medium Human Dev.
- Underdeveloped Countries: 146th to 177th or Low Human Dev.

The name of the last category piously changed into the more politically correct of:

- Developing Countries

This masks another assumption: that the path of development passes through the promotion of categories, as if they were sport leagues, and that therefore, the model of the upper categories is the **only** way to move up.

This, of course has raised the fear of the major players, that have come to think: "*Look out! If these fellows start to burn, fume, deplete and pollute water, even a tenth as much as we do, it will be the end of the world, we must put a* ***LIMIT TO THESE NAUGHTY DEVELOPMENTS."***

POLITICAL ASPECTS OF DEVELOPMENT

The needed full positive human development is hardly to occur through the means of mere business as usual. They must be brought by excellent governance that optimises all available resources.

We learned the hard way that the fertility of immense plains, the wealth of forests, oil and gas, and other mineral riches are not the most valuable resources a society can summon in order to attain the goals of human development. The most valuable capital is people, the human resource, in fact human intelligence, that should be made the best use of, in order to achieve excellent governance.

EXCELLENT GOVERNANCE

It rest on a tripod consisting of:

1. Best people at the helm (understanding by best the most capacitated, honest and committed to do the best for their governed), but how to attain this? The role of politics through democratic institutions should answer how.
2. A very strong density of the social contract between rulers and constituency.
3. As a consequence of 2, a day-to-day behaviour of the constituency that is as protagonic as that of the rulers if excellent governance is to be attained.

SOCIAL ASPECTS OF DEVELOPMENT

But if the conclusion is that the critical resource to be invested in order to achieve development is people, how do we match this with the fact that a sizable portion of that human resource is kept out of the game because of poverty.

This is the social conundrum of human development. The main goal is human development: that means elimination of poverty. But in order to eliminate poverty you need all hands on deck, and that means incorporating the poor, who precisely because of poverty are excluded to bring their best into play.

This is, precisely, the inescapable political challenge that we are facing.
A tall order indeed. The pessimists say: not possible, and intone Hobbes' mantra: *homo hominis lupus.*

The UN Millennium Project calls for "cutting extreme poverty by half by 2015, and end it altogether within the coming years. The world community has at its disposal the proven technologies, policies, financial resources, and most importantly, the human courage and compassion to make it happen."

**Table 1. Comparison Of Poverty Situations 1990 2003.**

| Rank | Country | 1 U.S.$/day90-03 | 2 U.S.$/day90-03 | National Poverty Line |
|---|---|---|---|---|
| 34 | Argentina | 3.3 | 14.3 | No data |
| 37 | Chile | below 2 | 9.6 | 17 |
| 53 | Mexico | 9.9 | 26.3 | 10.1 |
| 63 | Brazil | 6.2 | 22.4 | 17.4 |
| 113 | Bolivia | 14.4 | 34.3 | 62.7 |
| 139 | Bangladesh | 36 | 82.8 | 49.8 |

Source. Table 3 pages 227-28 UNDP/hdr05

## CONCLUSIONS

So far our PMP has concentrated on limits of development determined by:

- Depletion of resources such as soils, water, fossil fuels, etc.
- Environment pollution of air, water, etc,
- Solid waste and recycling
- Consequences of former in the greenhouse effect and global warming.

The Human Development Index adopted by WB and UNDP is formed as a function of GDPpp, life expectancy and education indexes.

HDI follows quite closely GDPpp; that may only indicate that GDPpp was given a decisive weight in the formula of the polynomic equation. However it has been said that there might be an increase in GDPpp without a proportional increase in HDI, but that the opposite seemed unlikely.

In the last 20 years, neither the *"trickle-down,"* nor the *"when the tide rises all boats go up"* effects have been particularly notorious in LA&Caribbean.

It is possible to improve several, if not most of the items, of the "wish list" menu without increasing that much the emission of $CO_2$ nor passing through the fad of 240HP powered SUVs common in the most developed economies.

## APPENDIX

Our PMP brought together architects/urbanists from Mexico, Brazil and Argentina concerned with the big issue of Latin American megacities. This particular geographic belonging opens the opportunity to some reflections about that part of the planet, that international institutions have brought together in the compartment "Latin America and the Caribbean." LA&C with 540M people represent 9.15% of world's population. At the same time the three countries, with 328M are 60.55% of the area (with Argentina the one of the three that is not among the 100M club). Our countries are not among the first 25, nor the 25 last ones, but in a sort of frontier line between them. The following table illustrates where they stand in between. Data of year 2003 Source Human Dev. Report 2005 UNDP.

**Table 2. Human Development Index 2003.**

| Rank | Country | GDPpc U.S.$ | GDPpcPPP U.S.$ | HD index |
|---|---|---|---|---|
| 19 | New Zealand | 19.847 | 22.582 | 0.933 |
| 34 | Argentina | 3.524 | 12.106 | 0.863 |
| 53 | Mexico | 6.121 | 9.168 | 0.814 |
| 63 | Brazil | 2.788 | 7.790 | 0.792 |
| 139 | Bangladesh | 376 | 1.770 | 0.520 |
| **37** | **Chile** | **4.591** | **10.274** | **0.854** |
| **113** | **Bolivia** | **892** | **2.587** | **0.687** |

Source table2: pages223-225 UNDP/hdr2005 and table14: pages266-269 ibid

New Zealand is the closest to the last of the first 25 and Bangladesh the closest to the first of the last 25 that seemed fit to compare.

Although not included in the lasts of the firsts nor the firsts of the lasts nor in the megacities club, Chile is included because it is currently considered (deservedly) the best of the grade, with over 20 years of stable economic growth under democratic government (nowadays through a coalition of Socialists and Christian Democrats, the two leading political parties). Bolivia was included because, sitting over fabulous oil, gas and mineral riches, is a society deeply split between an indigenous majority (very poor and with scarce political representation) and a ruling, illustrated minority. The country carries another split, a territorial one, between a rich Santa Cruz, in sharp contrast with extremely poor Lima suburbs, with constitutional implications. This situation has known strong social unrest and unseated several presidents. Recently-elected President Morales, himself a mine worker of proud Inca indigenous stock, has taken control of oil and mineral companies and promised to distribute these incomes to promote and include his majority constituency in social and political processes, and, so far, so good, although he is just inaugurating his presidential band.

Table 2 shows trends of human development between the years 1975 and 2003 and, on one part New Zealand, representing a last of the firsts and Bangladesh, as one of the first of the lasts, which for all five of them consistent data were available.

**Table 3. Trends Of Evolution Of HDI Between 1975 And 2003.**

| Country | Rank | 1975 | 1980 | 1985 | 1990 | 1995 | 2000 | 2003 | Improve 03/75 |
|---|---|---|---|---|---|---|---|---|---|
| New Zealand | 24 | 8440 | 8540 | 8680 | 8750 | 9050 | 9240 | 9330 | 10.54% |
| Argentina | 34 | 7840 | 7990 | 8080 | 8100 | 8330 | 8560 | 8630 | 10.07% |
| Mexico | 53 | 6890 | 7350 | 7550 | 7640 | 7820 | 8090 | 8140 | 11.81% |
| Brazil | 63 | 6450 | 6820 | 6980 | 7190 | 7470 | 7830 | 7920 | 12.28% |
| Bangladesh | 156 | 3450 | 3640 | 3890 | 4190 | 4520 | 5060 | 5200 | 15.07% |
| **Bolivia** | **113** | **512** | **549** | **580** | **604** | **636** | **672** | **687** | **34.18%** |
| **Chile** | **37** | **704** | **739** | **763** | **785** | **816** | **833** | **853** | **21.16%** |

Source table 2 pages 223-225 UNDP/hdr2005

It is interesting to note that the richer the country, the slowest trend they have in the improvement of HDIa 03/75. This motivated a little exercise to find out if this was consistent along the whole ranking of nations, with the following result.

**Table 4. Inequality of Income or Consumption.**

| **HDI Rank** | **Richest 20% Poorest 20%** | **Richest 10% Poorest 10%** | **Gini Index year** |
|---|---|---|---|
| 19 New Zealand | 6.8 | 12.5 | 36.2 1997 |
| 34 Argentina | 18.1 | 39.1 | 52.2 2001 |
| 53 Mexico | 19.3 | 45.0 | 54.6 2000 |
| 63 Brazil | 26.4 | 68.0 | 59.3 2001 |
| 139 Bangladesh | 4.6 | 6.8 | 31.8 2000 |
| **113 Bolivia** | **12.3** | **24.6** | **44.7 1999** |
| **37 Chile** | **18.7** | **40.6** | **57.1 2000** |

Source table15 pages270-272 UNDP/hdr2005

# LIMITS TO DEVELOPMENT: THE CULTURAL DIMENSION

ALBERTO GONZÁLEZ-POZO
Departamento de Teoría y Análisis, División CYAD
Universidad Autónoma Metropolitana – Xochimilco, MEXICO

## WHERE WE ARE

My colleagues in the working group focusing on the theme "Limits to Development" arrived with interesting ideas to our meeting. Most of them put questions to the original predictions of the Club of Rome conclusions in the 70s:

- Can environmental protection be greater than its damage, due to the increasing investment in protection measures? Is to get rich the best way to improve the environmental damage?
- Or is poverty the most important limit to development? And what about its relationship with democratic governance?
- Sustainable development is possible only if renewable resources are used, and if that is impossible, then what if they are used at a rate that gives time to find substitutes in the meantime?
- What is the difference in the indexes used the last 30 years to measure development, from plain GNP per capita to HDI or Human Development Index that includes life expectancy and education level? GNP per capita, for instance, doesn't take into consideration the unequal concentration of income of privileged elites in each nation.
- Which actions proposed by the Millennium Project have been achieved elsewhere? Do the developing economies face social and political limits?

It is a difficult task for me to add something more to these questions, but I'll try to take another approach: for a long time, the cultural dimension of development seems to be unlimited.

I use the term culture (and not only education or not only democracy) because it covers two "realities." Five decades ago, Melville Herskovits, a distinguished anthropologist and scholar, defined culture with two aspects: *culture is the man-made part of the environment* he said, *but it is the learned part of human behaviour, too.*[1] Both notions change the vulgar interpretation of "culture" only as something related to its "high" products such as science, arts and humanities. The latter are, of course, parts of culture, but language, education, social organization, beliefs, policies and conflicts are also included, as well as cultivated land, regional and urban infrastructures, buildings, clothes, machines, tools and artefacts, polluted environments, abused resources, war destruction and more…

I think that our Erice meetings since 2000 (when I joined the Limits to Development Workshop) have been dealing with both aspects equally:

- In 2000, we looked at the huge size of some megacities and asked us if there is a predictable limit to their growth. We concluded that each case shows us alternatives: decentralization of urbanites to a bigger region, technical improvement in the use of resources such as water and energy, and so on. If we observe more closely, some of these measures imply change in habits and costumes from the inhabitants and authorities, i.e., changes in the urban culture.

- In 2001, the theme of our group was related to water as a limit in megacities, something closely related to the depletion of water sources and the pollution caused by untreated sewages. Here the answer was of a double nature: better technologies to minimize waste and spill and to increase the recycling of used water on one hand, and on the other, how to increase a conscious use of urbanites of this vital resource.
- In 2002, our group worked upon waste disposal in megacities. I couldn't come that year but I read the proceedings edited yearly by Prof. Zichichi and Prof. Ragaini.[2] This time, the general discussion of the Erice Seminar focused on cultural aspects, such as Society and Structures and their Historical Perspective; Culture and Ideology, Economy and Culture; Psychology of Terrorism. In our group, the question had to do not only with technological measures to recycle garbage and avoid impacts to the environment, but with bad or good behavioural patterns in the handling and final disposal of solid waste too. All these aspects are part of a real urban culture.
- In 2003, we moved to urban mobility and transportation in megacities and examined several ways to reduce the need of endless road networks crowded with fuel-consuming personal cars and look for better alternatives in efficient public transportation systems. Again, movement of people and goods in a given territory, urban or rural, is linked to cultural patterns that are important to analyse.
- Then, in 2004 and again in 2005, our group focused its attention on migration aspects, starting with migration from impoverished rural population to megacities and later on with the same problem but from developing to developed countries. There are many approaches to this problem with demographic, socio-economic and even political consequences to regions and countries expelling or receiving migrants. And the cultural impact is significant for both: new languages, clothing, food habits, beliefs and other cultural features go back and forth from former territories alien to each other.

Of course there are big scientific and technological problems to overcome in these fields, and each one of them poses great questions to the limits of development. But the answers have been, always, a wiser use of applied science and technology to solve them combined with educational or social measures that imply a change in attitudes and behaviour of people. In other words, a better cultural response.

## WHY A CULTURAL DIMENSION?

Now I would like to focus on the cultural agenda of the Erice Seminars. In his presentation last year, our Chairman, Prof. Antonino Zichichi, gave a short history of the evolution of the Seminars through several decades and showed 15 transparencies, one for each theme discussed in the last decades, dividing them into sub themes. Our group, Limits of Development, is Number 6 (after 1 to 5: Water, Soil, Food, Energy and Pollution) and in Prof. Zichichi's presentation was subdivided into three sub themes: demographic growth, collapse of values in the megacities and water, soil, food and energy provisions.

Then come other 6 themes (Numbers 7 to 12: climatic change, global monitoring of the planet, new military threats, science and technology for developing countries to avoid north-south environmental holocaust, problems of organ substitutions and infectious diseases). The group Number 13, named "cultural pollution," comes next and is divided into 7 sub themes (political violence, education, the values of science, science and faith, science and technology, media and modern technology and democracy). The last two themes, Numbers 14 and 15, deal with a common response against cosmic objects and huge military investments.

Certainly, most of these themes and sub themes have been the subject, year after year, of interesting presentations inside the workshops and before the assembly gathering in the Paul Dirac Auditorium. But some of them had a real cultural dimension, for instance, the one dealing with "cultural planetary emergencies" in 2003 and 2004, even if it was restricted to cultural intolerance or information and communications.

A provisional conclusion is that all the themes of the Erice Seminars are connected totally or partially with cultural questions, because all imply different behavioural patterns of assimilation and response from scientists, decision-makers and social or political leaders. Most of them are related, too, with the dissemination of knowledge to wider audiences and public opinion in order to form, if possible, affirmative attitudes, common values and consensus. But common values and consensus are far from being universal. Cultural diversity is an asset, an advantage like biodiversity, but it is now endangered by globalisation, misunderstood as uniformity and adjustment to a single ideology or a single economic and political system.

## AN EXAMPLE: THE "CHINAMPAS" OF MEXICO CITY

Let me mention here a case of my own recent field of research: the ancient highly productive system of *chinampas* still living in the outskirts of Mexico City is an example that helps to understand the cultural dimension of the origin and persistence of this megacity.[3] The chinampas flourished in the shallow lakes of a closed basin, the Mexico Valley. Nobody knows exactly its origins, but some archaeologists think that it already existed eighteen centuries ago, near Teotihuacan, the first pre-Columbian metropolis. Its presence is detected more clearly since the 12$^{th}$ Century. Essentially, a chinampa is a small, elongated rectangle of rich, humid soil, surrounded by narrow waterways. In the pre-Columbian era, thousands of these small islets distributed in ca. 22,000 hectares formed a symbiotic unit with Tenochtitlan, the Aztec capital. This big city of more than 100,000 inhabitants (a huge one if compared with European standards of the Middle Ages and the Renaissance) provided a stable demand of food consumers of the orchard products harvested in the chinampas several times a year. And the city had in the chinampas a secure supply of food.[4]

That explains the prosperity of both the chinampas area and the Aztec city, capital of the big empire found and conquered by the Spaniards in the 16$^{th}$ Century. They destroyed the ancient city and replaced it with a new one, but were wise enough to allow the Indians to continue the cultivation of chinampas, thus assuring the prosperity of the new Spanish settlement, the forerunner of actual Mexico City. The system still exists, confined to the southern area of Xochimilco, and continues to play an important role in the economy of the Mexican capital, but is now endangered by the growth of the megacity whose problems I have tried to show you in previous meetings to this one.

A Mission of the United Nations' Food and Agriculture Organization (FAO) assessed, in 1984, the chinampas and strongly recommended their preservation as one of the best examples of sustainable high productivity agriculture in the world. And in 1976 another United Nations branch the Education, Science and Culture Organization (UNESCO) put the chinampas of Xochimilco in the List of Natural and Cultural Heritage of Mankind as one of the best examples of an ancient cultural landscape. This double way of the United Nations system of considering a single phenomenon in its economical as well as in its cultural dimension seems to me very appropriate.

The research we are doing now in my University has limited goals: to start an inventory of the thousands of chinampas still existing and understand the problems each one faces in order to organize better the preservation not only of their physical aspects but of their productivity too. The first findings are promising and allow the local authority to take adequate measures to keep alive this ancient culture.

HOW TO CONTINUE?

In my opinion, the future agenda of the Limits to Development group depends a lot on the research profile of its participants. We come from several fields of knowledge: chemistry, physics, journalism, education, but the majority is linked to urban development or urban management problems. If we keep the same composition we have now, the emphasis will fall inevitably upon the Megacities theme that grouped all of us initially. But if we really want to widen the discussion, then we must invite other Erice participants to join us who are now working in specialized groups such as food, water, energy, health and cultural emergencies. We must also include experts in the causes of poverty and in the real remedies to such condition. Democracy, even arms race and terrorism are themes vital to find real limits to development. And only a holistic approach that takes into account cultural diversity may link together such diverse fields in a coordinated common effort to understand the real situation of Limits of Development. I would like to work in such an enlarged group.

From this point of view, I add another suggestion: last year I suggested that our group could focus on the problems and measures to overcome Sudden Disasters like the flooding of many coastal regions around the Indian Ocean. Earthquakes and volcanic activity, too, periodically put human settlements and whole regions before the risk of damage and destruction. The **causes** of these risky phenomena are studied by scientists but their **vulnerability** and the **mitigation measures** that can be taken before, during and after a disaster are linked to cultural awareness, economic provision and democratic governance.

I think that we could reformulate the agenda of the Limits of Development Group in the Erice Seminars taking into account not only the territorial and economic aspects, but also the cultural dimension these realities.

REFERENCES

1. Melville Herskovits: *El hombre y sus obras*, México, Fondo de Cultura Económica, 1952. pp. 29,38. (The original version in English is *Man and his Works. The Science of Cultural Anthropology*, Alfred Knopf, New York, 1948.)
2. A. Zichichi (Series Editor and Chairman) and R. Ragaini (Editor), *International Seminar on Nuclear War and Planetary Emergencies*. 25th, 26th, 27th, 29th,30th, 31st and 32nd Sessions, World Scientific, Singapore, 2000, 2001, 2002, 2003, 2004 and 2005.
3. Alberto González Pozo, Salvador Díaz Berrio, Ignacio Armillas et al. *Catalogación de las Chinampas de Xochimilco: inicio de un proceso indispensable*, (projet report) Universidad Autònoma Metropolitana- Xochimilco, México, 2005.
4. Pedro Armillas, 1971, "Gardens on Swamps," *Science*, vol. **174**:653-661

# PANEL PARTICIPANTS

| | |
|---|---|
| Professor J.M. Borthagaray | Instituto Superior de Urbanismo<br>University of Buenos Aires<br>Buenos Aires, **Argentina** |
| Dr. Mbareck Diop | Science & Technology Advisor to the President of Senegal<br>Dakar, **Senegal** |
| Professor Christopher D. Ellis | Landscape Architecture and Urban Planning<br>Texas A&M University<br>College Station, **USA** |
| Dr. Bertil Galland | Writer and Historian<br>Buxy, **France** |
| Professor Alberto González-Pozo | Theory and Analysis Department<br>Universidad Autónoma Metropolitana,<br>Mexico D.F., **Mexico** |
| Professor Leonardas Kairiukstis | Laboratory of Ecology and Forestry<br>Kaunas-Girlonys, **Lithuania** |
| Professor Hiltmar Schubert | Fraunhofer-Institut für Chemische Technologie, ICT<br>Pfinztal, **Germany** |
| Professor Geraldo Gomes Serra | NUTAU<br>University of São Paolo<br>São Paulo, **Brazil** |
| Professor Wouter van Dieren | IMSA Amsterdam<br>Amsterdam, **The Netherlands** |

# 13. WORLD ENERGY MONITORING WORKSHOP

# THE FUTURE OF NUCLEAR ENERGY

AHMAD KAMAL
Senior Fellow, United Nations Institute for Training and Research
New York, USA

JEF ONGENA
Plasmaphysics Laboratory, Ecole Royale Militaire
Brussels, Belgium

The reputation of nuclear energy has swung wildly over the past decades. It was first touted as a saviour of the world under Eisenhower's Atoms for Peace programme. Major corporations competed with each other to spread the good word. Reactors of all shapes and sizes mushroomed around the world.

It was always known that the linkage between nuclear energy and nuclear weapons was intimate and integral. Action to de-link the two was not taken until the Permanent Five members of the United Nations Security Council (P-5) had all achieved nuclear weaponry themselves. Once that was so, negotiations on the Nuclear Non-Proliferation Treaty (NPT) were started in earnest in order to shut the door to all other outsiders. It was assumed that, since the nuclear weapons proliferation danger had been shut down for NPT signatories, the inherent right of access to nuclear energy could now be guaranteed to them under its Article 4.

The number of nuclear reactors then spread globally (Figures 1 and 2). In part, this was due to the imperative and growing need for energy, particularly in countries with limited or no reserves of fossil fuels. It was also due to the fact that non-signatories to the NPT understood the discriminatory nature of the game in which they were being denied the nuclear weapons technology that the P-5 wanted to retain only for themselves. If status and safety came from nuclear weapons capacity, then they wanted to have it too.

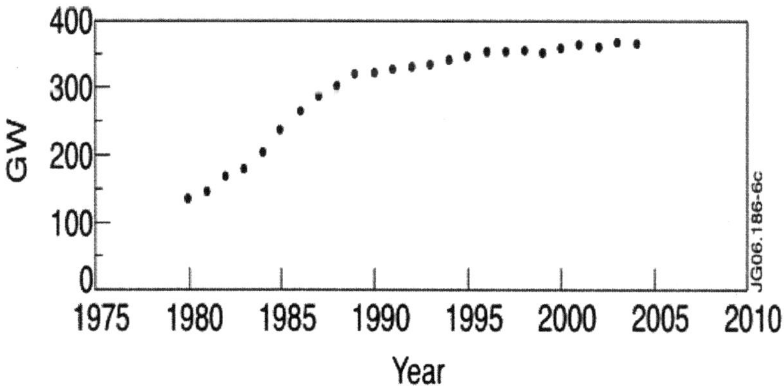


*Figure 1. Evolution of the installed nuclear capacity (in GW) in the world for the last 25 years. At this moment, 442 nuclear power plants are in operation, 27 are under construction and 6 in long term shutdown (data from IEA 2004[1] and IEA 2006[3]).*

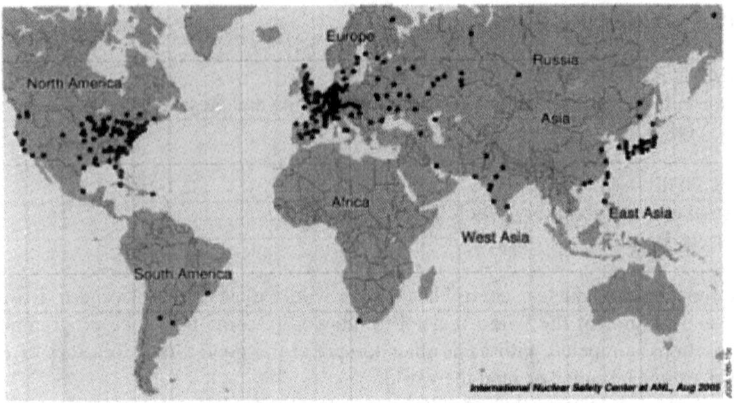


*Figure 2. Distribution of nuclear reactors around the world.*[2]

Meanwhile the safety and security aspects of nuclear reactors continued to exercise decision-makers and public opinion alike. Safety became a burning issue in the West after Three Mile Island, Chernobyl, and a host of other lapses came to light. Fear and panic in the public in most western countries turned them away from this source of clean energy, and this was then justified by arguments about the relative expensiveness of nuclear energy compared to classical fossil fuels. A partial list of some of the more serious incidents is annexed to this paper.

Despite the temerity of public opinion in most western countries, the shortage of access to fossil fuels led some of these countries, France and Japan in particular, to continue to hold up the banner of nuclear energy (Figure 3). They were among the lone exceptions in a new world in which the charm of nuclear energy began to tarnish, and the whole subject slowly became a taboo.

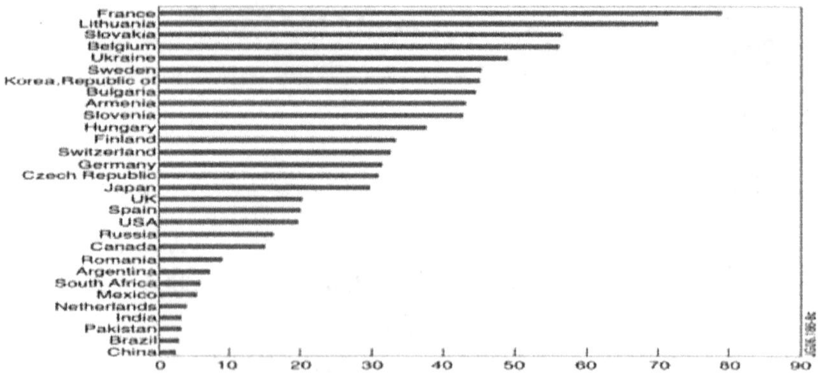


*Figure 3. Nuclear share in electricity generation for various countries in 2005.*[3]

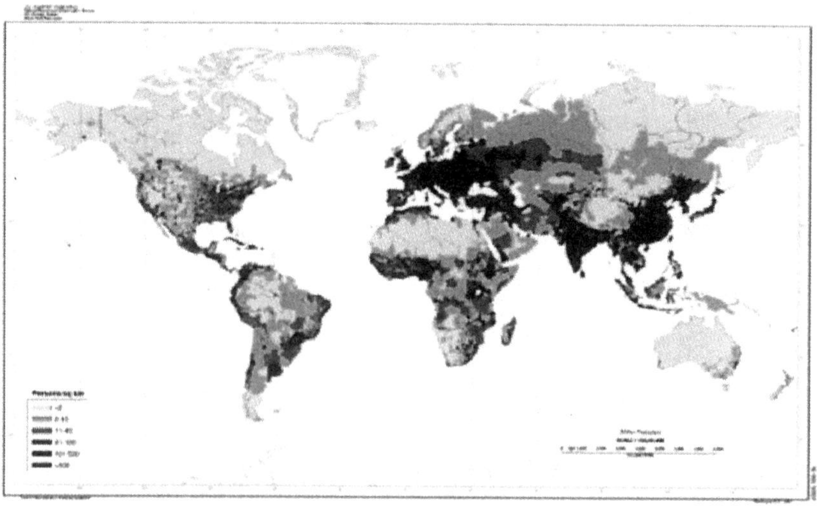

*Figure 4. Population density in the world.*[4]

The situation in the developing countries was substantially different. To begin with, these developing countries contained the overwhelming majority of the population of the world, as much as four-fifths of it (Figure 4). Some of these countries had the largest population masses in the world within their borders. Almost all of them were energy hungry (Figure 5). In fact, it was this extreme shortage of energy resources that was in great part responsible for their under-development.

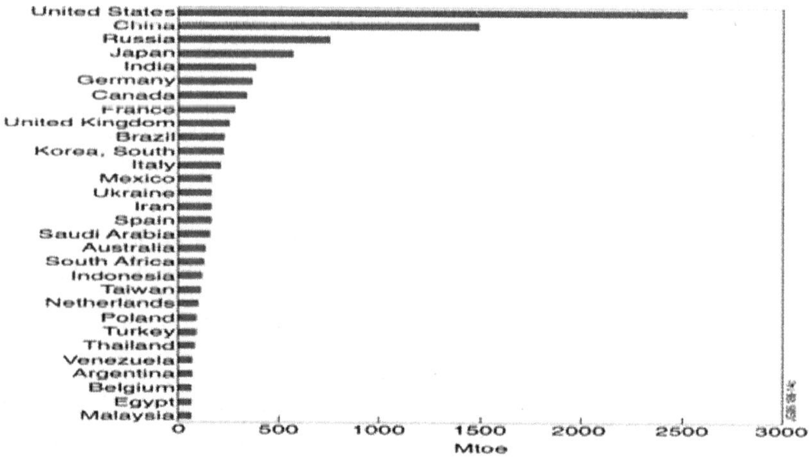


*Figure 5. Total primary energy consumption (in million tons of oil equivalent Mtoe) for the top 30 countries in the world.*[1]

The linkage between Per Capita Income and Energy Consumption is obvious (Figure 6). Developing countries could just not be straightjacketed into a future of continuing energy starvation.

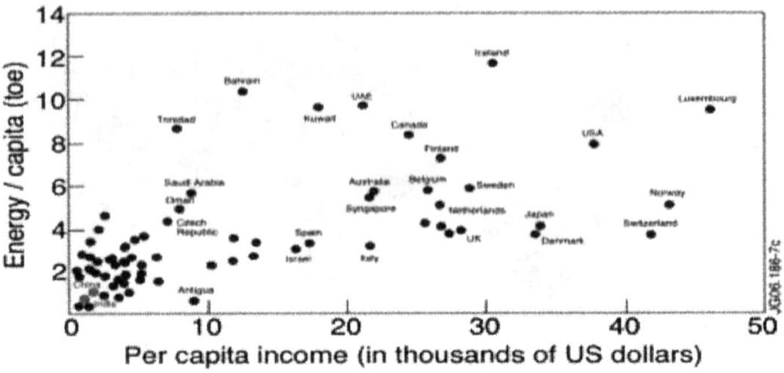


*Figure 6. Relation between energy use per capita (in tonnes of oil equivalent) and per capita income (in thousands of U.S. dollars) for various countries in the world.*[5] *Outliers are countries with a high income per capita (mostly states involved in banking) or high-energy use (mostly energy producing states). Most of the developing countries are in the bottom left corner of the graph, illustrating that low income is equivalent to low energy use.*

We thus have two powerful forces colliding frontally here. On the one side is the need for energy, without which vast populations of the world remain condemned to a wretched future. This is a tectonic force in a world which chatters *ad nauseam* about democracy and human centred development.

Opposing this force is the equally powerful force, generated this time by the developed countries, who just cannot allow their security interests to be threatened by upstarts in the developing world. The collision is fundamental, and its effects can be seen everywhere in the evolving geopolitics of the world today.

Unfortunately, the debate about nuclear energy is not really about security and safety in the immediate future. Its real parameters have to be defined scientifically, and that can be done only by projecting the trends in energy resources and energy requirements over a longer time frame.

Depending upon the increase in the world population and the increase in the energy use per capita, global energy needs could double or even quadruple in 2100. This is reflected in the three main energy scenarios studied by the IIASA-WEC working group[6] (Figures 7a, b, and c), showing three possible projections for the world primary energy needs: (a) high economic growth quadrupling our energy use in 2100; (b) middle course, with three times the current energy use in 2100; and (c) ecologically driven scenario, with maximum reduction in the use of fossil fuels and a mere doubling of energy consumption in 100 years. The difference between the projected total energy use and the energy provided by fossil fuels will have to be made up by non-fossil sources (indicated in yellow). At this moment about 6% of this is hydro, 6% nuclear, and 1-2 % is renewable energy. It is clear from these figures, that we face an enormous challenge:

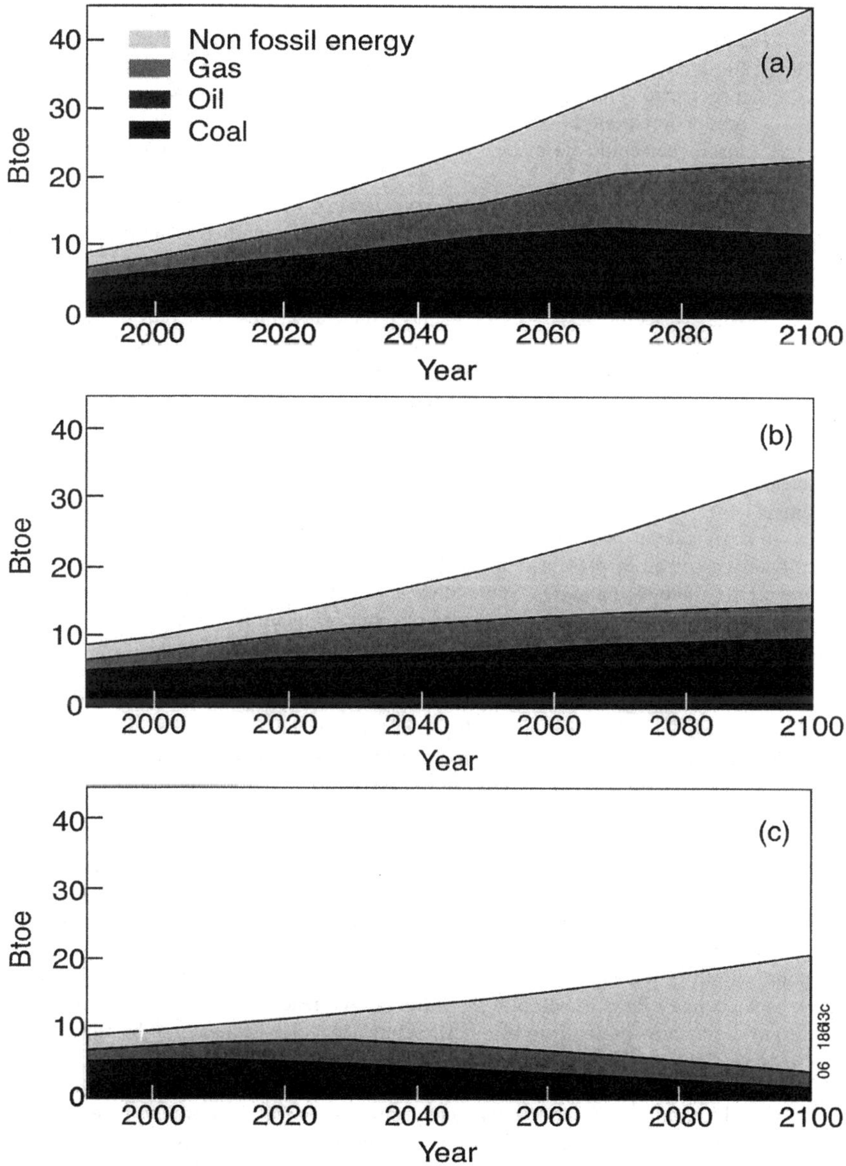


*Figure 7a, b, c. Energy scenarios for the projected world primary energy consumption (expressed in billions of tons of oil equivalent, Btoe) in the future.*[6]

- Scenario (a) assumes, in addition to high economic growth, an ample availability of oil and gas. While this reduces the amount of additional non-fossil power generation, but there are serious questions on the viability of such a scenario, certainly in view of ever stronger indications that the peak of oil production could be met at about this time. But even if such large quantities of fossil fuel could be made available, this is going to be challenging for our environment.
- Scenario (b) is more moderate on the use of fossil fuels, but still requires a large fraction of non-fossil fuel generation. Even this scenario could still overestimate the amount of oil and gas available or allowed to use.
- Scenario (c) probably underestimates future use of oil and gas, but shows what we could have to face in the future: a serious reduction in the use of fossil fuels due to resource depletion and/or restrictions because of environmental constraints. But even in this scenario a large fraction of non-fossil power generation is projected.

The conclusion is that even taking into account uncertainties in the future, as reflected by the three scenarios above, the fraction of non-fossil energy will be large in the future.

It is of course possible that there may be hidden variables which cannot be scientifically assessed at present. Part of this certainly will be generated by solar, wind and other renewable technologies. Some other new source of cheap energy may suddenly become available globally, nullifying the whole energy equation. All this is a possibility, and "hope springs eternal in the human breast."

The reality of life dictates otherwise. It appears inevitable then that we will have to look at currently available sources to fill the gap in energy requirements. But one wonders how one could generate a multiple of our current fossil energy production by renewables alone. Currently they contribute a mere 1-2%, and they all face the same basic problems limiting their potential: low energy density and/or intermittency. There is clearly a need for additional sources and nuclear techniques will have to be part of the energy mix in the future. That is where nuclear energy comes back into the picture.

Obviously, two types of safeguards will have to be crystallised as a precondition. The first relate to safety of nuclear reactors, in order to reassure public opinion in the West. The second relates to the safeguards which will have to be put into place to ensure that nuclear reactors do not continue to create leakages in the direction of nuclear weaponry. Both are feasible. Safety technology has increased dramatically in recent years, as we can see from the drop in incidents in civil reactors. As for safeguards, these are also available now in the form of modified nuclear fuels which cannot be reprocessed into weapons grade fissile materials.

The growing inevitability of nuclear energy will require the emergence of a consensus which can satisfy both safety and security concerns. The safety issue includes (a) the prevention of incidents by the wide dissemination of safety techniques; (b) cast-iron procedures for the storage of spent fuels, but not in the backyards of others; (c) strict control mechanisms to prevent illegal diversion of fissile material for military applications or terrorist actions.

Two important developments could deal very effectively with these problems in the future: so-called fourth generation nuclear reactors and nuclear fusion.

- Fourth generation fission reactors are being developed by a forum of 11 nations. Reactor designs have currently been selected on the basis of being clean, safe and cost-effective. They have a higher thermodynamic efficiency as their operating temperature is between 500°C and 1000°C depending upon the design, and could therefore also be used to directly generate hydrogen—a possible main energy carrier of the future. They are resistant to diversion of materials for weapons proliferation. Most of them employ a closed fuel cycle, maximising use of fuel and minimising high-level waste. Developers estimate that this type of reactor could be in commercial operation about 2030.
- With nuclear fusion, deuterium and tritium (two hydrogen isotopes) are fused together. As these are both positively charged, they repel each other. The nuclei have to be highly energetic in order to fuse, and this translates in extremely high fuel gas temperatures (100-200 million degrees). This process holds the promise to be environmentally friendly, safe, without the risk of run-away reactions, and there is no production of fissile material. Although large progress in fusion research is obtained in the last 20 years, there is still a considerable need for development. A big step forward, preparing further progress in fusion research is the ITER device. This large fusion machine will be built and operated by 7 nations (including India and China), representing in total more than half of the global population. The outcome of this experiment will define the parameters for a demonstration power plant based on fusion, DEMO. ITER and DEMO will take several years of construction and operation. The first commercially available fusion reactor will therefore not be available for the next decennia.

The security issues have been addressed partially by the Optional Protocol to the IAEA Safeguards, but this does not answer the whole problem of the security concerns of a number of countries who see themselves as under threat in their respective regions. That is a political problem which will have to find a political solution.

The basic debate between nuclear disarmament and nuclear non-proliferation will, of course, remain unaffected and inconclusive as long as some countries believe that their own security concerns do not allow them to accept nuclear disarmament. The question then is whether this debate should be allowed to continue to cloud our thinking on the need for nuclear energy as a scientific solution to an energy gap which jus cannot be resolved otherwise.

Compounding this problem is the rising costs of oil, and linked with that, of fossil fuels in general. This is only partly due to the unfolding unrest in the Middle East. It is much more due to the thirst for oil in a fuel hungry world. Almost all projections show that this will impact the price of oil in the future, taking it up to unsustainable levels. The comparative costs of nuclear energy will become consequently more and more attractive (Figure 8).

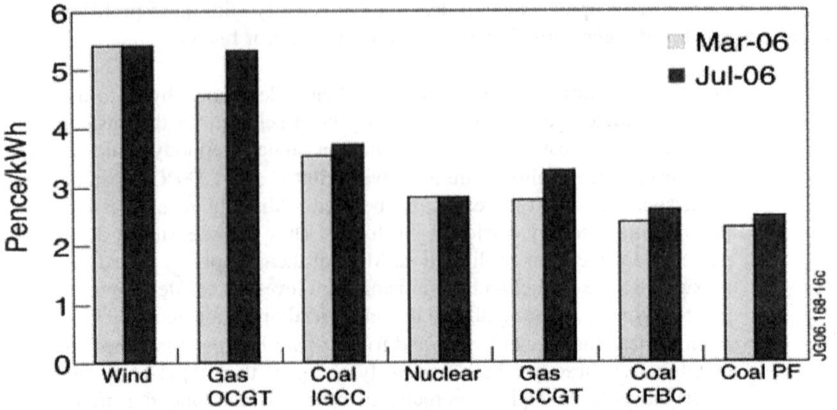


*Figure 8. Comparison of generation costs for electricity (in pence/kWh) for various options in the UK. Data based on Feb '06 fuel prices, in yellow and on Jul '06 fuel prices in green*[7]*. (OCGT: Open Cycle Gas Turbine; IGCC: Integrated Gasification Combined Cycle; CCGT: Combined Cycle Gas Turbine; CFBC: Circulating Fluidised Bed Combustion; :Pulverised Fuel).*

This paper does not attempt to suggest a way out of the political impasse which has bedevilled the whole thesis of nuclear energy for decades already. It does, however, point out emphatically that the energy gap is serious and growing, that fossil fuels have a limited life, that energy prices are not sustainable, and that all this will increasingly impact geopolitical security and stability in the world. Without access to energy, there can be no development, and without development there will be increasing unrest.

There is no easy solution to this problem of our future energy supplies. What is clear is that the debate needs to be addressed with much more urgency and care than is the case at present. The realization has already sunk in, with leading scientists now underlining the point that all energy production methods need to be developed simultaneously to help in bridging the gap.[8]

Unless there are some magic developments in energy availability, or some major new scientific discoveries, we will have to acknowledge that nuclear energy will be needed to fill the gap in future. The indicators are already there for all of us to see and ponder.

ANNEX

PARTIAL LIST OF NUCLEAR ACCIDENTS

**1952**

- **Dec. 12, Chalk River, near Ottawa, Canada:** a partial meltdown of the reactor's uranium fuel core resulted after the accidental removal of four control rods. Although millions of gallons of radioactive water accumulated inside the reactor, there were no injuries.

**1957**

- **Oct. 7, Windscale Pile No. 1, north of Liverpool, UK:** fire in a graphite-cooled military reactor spewed radiation over the countryside, contaminating a 200-square-mile area.
- **Sept. 29, near Kyshtym, South Ural Mountains:** explosion of radioactive wastes at Soviet nuclear weapons factory 12 mi from city of Kyshtym forced the evacuation of over 10,000 people from a contaminated area. No casualties were reported by Soviet officials.

**1959**

- **July 26, Santa Susana Field Lab, Simi Valley, California, USA :** a sodium cooled reactor suffered a partial core meltdown.

**1961**

- **Jan. 3, Idaho Falls, Idaho, USA :** A water vapor explosion occurred in the experimental military reactor SL-1, due to a criticality accident. At a distance of 30km from the reactor, the radiation from $^{131}I$ was about 100 times higher than the natural background.

**1964**

- **July 24, Wood River Junction facility in Charlestown, Rhode Island, USA.** A criticality accident occurred at the plant, designed to recover uranium from scrap material left over from fuel element production. An operator accidentally added a concentrated uranium solution to an agitated tank containing sodium carbonate, resulting in a critical nuclear reaction. The criticality exposed the operator to a fatal radiation dose of 10,000 rad (100 Gy). Ninety minutes later a second excursion happened, exposing two cleanup crew to doses of up to 100 rad (1 Gy) without ill effect.

**1966**

- **Oct. 5, Enrico Fermi demonstration nuclear breeder reactor on the shore of Lake Erie near Monroe, Michigan, USA.** A sodium cooling system malfunction caused a partial core meltdown. The accident was attributed to a piece of zirconium that obstructed a flow-guide in the sodium cooling system. Two of the 105 fuel assemblies melted during the incident, but no contamination was recorded outside the containment vessel.

**1967**

- **May, Chapelcross Magnox nuclear power station in Dumfries and Galloway, Scotland, UK.** Unit 2 of this power station suffered a partial meltdown when a fuel rod failed and caught fire after the unit was refuelled. Following the incident, the reactor was shut down for two years for repairs.

**1969**

- **Jan. 21, Lucens, Canton of Vaud, Switzerland.** A coolant malfunction in the experimental underground nuclear reactor. No injuries or fatalities resulted. The cavern was heavily contaminated and was sealed.

**1975**

- **Dec. 7, near Greifswald, DDR (now Germany):** radioactive core of reactor in the Lubmin nuclear power plant nearly melted down due to the failure of safety systems during a fire.

**1977**

- **Feb. 22, Jaslovske Bohunice, Czechoslovakia.** The nuclear power plant A1 in Jaslovske Bohunice experienced a serious accident during fuel loading. This INES level 4 nuclear accident resulted in damaged fuel integrity, extensive corrosion damage of fuel cladding and release of radioactivity into the plant area. As result the A1 power plant was shut down and is being decommissioned.

**1979**

- **March 28, Three Mile Island, near Harrisburg, Pennsylvania, USA:** one of two reactors lost its coolant, which caused overheating and partial meltdown of its uranium core. Some radioactive water and gases were released. This was the worst accident in U.S. nuclear-reactor history.

**1981**

- **March. Tsuruga, Japan**. More than 100 workers were exposed to doses of up to 155 millirem per day radiation during repairs of a nuclear power plant in Tsuruga, Japan, violating the company's limit of 100 millirems (1 mSv) per day.

**1982**

- **January 25, Rochester Gas & Electric Company's Ginna plant, Rochester, New York, USA.** A steam generator pipe broke, spilling radioactive coolant on the plant floor. Small amounts (about 80 Ci or 3 TBq) of radioactive steam escaped into the air.

**1983**

- **September 23, Buenos Aires, Argentina.** An operator error during a fuel plate reconfiguration led to a criticality accident at the RA-2 facility in an experimental test reactor. An excursion of $3x10^{17}$ fissions followed; the operator absorbed 2000 rad (20 Gy) of gamma and 1700 rad (17 Gy) of neutron radiation which killed him two days later. Another 17 people outside of the reactor room absorbed doses ranging from 35 rad (0.35 Gy) to less than 1 rad (0.01 Gy).

**1986**

- **April 26, Chernobyl, near Kiev, USSR (now Ukraine).** The worst accident in the history of nuclear power occurred at the Chernobyl nuclear power plant located near Kiev. This reactor type (RBMK) is dangerous without a well functioning control system, as it has a positive temperature coefficient. It is not used in the west. Remaining reactors of this type in the world are being phased out.
  Fire and explosions resulting from an experiment, carried out in an irresponsible way (control system was partially by-passed), left 31 dead in the immediate aftermath. Radioactive nuclear material was spread over much of Europe. Over 135,000 are evacuated from the areas immediately around Chernobyl and over 800,000 from the areas of fallout in Ukraine, Belarus and Russia. In 2005, a comprehensive study on the long-term health consequences of the accident was completed by the IAEA, World Health Organization and six other UN agencies, as well as the governments of Russia, Belarus and Ukraine.

**1986**

- **May 4, Hamm-Uentrop, Germany.** An experimental 300-megawatt THTR-300 HTGR released radiation after one of its spherical fuel pebbles became lodged in the pipe used to deliver fuel elements to the reactor. Operator actions to dislodge the obstruction during the event damaged the fuel pebble cladding, released radiation detectable up to two kilometers from the reactor.

**1993**

- **April 6, near Tomsk, Russian Federation.** At the Tomsk-7 Siberian Chemical Enterprise plutonium reprocessing facility, a pressure buildup led to an explosive mechanical failure in a 34 cubic meter stainless steel reaction vessel buried in a concrete bunker under building 201 of the radiochemical works. The vessel contained a mixture of concentrated nitric acid, uranium (8757 kg), plutonium (449 g) along with a mixture of radioactive and organic waste from a prior extraction cycle. The explosion dislodged the concrete lid of the bunker and blew a large hole in the roof of the building, releasing approximately 6 GBq of $^{239}Pu$ and 30 TBq of various other radionuclides into the environment. The accident exposed 160 on-site workers and almost two thousand cleanup workers to total doses of up to 50 mSv (the threshold limit for radiation workers is 100 mSv per 5 years). The contamination plume extended 28 km NE of building 201, 20 km beyond the facility property. The small village of Georgievka (pop. 200) was at the end of the fallout plume, but no fatalities, illnesses or injuries were reported.

**1999**

- **Sept. 30, Tokaimura, Japan:** Japan's worst nuclear accident to date takes place at a uranium reprocessing facility in Tokai-mura, Ibaraki prefecture, northeast of Tokyo, Japan. The direct cause of the criticality accident was workers putting uranyl nitrate solution containing about 16.6 kg of uranium, which exceeded the critical mass, into a precipitation tank. The tank was not designed to dissolve this type of solution and was not configured to prevent eventual criticality. Three workers were exposed to radiation doses in excess of allowable limits (two of these workers died); a further 116 received lesser doses of 1 msV or greater.

**2005**

- **April 19, 2005, Sellafield, UK**. Twenty metric tons of uranium and 160 kilograms of plutonium dissolved in 83,000 liters of nitric acid leaked undetected over several months from a cracked pipe into a stainless steel sump chamber at the Thorp nuclear fuel reprocessing plant.

REFERENCES

1. Energy Information Administration, International Energy Annual 2004, U.S. Department of Energy, Washington DC.
2. International Nuclear Safety Centre at Argonne National Laboratory, 2005, USA.
3. Power Reactor Information System, IAEA, Vienna, 2006
4. U.S. Deparment of Agriculture, Natural Resources Conservation Service, Washington DC, 2001
5. Worldbank, Washington DC, 2003.
6. Nakicenovic, N., Gruebler, A., McDonald, A. (eds), "Global Energy Perspectives", Joint IIASA-WEC Study, Cambridge University Press, Cambridge, UK (1998)
7. Ian Burdon and Dominic Cook, PB Power, Newcastle upon Tyne, UK, 2006
8. Dr. Martin Rees, Science 313, no. 5787, 591 (4 Aug 2006).
9. J.H.Fremlin, "Power Production: What are the risks?" 2nd Edition, Adam-Hilger, Bristol (1989)
10. R.F.Mould, "Chernobyl Record—The definitive history of the Chernobyl catastrophe", Institute of Physics Publishing, Bristol and Philadelphia (2000)
11. IAEA Bulletin 41, no.3, 2-17 (1999) and IAEA Safety Report Series No 4, Vienna (1998).

# WORKSHOP PARTICIPANTS

| | |
|---|---|
| Professor William A. Barletta | Accelerator & Fusion Research Division<br>Lawrence Berkeley National Laboratory<br>Berkeley, **USA** |
| Dr. Jacques Bouchard | French Atomic Energy Commission (CEA)<br>Paris, **France** |
| Dr. Carmen Difiglio | Office of Policy and International Affairs<br>U.S. Department of Energy<br>Washington, **USA** |
| Professor Steve Fetter | School of Public Policy<br>University of Maryland<br>College Park, **USA** |
| Professor William Fulkerson | Joint Institute for Energy and Environment<br>University of Tennessee<br>Knoxville, **USA** |
| Dr. Richard L. Garwin | Thomas J. Watson Research Center<br>IBM Research Division<br>Yorktown Heights, **USA** |
| Sir Brian Heap | St Edmund's College<br>University of Cambridge<br>Cambridge, **UK** |
| Professor Pervez Hoodbhoy | Physics Department<br>Quaid-e-Azam University<br>Islamabad, **Pakistan** |
| Dr. Richard Hoskins | International Atomic Energy Agency<br>Vienna, **Austria** |
| Dr. Jafar Dhia Jafar | Uruk Project Development Company<br>Dubai**, United Arab Emirates** |
| Professor Joachim Krause | Institute for Social Sciences<br>University of Kiel<br>Kiel, **Germany** |

| | |
|---|---|
| Professor Valery P. Kukhar | Institute for Bio-organic Chemistry<br>Academy of Sciences<br>Kiev, **Ukraine** |
| Dr. Kazuaki Matsui | The Institute of Applied Energy<br>Tokyo, **Japan** |
| Dr. Charles McCombie | Arius Association<br>Baden, **Switzerland** |
| Dr. Akira Miyahara | National Institute for Fusion Science<br>Tokyo, **Japan** |
| Dr. Jef Ongena | PlasmaPhysics Laboratory<br>Ecole Royale Militaire<br>Brussels, **Belgium** |
| Professor Donato Palumbo | World Laboratory Centre<br>Fusion Training Programme<br>Palermo, **Italy** |
| Professor Juras Pozela | Lithuanian Academy of Sciences<br>Vilnius, **Lithuania** |
| Professor Zenonas Rudzikas | Theoretical Physics & Astronomy Institute<br>Lithuanian Academy of Sciences<br>Vilnius, **Lithuania** |
| Dr. Bruce Stram | BST Ventures<br>Houston, **USA** |
| Ambassador Roland Timerbaev | Center for Policy Studies in Russia<br>Moscow, **Russia** |
| Professor François Waelbroeck | World Laboratory Centre<br>Fusion Training Programme<br>St. Amandsberg, **Belgium** |

# 14. SEMINAR PARTICIPANTS

| | |
|---|---|
| Dr. Giuseppe Tito Aronica | Department of Civil Engineering<br>University of Messina<br>Messina, **Italy** |
| Professor Aurelio Aureli | Department of Applied Geology<br>University of Palermo<br>Palermo, **Italy** |
| Professor William A. Barletta | Accelerator & Fusion Research Division<br>Lawrence Berkeley National Laboratory<br>Berkeley, **USA** |
| Dr. Michael J.S. Belton | Belton Space Exploration Initiatives, LLC<br>Tucson, **USA** |
| Dr. Antonio Boccafoschi | Depart. of Civil & Environmental Engineering<br>University of Catania<br>Catania, **Italy** |
| Professor J. M. Borthagaray | Instituto Superior de Urbanismo<br>University of Buenos Aires<br>**Buenos Aires,** Argentina |
| Dr. Jacques Bouchard | French Atomic Energy Commission (CEA)<br>Paris, **France** |
| Dr. Vladimir B. Britko | Information Systems Laboratory<br>Institute for Systems Analysis<br>Moscow, **Russia** |
| Colonel Christian Bühlmann | Research & Development<br>HQ Planification Services<br>Military Federal Department<br>Bern, **Switzerland** |
| Dr. Franco Buonaguro | Istituto Nazionale dei Tumori<br>"Fondazione G. Pascale"<br>Napoli, **Italy** |
| Dr. Philip Burgi | Environmental Water Resources Institute<br>Wheat Ridge, **USA** |

| | |
|---|---|
| Dr. Diego Buriot | Former Special Advisor to the Assistant Director<br>General Communicable Diseases<br>World Health Organisation<br>Geneva, **Switzerland** |
| Dr. Gina M. Calderone | EA Science and Technology<br>Newburgh, **USA** |
| Professor Antonino Cancelliere | Civil and Environmental Engineering<br>University of Catania<br>Catania, **Italy** |
| Dr. Angela Candela | Department of Hydraulic Engineering<br>University of Palermo<br>Palermo, **Italy** |
| Dr. Gregory R. Carmichael | Department of Chemical and<br>Biochemical Engineering<br>The University of Iowa<br>Iowa City, **USA** |
| Dr. Salvatore Carrubba | Department of Applied Geology<br>University of Palermo<br>Palermo, **Italy** |
| Dr. Nathalie Charpak | Kangaroo Foundation<br>Bogotá, **Colombia** |
| Dr. Sundar A. Christopher | Department of Atmospheric Sciences<br>University of Alabama<br>Huntsville, **USA** |
| Professor Robert Clark | Hydrology and Water Resources<br>University of Arizona<br>Tucson, **USA** |
| Dr. Socorro de Leon-Mendoza | Neonatology Unit<br>Jose Fabella Memorial Hospital<br>Manila, **Philippines** |
| Dr. Jean-François Debroux | Kennedy/Jenks Consultants<br>San Francisco, **USA** |

| | |
|---|---|
| Dr. Carmen Difiglio | Office of Policy and International Affairs<br>U.S. Department of Energy<br>Washington, **USA** |
| Dr. Mbareck Diop | Science & Technology Advisor to the President of Senegal<br>**Dakar,** Senegal |
| Professor Christopher D. Ellis | Landscape Architecture and Urban Planning, Texas A&M University<br>College Station, **USA** |
| Professor Andrew Curtis Elmore | Department of Geological Engineering<br>University of Missouri<br>Rolla, **USA** |
| Dr. Christopher Essex | Department of Applied Mathematics<br>University of Western Ontario<br>London, Ontario, **Canada** |
| Dr. Lorne Everett | Chancellor, Lakehead University<br>Thunder Bay, **Canada** and<br>Haley & Aldrich, Inc.<br>Santa Barbara, **USA** |
| Professor Steve Fetter | School of Public Policy<br>University of Maryland<br>College Park, **USA** |
| Dr. Philip J. Finck | Applied Science and Technology<br>Argonne National Laboratory<br>Argonne, USA |
| Dr. Lawrence Friedl | NASA Applied Sciences Program<br>NASA Headquarters<br>Washington, **USA** |
| Professor William Fulkerson | Joint Institute for Energy and Environment<br>University of Tennessee<br>Knoxville, **USA** |
| Dr. Bertil Galland | Writer and Historian<br>Buxy, **France** |

| | |
|---|---|
| Professor Richard L. Garwin | Thomas J. Watson Research Center<br>IBM Research Division<br>Yorktown Heights, **USA** |
| Professor Alberto González-Pozo | Theory and Analysis Department<br>Universidad Autónoma Metropolitana<br>Mexico D.F., **Mexico** |
| Professor Munther J. Haddadin | Former Minister of Water & Irrigation of the Hashemite Kingdom of Jordan<br>Amman, **Jordan** |
| Sir Brian Heap | St Edmund's College<br>University of Cambridge<br>Cambridge, **UK** |
| Dr. Udo Helmbrecht | Federal Office for the Security of Information Technologies<br>Bonn, **Germany** |
| Professor Pervez Hoodbhoy | Physics Department<br>Quaid-e-Azam University<br>Islamabad, **Pakistan** |
| Dr. Richard Hoskins | International Atomic Energy Agency<br>Vienna, **Austria** |
| Dr. Kembra Howdeshell | Reproductive Toxicology Division<br>NHEERL, ORD, EPA<br>North Carolina, **USA** |
| Professor Reiner K. Huber | Faculty of Information Technologies<br>German Armed Forces University München<br>Neubiberg, **Germany** |
| Professor Walter F. Huebner | Southwest Research Institute<br>San Antonio, Texas, **USA** |
| Dr. Christiane Huraux | Pédiatre-Neonatologue<br>Creteil, **France** |
| Dr. Jafar Dhia Jafar | Uruk Project Development Company<br>Dubai, **United Arab Emirates** |

| | |
|---|---|
| Professor Leonardas Kairiukstis | Laboratory of Ecology and Forestry<br>Kaunas-Girlonys, **Lithuania** |
| Dr. Ahmad Kamal | Ambassador (ret.) – United Nations Institute for Training and Research, New York Office<br>New York, **USA** |
| Dr. Hisham Khatib | World Energy Council<br>Amman, **Jordan** |
| Professor Pradeep Khosla | Carnegie Institute of Technology<br>Pittsburgh, Pennsylvania, **USA** |
| Professor Alexander Konovalov | Moscow State Institute of International Relations<br>Moscow, **Russia** |
| Professor Joachim Krause | Institute for Social Sciences<br>University of Kiel<br>Kiel, **Germany** |
| Professor Valery P. Kukhar | Institute for Bio-organic Chemistry<br>Academy of Sciences<br>Kiev, **Ukraine** |
| Professor Tsung-Dao Lee | Department of Physics<br>Columbia University<br>New York, **USA** |
| Professor Axel Lehmann | Institute for Technical Computer Sciences<br>German Armed Forces University München<br>**Neubiberg,** Germany |
| Dr. Kazuaki Matsui | The Institute of Applied Energy<br>Tokyo, **Japan** |
| Dr. Charles McCombie | Arius Association<br>Baden, **Switzerland** |
| Dr. Akira Miyahara | National Institute for Fusion Science<br>Tokyo, **Japan** |
| Captain Charles Moore | Algalita Marine Research Foundation<br>Long Beach, California, **USA** |

Professor Jörg Oehlmann
Department of Aquatic Ecotoxicology
Institute for Ecology, Evolution and Diversity
Johann Wolfgang Goethe University
Frankfurt, **Germany**

Dr. Jef Ongena
Plasmaphysics Laboratory
Ecole Royale Militaire
Brussels, **Belgium**

Professor Albert D.M.E. Osterhaus
Department of Virology
Erasmus Medical Center
Rotterdam, **The Netherlands**

Professor Paola Palanza
Evolutional and Functional Biology
University of Parma
Parma, **Italy**

Professor Donato Palumbo
World Laboratory Centre
Fusion Training Programme
Palermo, **Italy**

Professor Stefano Parmigiani
Evolutional and Functional Biology
University of Parma
Parma, **Italy**

Professor Margaret Petersen
Hydrology & Water Resources
University of Arizona
**Tucson,** USA

Professor Juras Pozela
Lithuanian Academy of Sciences
**Vilnius,** Lithuania

Professor Richard Ragaini
Department of Environmental Protection
Lawrence Livermore National Laboratory
Livermore, **USA**

Professor Ramamurti Rajaraman
School of Physical Sciences
Jawaharlal Nehru University
New Delhi, **India**

Professor Bartolomeo Reitano
Department of Civil and Environmental Engineering
University of Catania
Catania, **Italy**

| | |
|---|---|
| Professor Paolo Ricci | Department of Environmental Sciences<br>University of San Francisco<br>San Francisco, **USA** |
| Dr. Luca Rossi | Civil Protection Department<br>Planning Office for Risk evaluation and Prevention<br>Rome, **Italy** |
| Professor Giuseppe Rossi | Department of Civil and Environmental Engineering<br>University of Catania<br>Catania, **Italy** |
| Professor Zenonas Rudzikas | Theoretical Physics & Astronomy Institute<br>Lithuanian Academy of Sciences<br>**Vilnius,** Lithuania |
| Dr. Juan Ruiz | Department of Pediatrics<br>San Ignacio Hospital<br>Bogotá, **Colombia** |
| Professor Hiltmar Schubert | Fraunhofer-Institut für Chemische Technologie, ICT<br>Pfinztal, **Germany** |
| Professor Geraldo Gomes Serra | NUTAU<br>University of São Paolo<br>São Paulo**, Brazil** |
| Professor William A. Sprigg | Institute of Atmospheric Physics<br>University of Arizona<br>**Tucson,** USA |
| Dr. Graeme Stephens | Department of Atmospheric Science<br>Colorado State University<br>Fort Collins, **USA** |
| Dr. Bruce Stram | BST Ventures<br>Houston, **USA** |
| Dr. Shanna H. Swan | Center for Reproductive Epidemiology<br>University of Rochester<br>Rochester, **USA** |

| | |
|---|---|
| Dr. Chris Talsness | Department of Toxicology<br>Benjamin Franklin Medical Center<br>Charité Universitätsmedizin Berlin<br>Berlin, **Germany** |
| Ambassador Roland Timerbaev | Center for Policy Studies in Russia<br>Moscow, **Russia** |
| Professor Wouter van Dieren | IMSA Amsterdam<br>Amsterdam, **The Netherlands** |
| Dr. Frederick S. vom Saal | Division of Biological Sciences<br>University of Missouri<br>Columbia, **USA** |
| Professor François Waelbroeck | World Laboratory Centre<br>Fusion Training Programme<br>St. Amandsberg, **Belgium** |
| Dr. Henning Wegener | Ambassador of Germany (ret.)<br>Information Security Permanent Monitoring Panel<br>World Federation of Scientists<br>Madrid, **Spain** |
| Dr. Jody Westby | Global Cyber Risk LLC<br>Washington, **USA** |
| Professor Richard Wilson | Department of Physics<br>Harvard University<br>Cambridge, **USA** |
| Dr. Hajime Yano | Department of Planetary Science<br>The Graduate University for Advanced Studies<br>Kanagawa, **Japan** |
| Dr. John Zinn | Los Alamos National Laboratory<br>Los Alamos, **USA** |
| Professor Antonino Zichichi | CERN, Geneva, **Switzerland** and<br>University of Bologna, **Italy** |